6th International Conference on Biotechnology Engineering (ICBioE 2021)

IOP Conference Series: Materials Science and Engineering
Volume 1192

Kuala Lumpur, Malaysia
22 – 23 June 2021

Editors:

Sarina Sulaiman
Noor Illi Mohamad Puad
Mohd Firdaus Abd Wahab
Md Zahangir Alam

ISBN: 978-1-7138-4652-9
ISSN: 1757-8981

Printed from e-media with permission by:

Curran Associates, Inc.
57 Morehouse Lane
Red Hook, NY 12571

Some format issues inherent in the e-media version may also appear in this print version.

Institute of Physics
Dirac House, Temple Back
Bristol BS1 6BE UK

Phone: 44 1 17 929 7481
Fax: 44 1 17 920 0979

techtracking@iop.org

Additional copies of this publication are available from:

Curran Associates, Inc.
57 Morehouse Lane
Red Hook, NY 12571 USA
Phone: 845-758-0400
Fax: 845-758-2633
Email: curran@proceedings.com
Web: www.proceedings.com

TABLE OF CONTENTS

IOP Conf. Series: Materials Science and Engineering 1192 (2021) 011001 doi:10.1088/1757-899X/1192/1/011001

Preface

Published by IOP Conference Series: Materials Science and Engineering.

The International Conference on Biotechnology Engineering held its 6th Edition on the 22nd and 23rd June 2021. With the theme 'Nurturing Innovation for Sustainable Future', the conference revolved around scientifically rigor findings that emphasize the elements of sustainability and innovation.

ICBioE 2021 serves as a scientific platform for academic and industry researchers, entrepreneurs, and technologists to convene from around the world, and exchange their latest scientific findings in the field of Materials and Chemical Engineering. In addition, three keynote speeches by distinguished Professors from Malaysia and Singapore, as well as from a successful university spin-off entrepreneur, were delivered during this event.

Due to the global Covid-19 pandemic that took everyone by surprise, the initial plan to run ICBioE in 2020 was completely put on hold. After several readjustments and planning, the conference was revitalized into a virtual event in 2021. Every paper submitted were subjected to double-blind peer-review process by at least two field experts from the conference committee and international reviewers. A total of 64 papers were accepted and presented out of 126 submissions to the conference, with participations coming from Malaysia, Indonesia, Japan, Oman, Nigeria and United Arab Emirates. Papers were selected on their quality and relevance to the conference.

This volume of proceedings reflects the nature and field of study presented in the conference. It covers advances and study of materials and biomaterials from the perspective of Chemical and Biochemical Engineering, and ranging from the macro- to nano-sized molecules. It is hoped that this volume would find its use to readers from any discipline of science and technology, and opens up opportunities for collaboration with and among authors of this publication.

On behalf of the editors, we would like to thank all authors and contributors who have dedicated their time and effort to this volume, including the conference committee, the international advisory members and reviewers, keynote speakers, presenters, chairpersons and sponsors. We would also like to thank everyone who has participated in ICBioE 2021 for their support and interest.

EDITORIAL BOARD

IOP Conf. Series: Materials Science and Engineering 1192 (2021) 011001 doi:10.1088/1757-899X/1192/1/011001

- **Organizing Committee:**

Chairman
Nor Fadhillah Mohamed Azmin

Co-Chairman
Faridah Yusof

Secretary
Farah Ahmad
Mariyana Nawi

Technical Committee
Fazia Adyani Ahmad Fuad (Chairman)
Mohammed Saedi Jami
Raha Ahmad Raus
Fathilah Ali
Nurul Sakinah Engliman
Husna Ahmad Tajuddin

Publicity
Amanatuzzakiah Abd. Halim (Chairman)
Faridah Yusof
Ma'an Fahmi Al-Khatib
Noor Illi Mohamad Puad
Wan Wardatul Amani Wan Salim
Ricca Rahman Nasaruddin

Sponsorship
Azlin Suhaida Azmi (Chairman)
Maizirwan Mel
Dzun Noraini Jimat
Husna Ahmad Tajuddin
Munira Shahbudin
Mariatul Fazilah Mansor
Rashidah Rashid

Website
Wan Mohd Fazli Wan Nawawi (Chairman)
Mohd Iznan Ab. Hamid

Multimedia
Mohd Firdaus Abd Wahab (Chairman)
Dzun Noraini Jimat
Noor Illi Mohamad Puad
Yusilawati Ahmad Nor
Wan Mohd Fazli Wan Nawawi
Farah Ahmad
Mohd Iznan Ab. Hamid
Mohd Hafizul Shaibon
Nasaruddin Mohd Sulaiman

IOP Conf. Series: Materials Science and Engineering 1192 (2021) 011001 doi:10.1088/1757-899X/1192/1/011001

Keynote Speakers

Md. Zahangir Alam (Chairman)
Nassereldeen A. Kabbashi
Dzun Noraini Jimat
Wan Wardatul Amani Wan Salim

Registration

Yusilawati Ahmad Nor (Chairman)
Azlin Suhaida Azmi
Munira Shahbuddin
Mariatul Fadzillah Mansor
Rashidah Rashid
Aslan Sarif Asrimon
Anuar Ariffin
Siti Noradila Md Misri
Suharti Yahya

Journal

Dzun Noraini Jimat (Chairman)
Md. Zahangir Alam
Maizirwan Mel
Fathilah Ali
Nurul Sakinah Engliman

IOP Conf. Series: Materials Science and Engineering 1192 (2021) 011002 doi:10.1088/1757-899X/1192/1/011002

Peer review declaration

All papers published in this volume of *IOP Conference Series: Materials Science and Engineering* have been peer reviewed through processes administered by the Editors. Reviews were conducted by expert referees to the professional and scientific standards expected of a proceedings journal published by IOP Publishing.

- **Type of peer review: Single-blind / Double-blind / Triple-blind / Open / Other**

  Double Blind- Remove the author names and affiliations of the authors anywhere in the manuscript. Remove names and affiliations under the title within the manuscript. Do not include or remove any author names or affiliations in the Acknowledgements section of your manuscript. Author names and Funding information should be removed and can be included later in the peer review process. Do not sign your author response, rebuttals, or appeals with author names. Do not include or remove any names in any file names and ensure document properties are also anonymised. Check that your figures and schemes do not include author details.

- **Conference submission management system:**
  - https://easychair.org/conferences/submissions?a=26389499

- **Number of submissions received: 114 (Abstract)**

- **Number of submissions sent for review: 68**

- **Number of submissions accepted: 59**

- **Acceptance Rate (Number of Submissions Accepted / Number of Submissions Received X 100): 55.6%**

- **Average number of reviews per paper: 2**

- **Total number of reviewers involved: 83**

- **Any additional info on review process:**
  Turnitin was used to check plagiarism.

- **Contact person for queries:**

Assoc. Prof. Ir. Dr. Sarina Sulaiman, Department of Biotechnology Engineering, Kulliyyah of Engineering, International Islamic University Malaysia. Email: sarina@iium.edu.my

Assoc. Prof. Dr Noor Noor Illi, Department of Biotechnology Engineering, Kulliyyah of Engineering, International Islamic University Malaysia. Email: illi@iium.edu.my

Dr Mohd Firdaus Abd Wahab, Department of Biotechnology Engineering, Kulliyyah of Engineering, International Islamic University Malaysia. Email: firdaus@iium.edu.my

IOP Conf. Series: Materials Science and Engineering 1192 (2021) 011002 doi:10.1088/1757-899X/1192/1/011002

Prof. Dr. Zahangir Alam, Department of Biotechnology Engineering, Kulliyyah of Engineering, International Islamic University Malaysia. Email: zahangir@iium.edu.my

IOP Conf. Series: Materials Science and Engineering 1192 (2021) 012001 doi:10.1088/1757-899X/1192/1/012001

Extraction of Phenolic Compound using Natural Deep Eutectic Solvent from Biomass Waste

A H A Aziz[1], N S Engliman[1,2], M F Mansor[1] and R R Nasaruddin[1]

[1]Department of Biochemical-Biotechnology Engineering, Kulliyyah of Engineering, International Islamic University Malaysia (IIUM), P.O. Box 10, 50728, Gombak, Kuala Lumpur, Malaysia

[2]Corresponding author's email: sakinahengliman@iium.edu.my

Abstract. Phenolic compounds are aromatic compounds known for their bioactive substances which can be found in plants. It has been widely used in various applications due to its antibiotics, anti-inflammatory, anti-allergic, and other benefits. There are various methods to extract the phenolic compounds from plants including ionic liquid, liquid-liquid extraction as well as supercritical extraction. However, all of these methods requires energy extensive, laborious processes, advanced technology, and generate toxic waste. Therefore, there is a growing need to find an alternative green extraction method to reduce the environmental impact while improving the efficiency of the extraction process. Thus, natural deep eutectic solvents (NADES), a combination of two or more components that comes from primary metabolites like organic acids, choline chloride, or sugar, are able to form liquids upon mixing with lower melting point of individual constituents due to hydrogen bond interactions were proposed as alternatives to conventional extraction methods. Therefore, this research determined the suitable combination of NADES solvents (hydrogen bond acceptor/donor ratio) for extraction of phenolic compounds from biomass waste which included young and mature coconut shells and coconut husk, banana peel, empty fruit bunch, and palm oil fruit husk. The extracted compound was analysed using fourier-transform infrared spectroscopy (FTIR) and gas chromatography-mass spectrometry (GC-MS) to identify the functional groups and type of phenolic compounds present. The best biomass waste was coconut shell and the best NADES combination was choline chloride and ascorbic acid at 1:2 molar ratio. The FTIR analysis of coconut shell extracted by NADES showed peaks at 3404 cm^{-1} and 3523 cm^{-1} indicating OH stretching followed by 2915 cm^{-1}, 1388 cm^{-1} to 1473 cm^{-1}, and 1674 cm^{-1} showing C-H stretching, sp3 C-H band and C=C stretch respectively. As for GC-MS analysis, 26 compounds were detected and four phenolic compounds were identified at peaks 2, 13, 22, and 25. The research was successful in determining the best biomass waste and NADES combination for highest total phenol. The use of NADES was able to extract more phenolic compound from coconut shell than water due to the hydrogen bond between the choline chloride and ascorbic acid.

1. Introduction

Phenolic compounds are secondary metabolites that consist of aromatic rings with a hydroxyl group. They are readily available in the plant kingdom and possess beneficial properties such as antioxidant and anti-inflammatory. The use of phenolic compounds has been sought after for many years because of its benefits for health such as preventing cardiovascular diseases, diabetes, cancer, and obesity [1]. Phenols are also used in food technology as prevention from oxidation and bacterial growth [2]. In synthesis of metal nanoparticles, bioactive compounds like phenolic compounds are used as reducing agent, making it possible to produce the small material without using toxic chemicals [3]. The source of phenolic compounds can be found easily in nature such as in trees, leaves, or fruits, including biomass waste which the majority comes from agricultural and food waste. The abundance of biomass waste in an agriculture-rich industry like Malaysia has high potential to be utilized for other purposes instead of being disposed to the landfill. This will encourage the wealth generation from waste or known as the circular economy. In order to seize the opportunity, biomass waste can be utilized to extract the phenolic compounds for various applications.

There are many ways to extract phenolic compounds including liquid-liquid extraction, ionic liquids (ILs), and supercritical liquid. However, these techniques involve toxic chemicals, laborious and advance technology which might cause pollution and higher production cost [4]. At the same time, these techniques also have poor biodegradability, sustainability, and biocompatibility of the extraction product. Therefore, this study will use a green method in extracting the active compound from biomass waste. This research will use deep eutectic solvent (DES) to extract the active compound that is trapped inside the biomass waste. DES is a combination of two or more safe components to form a eutectic mixture through hydrogen bond interaction, which has a melting point lower than that of each component. DES combination with lower melting point developed strong hydrogen bond interaction between hydrogen bond donor and acceptor. The strength of hydrogen bond will increase the extraction yield of desired compound [5]. The low viscosity of DES helps in mixing and the high density helps with the separation of the phases in the matrix [6]. Most common DES are based on choline chloride (ChCl), carboxylic acids, urea, citric acid, and other hydrogen-bond donors. Another classification of DES is natural deep eutectic solvent (NADES) that uses natural compounds for both hydrogen donor and acceptor, such as primary metabolites like organic acids and sugars [4].

Therefore, this study focuses to determine the suitable combination ratio of hydrogen bond donor and acceptor that form the best NADES for maximum extraction of phenolic compounds from biomass waste. Then, the extract will undergo the qualitative analysis using GC-MS to identify the unknown active compounds that can be extracted using the selected extraction method. Then the comparison study will be done to differentiate between normal liquid-liquid extraction method with NADES solvent method based on the type of active compounds that can be identified from GC-MS. This result will be of guidance for other researchers to explore further on this type of extraction process so that many important active compounds can be extracted and to be used for various applications.

2. Materials and Methods

The lab grade chemicals used in this experiment were choline chloride, glucose, sucrose, lactic acid, ascorbic acid, hydrochloric acid, and Folin Ciocalteu's Reagent which were all purchased from BT Science Sdn. Bhd.

The types of equipment used in this experiment were UVILINE 9400 UV-Vis spectrophotometer, oven, centrifuge machine, magnetic stirrer, and pH indicator provided in the Bioprocess Engineering Laboratory in International Islamic University Malaysia (IIUM).

2.1. Screening of different biomass waste

The biomass waste used for screening were banana peel, empty fruit bunch (EFB), palm oil fruit husk, coconut shell, and coconut husk. The samples were dried overnight at 60 °C to ensure the removal of moisture without bioactive compounds degradation. Dried biomass was then grinded and stored in air-tight containers. Phenolic compounds were extracted by conventional solvent extraction, using 5 g of biomass waste and 100 mL distilled water, stirred continuously at 90 °C for 20 minutes. The Folin-Ciocalteu method [7] was used to measure the total phenol content. The biomass waste with the highest total phenol content is then used for the remaining of the experiments.

2.2. Screening of different NADES combinations

Different combinations of hydrogen bond donors and hydrogen bond acceptors were used to determine the best combinations for phenolic compound extraction. The heating method was used to develop the natural deep eutectic solvents with some modifications [8]. For screening, the NADES mixtures were placed in a shaking incubator at 65 °C for 30 minutes at 100 rpm. The NADES was then used to extract

IOP Conf. Series: Materials Science and Engineering 1192 (2021) 012001 doi:10.1088/1757-899X/1192/1/012001

phenols from 5 g of biomass waste at room temperature shaker for 20 minutes at 100 rpm followed by filtering. The extracted solution was measured for its total phenol content using the Folin-Ciocalteu method.

2.3. Fourier-transform infrared spectroscopy (FTIR) analysis

The NADES, its components, and the extracted solution were analyzed under FTIR to observe the functional groups present. The sample was analyzed in solid phase. FTIR analysis was carried out in transmission mode range of 650 cm^{-1} to 4000 cm^{-1}.

2.4. Gas chromatography-mass spectrometry (GC-MS) analysis

The extracted samples were observed using an Agilent 7890 gas chromatograph equipped with an Agilent MSD 5975C mass spectrometer. A capillary column HP-5 with 30 m × 0.25 μm and an internal diameter of 0.32 mm was used. Before analyzing the samples, the retention time was locked by changing column pressure using standard samples. A constant pressure model was then used for the entire analysis process. 1 % of sample to 99 % of hexane as a solvent in 1.5 mL vial was filtered from a 2 mm syringe filter was prepared beforehand. The gas chromatography (GC) oven temperature was programmed from 80 to 300 °C via a ramp of 5 °C min^{-1} and maintained at 80 °C for 2 min and at 300 °C for 5 min. The mass spectrum (MS) was operated in full-scan mode from m/z, 50–700 for qualitative analysis or selected ion monitoring (SIM) mode for quantitative analysis. The inlet and MS transfer line temperatures were maintained at 300 °C and the ion source temperature was 300 °C. Sample injection (1 μL) was done using splitless mode [9].

3. Results and Discussion

3.1. Screening of different biomass waste

The five different biomass wastes were screened by extraction using water at pH 3. The total phenol extracted was recorded in Table 1. The results show that water extracted young coconut shell had the highest total phenol content among the other five biomass waste. This is because of the abundance of polyphenolic compounds due to its natural property as protection in plants. The shell mainly consists of cellulose and lignin that is similar to wood, making it optimum for phenolic extraction [10].

Table 1. The total phenol extracted from biomass waste using water at pH 3.

No	Biomass waste	Total phenol (mg GAE/L)
1	Banana peel	739.00
2	Coconut shell	2427.50
3	Coconut husk	1199.50
4	Empty fruit bunch	102.00
5	Palm oil fruit husk	150.50

3.2. Screening of different NADES combinations

Since coconut shell was proven the best biomass waste with the highest total phenol, different combinations of chemicals with different molar ratios were screened. The eutectic mixture combinations and their total phenol extracted from the coconut shell are shown in Table 2. Based on the result, the best NADES combination was choline chloride with ascorbic acid at 1:2 molar ratio with 3715.67 mg GAE/L total phenol extracted. This is closely followed by glucose and ascorbic acid at 1:2 molar with 3682.67 mg GAE/L total phenol and sucrose with ascorbic acid at 1:2 molar ratio with 3670.00 mg GAE/L total phenol. The pattern here is the presence of ascorbic acid. Even though the use of NADES helps to produce hydrogen bond for optimum phenolic extraction, the use of ascorbic acid affects the polyphenols positively [11]. This is proven by the total phenol amount maintaining at high numbers when using NADES with ascorbic acid.

Table 2. The different chemical combinations and molar ratios of NADES and their total phenol extracted from coconut shell.

Acceptor	Donor	Molar Ratio	Total Phenol (mg GAE/L)	Acceptor	Donor	Molar Ratio	Total Phenol (mg GAE/L)
Water at pH 3		NA	454.00	ChCl	Ascorbic Acid	2:1	3678.00
Water at pH 7		NA	972.00	Glucose	Lactic Acid	1:1	602.00
ChCl	Glucose	1:1	628.00	Glucose	Lactic Acid	1:2	442.00
ChCl	Glucose	1:2	548.00	Glucose	Lactic Acid	2:1	515.00
ChCl	Glucose	2:1	543.00	Glucose	Ascorbic Acid	1:1	3611.00
ChCl	Sucrose	1:1	528.00	Glucose	Ascorbic Acid	1:2	3682.67
ChCl	Sucrose	1:2	583.00	Glucose	Ascorbic Acid	2:1	3557.00
ChCl	Sucrose	2:1	658.00	Sucrose	Lactic Acid	1:1	587.50
ChCl	Lactic Acid	1:1	1510.50	Sucrose	Lactic Acid	1:2	537.00
ChCl	Lactic Acid	1:2	1497.50	Sucrose	Lactic Acid	2:1	434.00
ChCl	Lactic Acid	2:1	2353.00	Sucrose	Ascorbic Acid	1:1	3558.50
ChCl	Ascorbic Acid	1:1	3624.00	Sucrose	Ascorbic Acid	1:2	3649.50
ChCl	Ascorbic Acid	1:2	3715.67	Sucrose	Ascorbic Acid	2:1	3670.00

**NA: not applicable*
**ChCl: choline chloride*

3.3. Fourier transform infrared spectroscopy (FTIR) analysis

Figure 1 shows the FTIR analysis of a) choline chloride, b) ascorbic acid and c) NADES of choline chloride and ascorbic acid at 1:2 molar ratio. The spectra of choline chloride show a very strong band at 3223 cm^{-1} that indicates intramolecular hydrogen bond since both hydrogen donor and hydrogen acceptor is present in the compound [12]. The weak peak at 2845 cm^{-1} also proves the presence of hydrogen bond. The peak at 1482 cm^{-1} refers to the alkyl group that is present in choline chloride [13]. In figure 1 b), the peaks at 3524 cm^{-1} and 3407 cm^{-1} shows the OH stretching that is present in the ascorbic acid. Followed by the peak at 2916 cm^{-1} for C=C stretching and peak at 3032 cm^{-1} for sp2 C-H stretching. The band at 1754 cm^{-1} and 1672 cm^{-1} corresponds to C=O stretching and C=C stretching which are in line with ascorbic acid properties. From figure 1 c), it shows that NADES is an overlap of choline chloride and ascorbic acid. Changes of vibrations on wavenumbers correspond to the change in bond length which can depict the establishment of hydrogen bond [14]. The OH stretching from choline chloride and ascorbic acid broadens and widens in DES at 3344 cm^{-1}, the shift indicates the formation of hydrogen bond between the two components. The peaks at 2883 cm^{-1} and 2774 cm^{-1} shows the presence of C-H stretching and the many peaks at 1709 cm^{-1} to 1787 cm^{-1} shows the presence of C=O stretching.

Figure 2 shows FTIR analysis between extract from coconut shell using a) water at pH 3 and b) NADES. Between the two spectra, coconut shell extracted with water (CSW) has fewer peaks than coconut shell extracted with NADES (CSNADES). This is due to the lower amount of different functional groups detected in the extract. Meanwhile, the NADES extracted more compounds from a wider range of functional groups. In figure 2 a), the peak at 3304 cm^{-1} indicates OH stretching. Since this extract used water as a solvent, large OH stretching may also cause noise and inaccuracy to detect other functional groups. Apart from that, the peak at 1077 cm^{-1} corresponds to C-O stretching and peak at 1651 cm^{-1} indicates C=O vibration. Not many peaks are enough to make deductions on functional groups present in the extract. Nonetheless, figure 2 b) shows more characteristic peaks. The OH stretching is evident between peaks 3404 cm^{-1} and 3523 cm^{-1}. The peak at 2915 cm^{-1} indicates a C-H stretch and peaks between 1388 cm^{-1} to 1473 cm^{-1} corresponds to the sp3 C-H band. A C=C stretch is proven at 1674 cm^{-1}. All of these peaks help to indicate the presence of phenol in the extract which consists of an aromatic ring with a hydroxyl group.

IOP Conf. Series: Materials Science and Engineering 1192 (2021) 012001 doi:10.1088/1757-899X/1192/1/012001

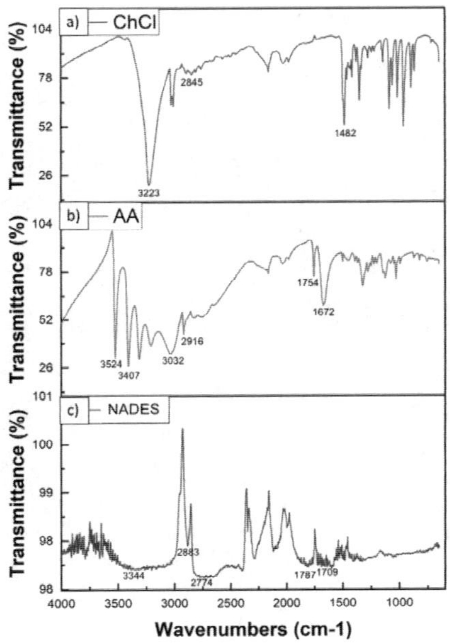

Figure 1. FTIR analysis of a) choline chloride, b) ascorbic acid, and c) NADES.

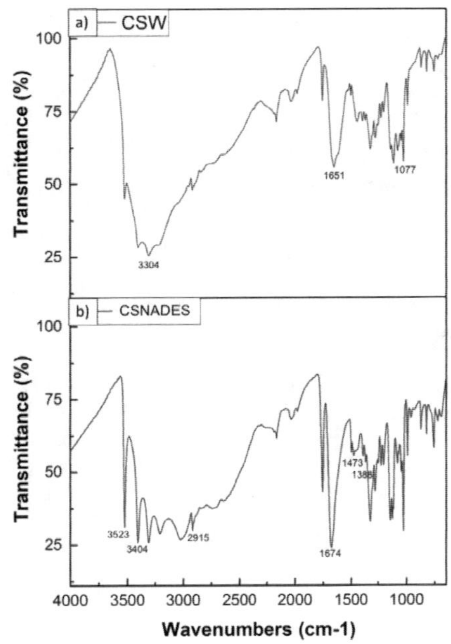

Figure 2. FTIR analysis of extracted coconut shell using a) water at pH 3 and b) NADES.

3.4. Gas chromatography-mass spectrometry (GC-MS) analysis

GC-MS analysis was used to detect and compare the compounds extracted from coconut shell using water and NADES. Table 3 shows the compounds present in the extract after going through liquid-liquid extraction with water at pH 3 and NADES. A visual representation is shown in Figure 3. From the table, a total of 26 different compounds from varying chemical groups were detected. From the compounds present in both extracts, carotenoids were more prominent among the other chemical groups. Lycoxanthin, astaxanthin, lycopene, and α-Carotene have dietary benefits, helps reduces blood pressure, and positively affect neurodegenerative diseases [15]. Alkaloids, acids, esters, aldehydes, and ketones with similar antioxidant properties were also present within the extracts.

For liquid-liquid extraction, only one phenolic compound is present, at peak number 13, the 2-Nonaprenyl-6-methoxyphenol. In contrast, NADES was able to extract four phenolic compounds from coconut shell at peak numbers 2, 13, 22, and 25 which are the 1,2,4-Benzenetricarboxylic acid, 1,2-dimethyl ester, the 2-Nonaprenyl-6-methoxyphenol, the Trimethyl[4-(1,1,3,3,-tetramethylbutyl)phenoxy]silane, and the 2',6'-Bis(trimethylsiloxy)acetophenone respectively. Both extraction methods were able to extract a good amount of antioxidants that were trapped within the biomass waste. Nonetheless, NADES was more superior in extracting phenolic compounds than water. This is due to the difference in polarity between water and NADES with phenolic compounds. Even though water is a green solvent that can be utilized for extraction, the hydrogen bond network in water is easily disintegrated with high temperature and pressure [16]. NADES, utilizing the hydrogen bond between choline chloride and ascorbic acid, helps to improve the extraction of phenolic compounds from coconut shell.

Table 3. List of compounds detected in coconut shell extract using GC-MS.

	Liquid-liquid extraction method with water at pH3		Natural deep eutectic solvent extraction method	
No	Compound Name	R.T. (s)	Compound Name	R.T. (s)
1	1-Methyl-3-phenylindole	3351.33	Tungsten, pentacarbonyl(4,5-diethyl-2,2,3-trimethyl-1-phenyl-1-phospha-2-sila-5-boracyclohex-3-ene-P1)-, (oc-6-22)-	180.232
2	1,2,4-Benzenetricarboxylic acid, 1,2-dimethyl ester	3429.75	1,2,4-Benzenetricarboxylic acid, 1,2-dimethyl ester	3397.45
3	D-Glucopyranosiduronic acid, 3-(5-ethylhexahydro-1,3-dimethyl-2,4,6-trioxo-5-pyrimidinyl)-1-methylbutyl 2,3,4-tris-O-(trimethylsilyl)-, methyl ester	3451.79	1H,15H-Hexadecamethyloctasiloxane	3450.86
4	2,4,6-Cycloheptatrien-1-one, 3,5-bis-trimethylsilyl-	3464.34	2,2-Bis[4-[(4,6-dichloro-1,3,5-triazin-2-yl)oxy]phenyl]-1,1,1,3,3,3-hexafluoropropane	3480.31
5	Decaethylene glycol, bis(pentafluoropropionate)	3474.14	Lycoxanthin	3488.45
6	Zeaxanthin	3498.24	Pregnane-11,20-dione, 3,17,21-tris[(trimethylsilyl)oxy]-, 20-[O-(phenylmethyl)oxime], (3α,5α)-	3497.71
7	Rhodovibrin	3521.91	5-Aminododecahydro-as-indacen-4-ol	3543.59
8	2-(5-(5-[Cyano-(9,9-dimethyl-1,4-dioxa-7-aza-spiro[4.4]non-7-en-8-yl)-methylene]-3,3-dimethylpyrrolidin-2-ylidenemethyl)-3,3-dimethyl-ё1-pyrrolin-5-ylidenemethyl-4,4,5-trimethyl-ё1-pyrroline-5-carbonitrile]	3584.83	3-Pyridinecarboxylic acid, 2,7,10-tris(acetyloxy)-1,1a,2,3,4,6,7,10,11,11a-decahydro-1,1,3,6,9-pentamethyl-4-oxo-4a,7a-epoxy-5H-cyclopenta[a]cyclopropa[f]cycloundecen-11-yl ester, [1aR-(1aR*,2R*,3S*,4aR*,6S*,7S*,7aS*,8E,10R*,11R*,11aS*)]-	3544.69
9	Lycoxanthin	3594.89	Tris(cyclopentadienyl-cobalt)-hexapropenylbenzene	3576.96
10	Cyclotrisiloxane, hexamethyl-	3595.12	5H-Indeno[1,2-b]pyrazin-5-one, 7,8,9-tribromo-N,N'-diethyl-1,2,3,4-tetrahydro-6-(phenoxycarbonyl)-	3591.53
11	α-Carotene	3599.8	2α,4a-Epoxymethylphenanthrene-7-methanol, 1,1-dimethyl-2-methoxy-8-(1,3-dithiin-2-ylidene)methyl-1,2,3,4,4a,4b,5,6,7,8,8a,9-dodecahydro-, acetate	3597.18
12	2-Fluoro-5-trifluoromethylbenzoic acid, ethyl ester	3601.59	Dimethoxylycopene	3617.3
13	2-Nonaprenyl-6-methoxyphenol	3609.56	2-Nonaprenyl-6-methoxyphenol	3653.35
14	Arsenous acid, tris(trimethylsilyl) ester	3699.27	Pregn-4-ene-3,11,20-trione, 6,17,21-tris[(trimethylsilyl)oxy]-, 3,20-bis(O-methyloxime), (6α)-	3657.47
15	Rhodoviolascin	3709.56	Spheroidenone	3682.4
16	Silane, [[(3α,5α,11α,20S)-pregnane-3,11,17,20,21-pentayl]pentakis(oxy)]pentakis[trimethyl-	3726.86	Lycopene	3685.69
17	Astaxanthin	3749.56	2,2-Bis[4-[[4-chloro-6-(3-ethynylphenoxy)-1,3,5-triazin-2-yl]oxy]phenyl]propane	3688.87
18	4-Methyl-2,4-bis(4'-trimethylsilyloxyphenyl)pentene-1	3754.18	Milbemycin B, 5-demethoxy-5-one-6,28-anhydro-25-ethyl-4-methyl-13-chloro-oxime	3693.46
19	2,4-Imidazolidinedione, 5-[3,4-bis[(trimethylsilyl)oxy]phenyl]-3-methyl-5-phenyl-1-(trimethylsilyl)-	3755.57	1,N-Dimethyl-N-propargyl-2-phenylethylamine	3698.77
20	1,3-Dichloro-1,3-bis(norbornadien-2-yl)-1,3-bis(3-trimethylsilylpropyl)disiloxane	3756.57	2,4-Imidazolidinedione, 5-[3,4-bis[(trimethylsilyl)oxy]phenyl]-3-methyl-5-phenyl-1-(trimethylsilyl)-	3710.65
21	Tetrahydrospirilloxanthin	3758.73	7aH-Cyclopenta[a]cyclopropa[f]cycloundecene-2,4,7,7a,10,11-hexol, 1,1a,2,3,4,4a,5,6,7,10,11,11a-dodecahydro-1,1,3,6,9-pentamethyl-, 2,4,7,10,11-pentaacetate	3712.91
22	1,4-Dibromo-2-butene	3774.27	Trimethyl[4-(1,1,3,3,-tetramethylbutyl)phenoxy]silane	3720.95
23	2,2-Bis[4-[[4-chloro-6-(3-ethynylphenoxy)-1,3,5-triazin-2-yl]oxy]phenyl]propane	3791.46	Dithiocarbonic acid, S-(2,3-diphenyl-1-p-tolyl-cycloprop-2-enyl)ester-O-ethyl ester	3721.64
24	Pregn-4-ene-3,11,20-trione, 6,17,21-tris[(trimethylsilyl)oxy]-, 3,20-bis(O-methyloxime), (6α)-	3792.62	Prosta-5,13-dien-1-oic acid, 9,11,15-tris[(trimethylsilyl)oxy]-, trimethylsilyl ester, (5Z,9α,11α,13E,15S)-	3756.77
25	Milbemycin B, 5-demethoxy-5-one-6,28-anhydro-25-ethyl-4-methyl-13-chloro-oxime	3818.85	2',6'-Bis(trimethylsiloxy)acetophenone	3759.26
26	1-Methyl-2-phenylbenzimidazole	3820.75	Astaxanthin	3814.44

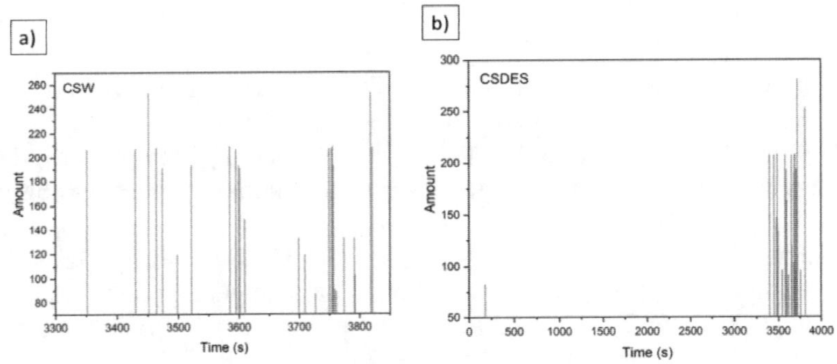

Figure 3. GC-MS analysis of extracted coconut shell using a) water at pH 3 and b) NADES.

4. Conclusion

This study has been successful in screening the best biomass waste and the best NADES combinations with the highest total phenol. It can be concluded that the coconut shell consists of the highest phenol among the biomass waste and that choline chloride with ascorbic acid at 1:2 molar ratio has the highest phenol extraction capacity. The use of NADES improves the phenolic extraction as compared to using water at pH 3. This is due to the hydrogen bonding between the components and the use of ascorbic acid that positively affects the phenol amount.

Acknowledgments

The author would like to thank the Ministry of Higher Education Malaysia for granted the fund for this research under the Fundamental Research Grant Scheme, FRGS19-105-0714 and to International Islamic University for research facilities.

References

[1] Vu H T, Scarlett C J and Vuong Q V 2017 Phenolic compounds within banana peel and their potential uses: A review *J. Funct. Foods* **40** pp 238–248

[2] Dai J and Mumper R J 2010 Plant Phenolics: Extraction, Analysis and Their Antioxidant and Anticancer Properties *Molecules* **15** pp 7313–7352

[3] Salgado P, Márquez K, Rubilar O, Contreras D and Vidal G 2019 The effect of phenolic compounds on the green synthesis of iron nanoparticles (Fe_x0_y-NPs) with photocatalytic activity *Appl. Nanosci.* **9** pp 371–385

[4] Paiva A, Craveiro R, Aroso I, Martins M, Reis R L and Duarte A R C 2014 Natural deep eutectic solvents - Solvents for the 21st century *ACS Sustain. Chem. Eng.* **2** pp 1063–1071

[5] Smith E L, Abbott A P, and Ryder K S 2014 Deep Eutectic Solvents (DESs) and Their Applications *Chem. Rev.* **114** pp 11060–11082

[6] Cunha S C and Fernandes J O 2018 Extraction techniques with deep eutectic solvents," *TrAC - Trends Anal. Chem.* **105** pp 225–239

[7] Waterhouse A L 2002 Determination of Total Phenolics *Curr. Protoc. Food Anal. Chem.* **6** p I1.1.1-I1.1.8

[8] Dai Y, van Spronsen J, Witkamp G J, Verpoorte R and Choi Y H 2013 Natural deep eutectic solvents as new potential media for green technology *Anal. Chim. Acta* **766** pp 61–68

[9] Zhong W J, Wang D H, Xu X W, Wang B Y, Luo Q, Kumaran S S and Wang Z J 2011 A gas chromatography/mass spectrometry method for the simultaneous analysis of 50 phenols in wastewater using deconvolution technology *Chinese Sci. Bull.* **56** pp 275–284

[10] Rodrigues S, Pinto G A S and Fernandes F A N 2008 Optimization of ultrasound extraction of phenolic compounds from coconut (Cocos nucifera) shell powder by response surface methodology *Ultrason. Sonochem.* **15** pp 95–100

[11] Kolniak-Ostek J, Oszmiański J and Wojdyło A 2013 Effect of l-ascorbic acid addition on quality, polyphenolic compounds and antioxidant capacity of cloudy apple juices *Eur. Food Res.*

Technol. **236** pp 777–798
[12] Troter D Z , Zlatkovic M Z, Dordevic B S, Dokic-Stojavonic D R, Konstantinovic S S, Todorovic Z V and Veljkovic V B 2016 Preparation and FTIR Characterization of Choline Chloride-based Deep Eutectic Solvents with Thiourea and Dimethylurea *Phys. Chem.* **1**
[13] Ling J K U, Chan Y S, Nandong J, Chin S F and. Ho B K 2020 Formulation of choline chloride/ascorbic acid natural deep eutectic solvent: Characterization, solubilization capacity and antioxidant property *Lwt* **133** p 110096
[14] Gonjo T, Futami Y, Morisawa Y, Wojcik M J and Ozaki Y 2011 Hydrogen bonding effects on the wavenumbers and absorption intensities of the OH fundamental and the first, second, and third overtones of phenol and 2,6-dihalogenated phenols studied by visible/near-infrared/infrared spectroscopy *J. Phys. Chem. A* **115** pp 9845–9853
[15] Przybylska S 2019 Lycopene – a bioactive carotenoid offering multiple health benefits: a review *Int. J. Food Sci. Technol.* **55** pp 11-32
[16] Chemat F, Vian M A, Ravi H K, Khadraoui B, Hilali S, Perino S, Tixier A F 2019 Review of Alternative Solvents for Green Extraction of Food and Natural Products: Panorama, Principles, Applications and Prospects *Molecules* **24** pp 1–26

IOP Conf. Series: Materials Science and Engineering 1192 (2021) 012002 doi:10.1088/1757-899X/1192/1/012002

The effect of chemically treated all-cellulose composites (ACCs) with dodecyltriethoxysilane (DTES) solution on the structural-property relationship

M. Mat Salleh[1, 2], M. F. Mohd Fauzi[1], S. S. Md Nor [1]

[1]Faculty of Mechanical Engineering Technology, Universiti Malaysia Perlis, Kampus Pauh Putra, 02600 Arau, Perlis, Malaysia.

[2]Corresponding author's e-mail: mokhtarms@unimap.edu.my

Abstract. Cellulose is the most abundant natural polymer on the Earth that is widely used in bio-based composites due to its high mechanical properties, availability and biodegradability. All-cellulose composites (ACCs) are known as a new class monocomponent of biocomposites due to both reinforcing and matrix phases that are based on cellulose. However, a technical challenge for ACCs is observed due to their propensity for high moisture absorption (water uptake), leading to the instability and deterioration of the mechanical properties. Therefore, this research focussed towards the improvement of the surface of ACCs in order to increase the resistance to water absorption. Prior to the characterisations, ACCs were chemically treated using dodecytriethoxysilane (DTES) coating solution by dip coating method. In this present study, the effects of two control factors: (i) DTES concentration (1.5, 7.5, and 12.5 vol%), and (ii) heating temperature (50, and 100 °C), were investigated on the ACCs. Upon completion of this treatment, three possible characterisations were conducted including of Fourier Transform Infrared (FTIR) spectroscopy analysis, scanning electron microscopy (SEM), and water absorption (WA) testing. Creation of polysiloxane layer was expected to reduce the tendency to absorb water in ACCs while being applied in the outdoor applications.

1. Introduction

Cellulose was the most abundant natural polymer on the Earth that was widely used in bio-based composites due to its high mechanical properties, availability and biodegradability [1, 2]. All-cellulose composites (ACCs) materials composed of a cellulose for the reinforcement and matrix phases. The concept of using the same polymer for both phases was aimed at facilitating the recyclability and improving mechanical performance of the final composite [3]. Nishino *et al.* first reported on the terminology, processing and mechanical properties of ACCs [4]. Gindl *et al.* reported that ACCs may be manufactured by the partial (or selective) dissolution of cellulose I with a suitable solvent, where the portion of dissolved cellulose was then subsequently regenerated (*via* solvent exchange with a coagulant) to form a cellulose II matrix phase *in-situ* [5]. The advantages of ACCs had been reported previously, including of its superior mechanical properties and more rapid biodegradability as compared to other bio-based composites [6, 7]. Previous research also showed that ACCs were intended to overcome the problem with chemical incompatibility due to the use of cellulose, for both phases, with the expectation of producing of interfaceless ACCs [8].

However, a technical challenge for ACCs was observed due to their propensity for high water absorption, leading to the instability and deterioration of the mechanical properties. These results had been shown by the previous studies of other bio-based composite materials [9, 10]. It happened due to the hydrophilic behaviour of cellulose which was attributed by hydroxyl (OH) groups that are located on the surface of cellulose. The OH groups tended to bind with water molecules which rendered to a higher water absorption. Therefore, ACCs surfaces had to be chemically treated in order to prevent the attachment of water molecules with the OH groups.

Several approaches had been studied in order to reduce the hydrophilic character of the cellulose surface. Previous studies reported that the use of fluorine-based or alkyl-based silane coupling agents was found to be efficient due to their high water resistance and ability to couple both hydrophilic and hydrophobic groups, which had been widely applied in various fields [11, 12]. Based on the previous study by Xie *et al.*, silanes were classified as extensively used coupling agents in composites and adhesive formulations [11]. The advantages of silanes included of the large-scale commercial availability with several chemical functional groups, and the ability to couple with any polymer or mineral in composites. The organofunctional group was responsible for bonding with any polymer through a covalent reaction with other phases. In addition, silane coupling agents were also used to enhance the compatibility and dispersion of nanocellulose in the polymer matrix [13]. These silane coatings also had excellent compatibility and high adhesion to the wide range of inorganic and organic interfaces [14, 15]. The silane compounds used for coating applications contain at least one non-hydrolysable functional group (amino, epoxy, etc) that chemically interacted with overlying organic coating and also hydrolysable portions that were responsible for forming covalent bonds with the substrate surface. Besides, a hybrid silane coating's primary protective mechanism was to build a barrier siloxane network, which was obtained by condensing silanol (Si-OH) groups [16–18].

In this present study, DTES was used to impart water repellency to ACCs. Upon completion of this treatment, the structural-property relationship of chemically-treated ACCs was examined with FTIR spectroscopy, SEM, and WA testing. To the best of our knowledge, no report related to the chemically-treated ACCs produced via sodium hydroxide (NaOH)/urea with DTES had been published previously.

2. Experimental procedures

A rayon fiber textile (Cordenka GmbH, Obernburg Germany) in the form of a K2/2 twill weave was used. The textile was based on a multifilament yarn (Cordenka 700, 1840 dtex, f1000) with a filament diameter of 12 μm and final areal weight of 450 g/m^2. The NaOH, urea, DTES, and ethanol were commercially available and used without further purification.

The solvent was prepared by mixing 7/12/81 wt.% of NaOH/urea/distilled water in a glass beaker. The mixture is stirred by using magnetic stirrer at room temperature until clear. The solvent is referred to as NaOH/urea.

The procedures and solvent infusion processing were used to manufacture ACCs produce *via* NaOH/urea as described by Jan *et al.* [1]. The rayon textile was cut in strip (120 mm x 160 mm) and conditioned at 20% relative humidity (RH) for 24 h. Then, a layer of conditioned textile was placed accordingly on flat plate and sealed with vacuum bag. This sealed conditioned textile was connected to vacuum infusion setup with an inlet and outlet valve to control the flow of solvent. Then, infused setup was placed in 101 cooling baths at -12 °C for designated dissolution time. The partially dissolved cellulose was pressed at 1 MPa for 15 min at 21 ± 2 °C. Then, the partially dissolved cellulose was regenerated and washed in 500 ml bath of distilled water with at least 72 h (two times water change daily). Finally, the regenerated cellulose was dried in a vacuum furnace at pressure of 0.2 MPa and a temperature of 60 °C. This sample is called as ACCs produced *via* NaOH/urea. The sample is placed in a sealed plastic bag prior to the characterisations.

The procedures to the chemically treated ACCs with DTES were followed, as described by Yousefi *et al.*[12]. A mixture solution of DTES and the ratio 80:20 of ethanol/ distilled water was prepared by vigorous stirring for 2 h. The ACCs specimen was immersed into three different concentrations of

DTES (1.5, 7.5, and 12.5 vol%) using dip coating method. The ACCs specimen was immersed in DTES solution for 1 h. Then, the samples were dried at room temperature, followed by further heating temperature of 50, and 100 °C for 1 h each different temperature. All treated samples were stored in sealed plastic bag prior to the characterisations.

Both treated and untreated ACCs were characterized using a Perkin Elmer (Waltham, MA, USA) FT-IR spectrometer Frontier with universal ATR sampling accessory. Wavenumbers ranging from 4000 to 650 cm^{-1} were scanned 24 times at 0.4 cm^{-1} resolution. Microscopy samples were cut from ACCs laminae into a rectangular shape with dimensions of (10 (l) × 10 (w) mm). The specimen was firstly dried under 0.2 MPa external pressure in vacuum state at 60 °C temperature. Then the sample surface was sputtered-coated with gold film at 25 mA for 180 s. Gold sputter-coated (180 s, 25 mA) samples were imaged using a TM3000 scanning electron microscopy (SEM) at an accelerating voltage of 5 kV.

Water absorption testing were conducted in accordance with ASTM D570-98 in which the specimens were immersed in water for 7 days at a room temperature. The weight gain was then measured until the samples attained equilibrium. The water absorption was measured as the weight gain over the oven-dry of samples, and calculated by using the following equation [10]:

$$WA\ (\%) = \frac{m_t - m_o}{m_o} \times 100 \quad (1)$$

where m_o and m_t are the oven-dry weight and weight after time, respectively. The equilibrium moisture content (*EMC*) of the specimens is the moisture content when the daily weight change of sample was less than 0.01% and thus to be reached the equilibrium state.

3. Results and discussion

The effect of various DTES concentrations (1.5, 7.5, and 12.5 vol%) and heating temperatures (50 and 100 °C) on ACCs were presented. These results were examined using several characterisations, including of FTIR, SEM, and WA testing.

The FTIR spectrums of untreated and treated ACCs with various DTES concentration are presented in Figure 1. The untreated ACCs shows several typical characteristic bands for cellulose, including of the hydrogen bonded OH stretching, CH stretching, OH bending of absorbed water, CH bending, CH_2 stretching, and OH out of plane bending at 3420.7, 2900, 1632.3, 1371, 900, and 791.68 cm^{-1}, respectively (Figure 1). The spectrum for untreated ACCs is labelled as raw ACCs, as shown in Figure 1. The observation of the pristine cellulose characteristic band in untreated ACCs is expected due to no changes occur in the ACCs structure prior to the DTES treatment.

The narrow range of FTIR spectrum is carried out in order to clearly observe the presence of silane on the cellulose structure following the DTES treatment. It was observed that the strong peak at 1061.5 cm^{-1} in 1.5 vol% DTES (50 °C), 1059.3 cm^{-1} in 1.5 vol% DTES (100 °C), 1077 cm^{-1} in 7.5 vol% DTES (100 °C), and peak at 1077 cm^{-1} in 7.5 vol% DTES (50 °C), 12.5 vol% DTES (50 °C), and 12.5 vol% DTES (100 °C) was attributed to the Si-O-C bands. This result indicated that the silane was completely attached with the carbon group in the cellulose structure, resulting in a formation of covalent bond in treated ACCs. The Si-O-C peaks for the treated ACCs with 7.5 vol% DTES (50 °C) and 12.5 vol% DTES (100 °C) showed a significant increase in their intensity due to the increasing DTES concentration. It was exhibited by the narrower and sharper peak in the wavenumber range of 1077-1078 cm^{-1}. This finding indicated that the silane has successfully attached to the carbon component in ACCs structure, as aforementioned previously. Meanwhile, for Si-O-Si peak, the peak appears after the DTES treatment in ACCs at 1100 cm^{-1}, resulting in a formation of polysiloxane layer on ACCs surface structure. However, these strong two bands that occurred between regions 1000-1200 cm^{-1} make it difficult to completely assigned Si-O-Si (siloxane) and Si-O-C band which might be due to the overlapping of Si-O-Si band and C-O stretching of cellulose.

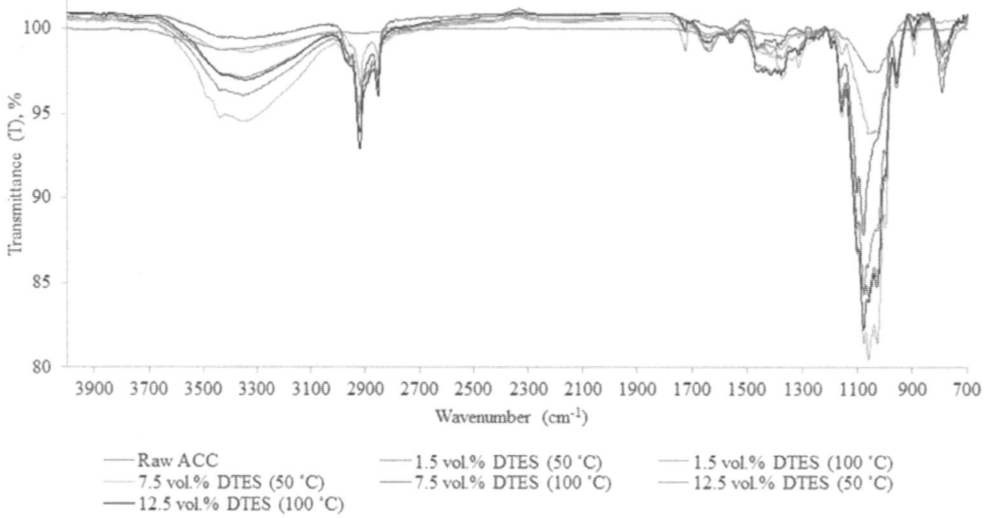

Figure 1. FTIR spectrum for untreated ACCs and treated ACCs.

Besides, the effect of heating temperature on the formation of Si-O-C was considerably varied for each DTES concentration. It was observed that the heating temperature of 50°C produced a broader peak as compared to heating temperature of 100°C at DTES concentration up to 7.5 vol%. However, the sharper peaks were found at DTES concentration of 12.5 vol% for both temperatures, as compared to the others. These results showed that the increasing Si-O-C formation was evidenced with the increasing DTES concentration rather than that of the heating temperature.

Figure 2 show the SEM images of surface for treated and untreated ACCs with DTES. Based on Figure 2 (a), the surface for untreated ACCs was considerably smooth, and large gap between the fiber and matrix phases was also observed. The large gap between fiber and matrix indicated that the poor interfacial bonding in ACCs structure, resulting in the existence of the intermolecular crack. This may be due to the insufficient pressure at the pressing machine, leading to the poor arrangement of fiber and matrix phases in ACCs. The untreated ACCs also exhibited a large void in the cross-sectional view. These voids were produced due to the small volume of air that trapped during the SIP of ACCs laminates, as also discussed by Jan *et al.* [1].

For the treated ACCs with DTES which are presented in Figure 2 (b-g), it was observed that the surface of ACCs significantly changed after the silane treatment, resulting in a rougher surface of treated ACCs, as compared to the untreated ACCs. This morphology changes can be attributed to the removal of impurities on the surface of ACCs after the formation of silane treatment. In addition, the formation of thin silane layer and granular-like structure on the ACCs surface were evidenced upon completion of DTES treatment. However, a larger void was not completely removed after the silane treatment, whereas a thin layer of DTES was observed due to the presence of a polysiloxane layer, as presented by the FTIR analysis.

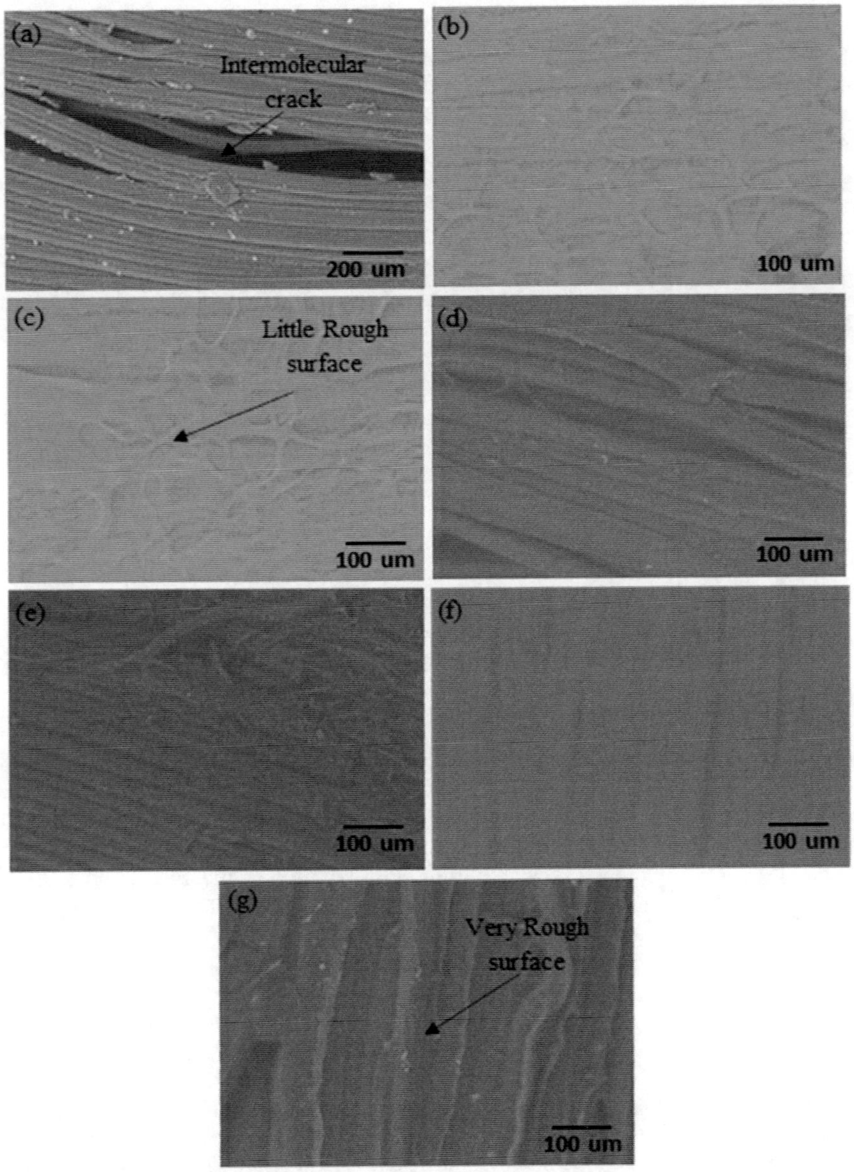

Figure 2. SEM images for (a) untreated and (b-g) treated ACCs (surface view).

IOP Conf. Series: Materials Science and Engineering 1192 (2021) 012002 doi:10.1088/1757-899X/1192/1/012002

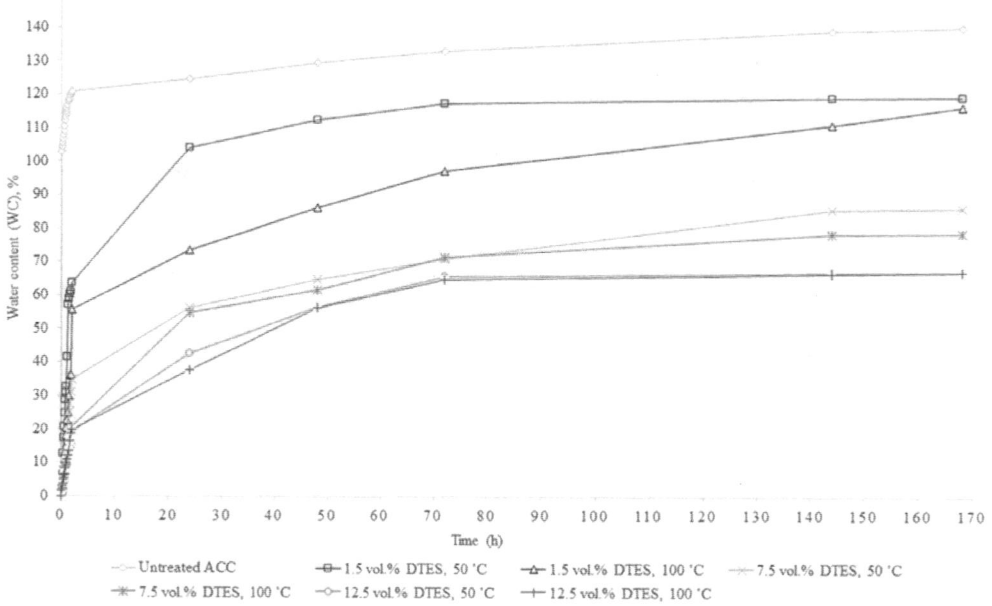

Figure 3 Water absorption results for untreated and treated ACCs.

Figure 3 demonstrates the effect of untreated and treated ACCs with DTES concentration of 1.5, 7.5, and 12.5 vol% and heating temperatures of 50 and 100 °C of the WA. It is observed that the WA increased with immersion time until equilibrium states were reached. The WA of untreated ACCs increased rapidly with immersion time in the initial state. This result was expected due to the hydrophilicity of ACCs that consisted of cellulose for both fiber and matrix phases. It has been reported that the numerous accessible hydroxyl groups in cellulose structure led to the binding of water molecules with the hydroxyl groups [9, 10]. Thus, a higher WA was significantly found in untreated ACCs. In addition, the increasing of WA for untreated ACCs also may be also attributed due to the occurrence of voids and microcracks in ACCs. A higher void content and microcracks of untreated ACCs has resulted in a higher water absorption within the more confined free space of the microstructure defects.

For the DTES treated ACCs, the WA decreased significantly due to the OH-rich surface that was covered by a layer of polysiloxane, thus preventing the water penetration into the cellulose structures. By increasing the DTES concentration, the decreasing water absorption was observed in the treated ACCs. The maximum of WA after 7 days of the testing for the untreated ACCs was 140 % It was followed by the treated ACCs of 1.5 vol% DTES (50 °C) and 1.5 vol% DTES (100 °C) with WA of 119.58%, and 116.67%, respectively. It was found that the maximum WA exceeded 100% for untreated and treated ACCs with 1.5 vol% DTES (50 °C) and 1.5 vol% DTES (100 °C). This is probably due to the larger void that was existed in untreated ACCs and insufficient amount of DTES concentration for both treated ACCs with 1.5 vol% DTES at 50 and 100 °C.

Apart from that, a significant decrease in water absorption was observed for treated ACCs with 7.5, and 12.5 vol% DTES. The consumption of sufficient DTES can enhance the bonding in ACCs, resulting in the formation of multilayer silanol groups that led to the existence of polysiloxane layer on the ACCs surface. The maximum WA for treated ACCs with 12.5 vol% DTES (50 °C) was 67.11%. Meanwhile, it was reduced to 67.24% for 12.5 vol% DTES (100 °C). It can be concluded that a higher

DTES concentration resulted in a lower the percentages of water absorption. However, the effect of heating temperature on the WA seemed to be comparable in each DTES concentration. It is difficult to observe any significant changes of various heating temperature on WA. Therefore, it could be concluded that the heating temperature has little effect on the WA at specified DTES concentration.

4. Conclusion

As a conclusion, ACCs were successfully treated with DTES in this present study. Several findings can be drawn as discussed below.

FTIR spectroscopy was used to examine the peak of functional group that appeared through this silane treatment which was expected in the formation of covalent bonds. The presence of silane functional groups (Si-O-C bonds) on the ACCs was observed at the range 1059.3-1078 cm^{-1}. The silanol groups was formed a multilayer on ACCs, thus the formation of rigid polysiloxane layer on the surface. The increasing DTES concentration showed a significant increase in the peak associated with the Si-O-C functional group, representing by a narrower and sharper peak at the specified wavenumber 1077-1078 cm^{-1}. This result indicated that the presence of silane was successfully attached on ACCs surface structure, thus preventing the water absorption into the ACCs structures.

SEM images showed that the microstructural analysis of untreated and treated ACCs with DTES. The surface of ACCs that treated with 12.5 vol% DTES concentration produced rougher surface structures as compared to the others. In addition, the presence of DTES were also evidenced by the observation of granule-like structure and thin layer of DTES on ACCs structures.

WA results exhibited that the treated ACCs with 12.5 vol% DTES provided a lower percentage of water absorption, as compared to the others with the increasing DTES concentration. This finding showed that the water barrier was formed upon the completion of DTES surface treatment, indicating the successful implementation of coating procedures in this present study.

The effect of DTES concentration was rather significant as compared to the effect of heating temperature. It was showed by an improvement in reducing the WA rather than that of heating temperature. However, several future works are required in order to examine the hydrophobic behaviour of treated ACCs in the future.

Acknowledgment

The authors are grateful to technical staffs and for technical assistance at Universiti Malaysia Perlis. One of the authors (MMS) gratefully acknowledges financial support, FRGS/1/2019/TK10/UNIMAP/02/3, provided by the Ministry of Higher Education Malaysia (MOHE) and Universiti Malaysia Perlis (UniMAP).

References

[1] Faruk O, Bledzki AK, Fink HP, et al. Biocomposites reinforced with natural fibers: 2000-2010. *Prog Polym Sci* 2012; 37: 1552–1596.

[2] Huber T, Müssig J, Curnow O, et al. A critical review of all-cellulose composites. *J Mater Sci* 2012; 47: 1171–1186.

[3] Nishino T, Matsuda I, Hirao K. All-cellulose composite. *Macromolecules* 2004; 37: 7683–7687.

[4] Qin C, Soykeabkaew N, Xiuyuan N, et al. The effect of fibre volume fraction and mercerization on the properties of all-cellulose composites. *Carbohydr Polym* 2008; 71: 458–467.

[5] Huber T, Bickerton S, Müssig J, et al. Solvent infusion processing of all-cellulose composite materials. *Carbohydr Polym* 2012; 90: 730–733.

[6] Kalka S, Huber T, Steinberg J, et al. Biodegradability of all-cellulose composite laminates. *Compos Part A Appl Sci Manuf* 2014; 59: 37–44.

[7] Mohanty AK, Misra M, Dreal LT. Surface modifications of natural fibres and peformance of

the resulting biocomposite. *Compos Interfaces* 2001; 8: 313–343.
[8] Dormanns JW, Schuermann J, Müssig J, et al. Solvent infusion processing of all-cellulose composite laminates using an aqueous NaOH/urea solvent system. *Compos Part A Appl Sci Manuf* 2016; 82: 130–140.
[9] Espert A, Vilaplana F, Karlsson S. Comparison of water absorption in natural cellulosic fibres from wood and one-year crops in polypropylene composites and its influence on their mechanical properties. *Compos Part A Appl Sci Manuf* 2004; 35: 1267–1276.
[10] Adhikary KB, Pang S, Staiger MP. Dimensional stability and mechanical behaviour of wood-plastic composites based on recycled and virgin high-density polyethylene (HDPE). *Compos Part B Eng* 2008; 39: 807–815.
[11] Xie Y, Hill CAS, Xiao Z, et al. Silane coupling agents used for natural fiber/polymer composites: A review. *Compos Part A Appl Sci Manuf* 2010; 41: 806–819.
[12] Yousefi H, Nishino T, Shakeri A, et al. Water-repellent all-cellulose nanocomposite using silane coupling treatment. *J Adhes Sci Technol* 2013; 27: 1324–1334.
[13] Sheltami RM, Kargarzadeh H, Abdullah I. Effects of silane surface treatment of cellulose nanocrystals on the tensile properties of cellulose-polyvinyl chloride nanocomposite. *Sains Malaysiana* 2015; 44: 801–810.
[14] Alibakhshi E, Ghasemi E, Mahdavian M, et al. Active corrosion protection of Mg-Al-PO43− LDH nanoparticle in silane primer coated with epoxy on mild steel. *J Taiwan Inst Chem Eng* 2017; 75: 248–262.
[15] Kong G, Lu J, Zhang S, et al. A comparative study of molybdate/silane composite films on galvanized steel with different treatment processes. *Surf Coatings Technol* 2010; 205: 545–550.
[16] Palomino LEM, Suegama PH, Aoki I V., et al. Investigation of the corrosion behaviour of a bilayer cerium-silane pre-treatment on Al 2024-T3 in 0.1 M NaCl. *Electrochim Acta* 2007; 52: 7496–7505.
[17] Chakraborty Banerjee P, Singh Raman RK. Electrochemical impedance spectroscopic investigation of the role of alkaline pre-treatment in corrosion resistance of a silane coating on magnesium alloy, ZE41. *Electrochim Acta* 2011; 56: 3790–3798.
[18] Palomino LM, Suegama PH, Aoki I V., et al. Electrochemical study of modified cerium-silane bi-layer on Al alloy 2024-T3. *Corros Sci* 2009; 51: 1238–1250.

Statistical Optimization of Adsorption Processes for The Removal of Phenol by Activated Carbon Derived from Baobab Fruit Shell

R Nedjai[1], N A Kabbashi[1,2], M Z Alam[1], and M F R Al-Khatib[1]

[1]Department of Biotechnology Engineering, Faculty of Engineering, International Islamic University Malaysia, Kuala Lumpur, Malaysia.

[2]Corresponding Author's E-mail: nasreldin@iium.edu.my

Abstract. Baobab fruit shell (BFS) biomass was used as an alternative precursor for producing high surface area and microporous baobab fruit shell activated carbon (BFS-AC) by chemical activation using KOH. Scanning electron microscope (SEM) was performed for the characterization of baobab fruit shell activated carbon. The adsorption property of BFS-AC for the removal of phenol from aqueous solution was evaluated. The effect of key adsorption parameters such as the contact time (10-20 min), BFS-AC dose (2.5-3.5 g/L), pH (1-3), and agitation speed (150-250 rpm) were optimized using a response surface methodology (RSM) with faced centered central composite design (FCCCD). Consequently, a maximum adsorption capacity (196.86 mg/g) was achieved at 15 min of contact time, 2 of pH, 3 g/L of adsorbent dosage, and 250 rpm of agitation speed. The results reveal that BFS-AC has an efficient performance for the removal of phenol from aqueous media.

1. Introduction

Phenol has been classified as a hazardous substance by the harm may cause when entering the body of living organisms. Although phenol can cause serious damage to living organs, it is not listed as a carcinogen [1]. Since 1860, phenol has been utilized in the industry and was mainly used as an antiseptic. The use of phenol has been increased to include the dyes, aspirin, plastics, pharmaceutical, petrochemical, and pesticide chemical industries at the early of the 19th century and afterward [2]. Consequently, phenol-contaminated water may be released into the existing water sources without treatment. This devastates the whole ecosystem enormously. Considering the toxicity of phenol, the U.S. Environmental Protection Agency (US-EPA) has listed phenols as a priority pollutant, with a permissible limit of 0.1 mg/L in wastewater and 0.001 mg/mL in water supplies [3,4].

Various techniques have been applied to eliminate phenolic compounds from aqueous solutions including ion-exchange [5], absorption, distillation [6], chemical oxidation [7], coagulation [8], extraction [9], electrochemical oxidation [10], and adsorption [11]. Many studies have revealed that adsorption to be an extremely efficient technique for the removal of phenol from the wastewater and activated carbons (ACs) have been widely utilized as adsorbents. This technique gained much attention owing to the advantages such as the convenience of operation and selectivity, high performance, superior design flexibility as well as no formation of harmful by-products [12].

In general, the mode of preparation of activated carbon involves either by physical or chemical activation. A chemical activation presents some advantages over physical activation such as product with larger surface area and greater yield with lower process temperature and shorter treatment time [13]. The chemical activation process involves pyrolytic heating in the existence of activating agent potassium hydroxide (KOH), zinc chloride ($ZnCl_2$), sodium hydroxide (NaOH), potassium carbonate (K_2CO_3), phosphoric acid (H_3PO_4), and ferric chloride ($FeCl_3$), which act as dehydrating agents and oxidants. Therefore, improving pore development and increasing carbon yield [14]. Among various chemical activating agents, KOH is often preferred for the preparation of AC owing to its advantages such as high surface area of produced AC, eco-friendly, and formation of oxygen functional groups [15].

Due to the relatively high cost of activated carbons, low-cost alternative precursors have been attempted for the production process. In recent years, there has been increasing interest in using lignocellulosic biomass as an alternative source to produce activated carbon that was used for the removal of different pollutants from wastewaters.

Baobab is a massive tree and may live for serval hundred or thousand years [16]. The potential average yield of whole baobab fruit of 673,000 tonnes per year [17]. The abundance and low cost of baobab fruit shell (BFS) makes it a good candidate as an adsorbent for the removal of many pollutants from wastewaters. The BFSs were demonstrated to be effective as precursor for production of activated carbon, which have been used for the removal of Cu(heavy metal) [18] and dye (methylene blue) [19]. Its high efficiency removal rate and its properties make the baobab fruit shell based activated carbon a feature potential candidate for the removal of phenol from wastewater.

This study aims to demonstrate the production of activated carbon derived from baobab fruit shell by chemical activation using KOH and their adsorption capacities on phenol removal. The optimal values of contact time, adsorbent dosage, pH, and agitation speed as the process parameters for effective phenol removal were established using face-centered central composite design (FCCCD). Moreover, the morphological structure was investigated.

2. Materials and Methods

2.1. Materials

Baobab fruit shells (BFS) were collected from North of Sudan. Potassium hydroxide (KOH) and hydrochloric acid (HCl) were purchased from R&M Chemicals. Nitrogen gas (99.95%) was acquired from Fuelink Marketing Sdn, Bhd. Phenol of reagent grade was obtained from Sigma-Aldrich.

2.2. Sample Preparation

Dried BFS sample (20 g) was mixed in a beaker with potassium hydroxide (KOH), as an activating agent, at impregnation ratio of 1:2. A volume of 160 ml of distilled water was added. The mixture was stirred for 60 min at 50°C in the rotary shaker and was placed to oven overnight at 100°C to obtain a thick paste. The produced thick bulk was inserted into the reactor tube of the pyrolysis furnace for carbonization under high-purity nitrogen gas, which was afterward heated at 10°C/min at 700°C for designated time (60 min). The activated product was subsequently taken out from the furnace to the desiccator to cold. The carbon was washed with an amount of hydrochloric acid solution (0.5M) and then with warm distilled water until a constant pH was achieved. The activated carbon was inserted in the oven to dry for 48 h at 110°C [17]. The activated carbon samples were stored away in a desiccator for additional utilization.

2.3. Scanning Electron microscope (SEM)

The morphological structure of prepared BFS-AC was observed using a JEOL-IT 100 SEM instrument.

2.4. Design Experiments

The optimal conditions for the phenol adsorption were determined by examining the effects of four parameters (contact time, BFS-AC dose, pH, agitation speed) using Face-Centered Central Composite Design (FCCCD), which was selected for a minimum number of experiments to fit a quadratic surface for the response. A range of the variables are reported in Table 1.

Table 1. Experimental factors and their high and low level for central composite design (CCD)

Factor	Name	Low level (-)	High level (+)
A	Contact time (Minutes)	10	20
B	Dose of BFS-AC (g/L)	2.5	3.5
C	pH	1	3
D	Agitation speed (rpm)	150	250

2.5. Batch Adsorption
A series of experiments were carried out batch-wise in several conical flasks contained BFS-AC samples. They were then placed on the rotary shaker and an amount of phenol solution of 600 mg/L was added to each conical flask. The mixtures were agitated under different agitation speeds at 27±1°C. The solution was filtered, and the residual concentration of phenol was determined at 270 nm using a UV/vis spectrophotometer. The adsorption capacity (q_e) was determined from the difference between the initial concentration (C_i) and equilibrium phenol concentration (C_f), which is as follows:

$$q_e = \frac{(C_i - C_f)\mathrm{V}}{\mathrm{W}} \quad (1)$$

where V: the phenol solution volume (L) and W: the weight of BFS-AC used (g) [20].

2.6. Statistical Analysis
The data obtained from experimental should be studies and interpreted using the statistical software Design-Expert version 12.0.12.0 (STAT-EASE, Inc., Minneapolis, US) for the regression analysis in accordance with the second-degree polynomial equation and to determine the statistical value of the formed equations.

3. Discussion and Results

3.1. Scanning Electron Microscope (SEM)
The SEM technique was applied to observe the morphological surface of BFS and the produced BFS-AC. Figure 1 shows a significant variance in the morphological surface of the raw material and BFS-AC.

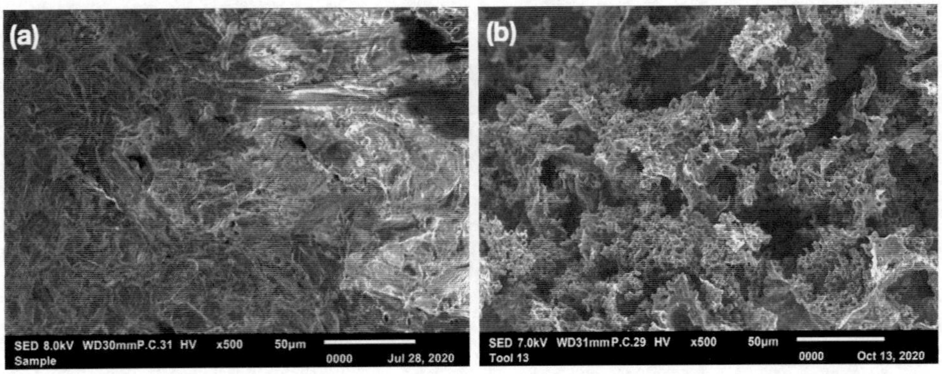

Figure 1. Morphological structure of BFS (a) and BFS-AC produced (b) (magnification scale ×500).

A micrograph of the raw BFS (Figure 1a) contains the lowest original pore structures. While the BFS-AC micrograph (Figure 1b) showed more porous structures occured. This morphological change signifies that the chemical activation process was successful.

3.2. Determination of the Optimum Conditions for Adsorption of Phenol by BFS-AC Using DoE

To determine the optimum conditions for phenol adsorption by the AC based-BFS produced, an experimental design was conducted. Statistical analysis of the data collected for phenol adsorption of the 30 different samples, prepared according to the design matrix, was analyzed with the Design of Expert software. The different parameters which were optimized using FCCCD are contact time, Selected BFS-AC dose, pH, and agitation speed. The response was phenol adsorption capacity. The complete design matrixes and the response values are presented in Table 2. The results showed that the adsorption capacity (q_e) of BFS-ACs was found to range from 177.38 to 196.86 mg/g. The highest adsorption capacity of 196.86 mg/g was found at contact time of 15 min, AC dose of 3g/L, and pH of 2 at 250 rpm.

Table 2. The Adsorption Capacity Results of Phenol Adsorption for its optimum Adsorption Process Conditions by BFS-AC.

Run	Variables				Response
	A: Contact Time (min)	B: BFS-AC Dose (g/L)	C: pH	D: Agitation Speed (rpm)	Adsorption Capacity (q_t) (mg/g)
1	10	2.5	3	150	177.92
2	15	3	3	200	189.72
3	15	3	2	200	193.74
4	10	2.5	3	250	178.20
5	10	3.5	3	150	183.96
6	15	3	2	250	196.86
7	15	3	1	200	188.36
8	20	3.5	3	250	190.0
9	20	3.5	3	150	186.36
10	15	3.5	2	200	191.92
11	10	3.5	1	150	180.4
12	20	2.5	1	250	184.78
13	15	3	2	150	191.64
14	10	3.5	1	250	185.06
15	10	3.5	3	250	184.12
16	15	3	2	200	195.22
17	20	3.5	1	250	192.46
18	10	3	2	200	188.08
19	20	3.5	1	150	187.26
20	10	2.5	1	150	177.38
21	15	2.5	2	200	188.36
22	15	3	2	200	195.22
23	20	2.5	3	150	180.4
24	20	2.5	1	150	179.84
25	20	2.5	3	250	182.86
26	20	3	2	200	191.92
27	15	3	2	200	195.22
28	15	3	2	200	195.22
29	15	3	2	200	195.22
30	10	2.5	1	250	177.92

The suitability of the models is justified by the variance analysis (ANOVA) for the adsorption capacity of phenol by BFS derived activated carbon. As shown in Table 3, the quadratic model to be highly significant for assessing the phenol removal, where the *F*-value was 63.84. The *p*-values were less than 0.0500 indicating the model terms are significant. The results also showed that the contact time(A), BFS-AC dose (B), and the agitation speed (D) were highly significant since their *p*-values were <0.0001. Consequently, these factors had a great effect on the adsorption properties of the baobab fruit shell-based activated carbon produced. The "Lack of Fit *F*-value" of 4.43 implies the Lack of Fit is not significant relative to the pure error. There was a 5.7% chance that a "Lack of Fit *F*- value" this large might occur owing to noise. A non-significant lack of fit is good as it shows that the model is in a good fit. A non-significant lack of fit is good because it indicates that the model is fit.

The quality of the model developed was assessed according to the correlation coefficient value. The model equation showed high R^2 statistics of 0.9835 which showed 98.35% of the variations in adsorption capacity of phenol (q_m) can be explained by the independent variables, contact time, BFS-AC dose, pH, and agitation speed. The obtained R-value was relatively high, which indicates that the experimental values were well agreed with the theoretical (predicted) values for the model. Furthermore, the proposed model showed that R^2 is compatible with the adjusted R^2 values of 0.9681. The best model can be identified according to these parameters. The optimum conditions selected by the software as the first choice were contact time of 15 min, BFS-AC dose of 3 g/L, pH of 2, and agitation speed of 250 rpm.

Table 3. Analysis of variance (ANOVA) for the adsorption capacity using BFS-AC.

Source	Sum of Squares	DF	Mean Square	*F*-value	*p*-value	
Model	1103.48	14	78.82	63.84	< 0.0001	significant
A-Contact time	99.58	1	99.58	80.65	< 0.0001	
B-BFS-AC dose	159.23	1	159.23	128.96	< 0.0001	
C-pH	0.0595	1	0.0595	0.0482	0.8293	
D-agitation speed	45.46	1	45.46	36.81	< 0.0001	
AB	3.32	1	3.32	2.68	0.1221	
AC	3.21	1	3.21	2.60	0.1280	
AD	5.80	1	5.80	4.70	0.0467	
BC	0.0521	1	0.0521	0.0422	0.8400	
BD	1.21	1	1.21	0.9809	0.3377	
CD	6.30	1	6.30	5.10	0.0392	
A^2	36.05	1	36.05	29.20	< 0.0001	
B^2	33.36	1	33.36	27.02	0.0001	
C^2	57.33	1	57.33	46.43	< 0.0001	
D^2	0.8736	1	0.8736	0.7075	0.4135	
ABC	1.54	1	1.54	1.27	0.2781	
Residual	18.52	15	1.23			
Lack of Fit	16.64	10	1.66	4.43	0.0570	not significant
Pure Error	1.88	5	0.3756			
Cor Total	1122.01	29				
R-Squared	0.9835					
Adj R-Squared	0.9681					
Pred R-Squared	0.9299					
Adeq Precision	25.6515					

3.3. Effects of Optimization Variables on Phenol Removal

Model equation generated by Design Expert 12.0.12.0 software can be represented graphically, which could facilitate the evaluation of the effects of the optimization factors on adsorption of phenol. Figure 2 shows the three-dimensional surface plots between the optimization variables obtained by Design-Expert software.

The 3D surface plots presented in Figure 2a and 2d display the effect of different BFS-AC dosage on the phenol adsorption capacity. As can be seen that the phenol adsorption capacity was increased by increasing the dose of BFS-AC. When activated carbon was added to the phenol solution, more surfaces are allocated to remove the phenol, and consequently, increase the adsorption capacity of phenol. Figure 2b shows the interactive effect of contact time and pH on the removal capacity, which implies the important pH role in the phenolic compound adsorption process. It is obvious that the phenol removal augmented with the rise of pH up to 1, and very high adsorption capacity of 196.86 mg/g was attained with 98.43% of phenol removal. While the efficiency of phenol removal reduced with pH increase up to the optimum point, which was 1.8. At low pH, the surface of the catalyst is positively charged and induces the phenolate anion with a negative charge to further the removal of phenol [21].

Figure 2c is the illustration of the interactive effect of contact time and agitation speed on the removal capacity. It was observed that the phenol adsorption capacity increased with the increase of time. This result may be owing to the availability of several empty sites that saturated with time. With the increase of time, the adsorption capacity phenol tends to slow. A similar trend have been reported by other researchers [22,23]. From Figure 2c and 2d, it can be seen that the agitation speed has a minor effect on the phenol adsorption.

Generally, the response in all figures shows a curvature, indicating good interaction effect between contact time (10-15 min), activated carbon dose (2.5-3.5 g/L), pH (1-3), and agitation speed (150-250 rpm) on adsorption of phenol by BFS-AC. The optimum conditions for best phenol adsorption selected by FCCCD design as the first choice were contact time of 15 minutes, BFS-AC dose of 3.2 g/L, pH of 2, and agitation speed of 242 rpm. The mid-level area of pH (pH=2) leads to achievable higher phenol adsorption. Moreover, the removal of phenol is increased with the increase of contact time, AC dose, and agitation speed. These findings reflect the results that several authors have published [24].

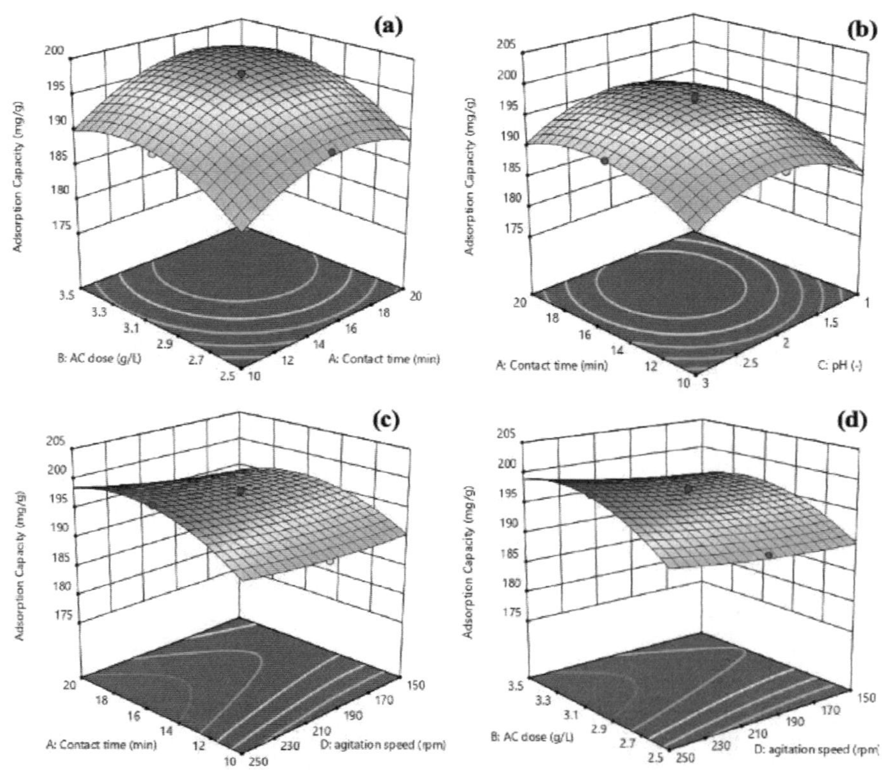

Figure 2. The 3D surface plot of the interaction between (a) BFS-AC dose and contact time, (b) contact time and pH, (c) contact time and agitation speed, and (d) BFS-AC dose and agitation speed.

4. Conclusion

From this study, it can conclude that the baobab fruit shell is good a precursor for the production of activated carbon by chemical activation method using KOH. The results indicated that the maximum adsorption capacity of phenol was achieved at 196.86 mg/g at optimum values of contact time 15 min, 2 of pH, 3 g/L of adsorbent dosage, and 250 rpm of agitation speed. Phenol adsorption achieved its higher level on the area of the mid-level of pH (pH=2). Furthermore, the phenol removal increases with the increase of contact time, AC dose, and agitation speed. Statistical analysis modeling indicated that the model was significant and thus confirm the effect of the combination parameters and conditions for phenol removal by produced BFS-AC from the aqueous solution.

5. References

[1] Karunarathne H D S S and Amarasinghe B M W P K 2013 *Energy Procedia* **34** 83–90.

[2] Abd Gami A, Abd. Shukor M, Abdul Khalil K, Dahalan F, Khalid A and Ahmad S 2014 *J. Environ. Microbiol. Toxicol.* **2** 11–23.

[3] Abd-Hadi G M and Salman S D 2020 *Sains Malaysiana* **49** 57–67.

[4] Xie B, Qin J, Wang S, Li X, Sun H and Chen W 2020 *Int. J. Environ. Res. Public Health* **17** 789. https://doi.org/10.3390/ijerph17030789.

[5] Ochando-Pulido J M, González-Hernández R and Martinez-Ferez A 2018 *Purif. Technol.* **195** 50–59. https://doi.org/10.1016/j.seppur.2017.11.075.

[6] Mohammadi S, Kargari A, Sanaeepur H, Abbassian K, Najafi A and Mofarrah E 2015 *Desalin. Water Treat.* **53** 2215–2234. https://doi.org/10.1080/19443994.2014.883327.

[7] Peings V, Frayret J and Pigot T 2015 *J. Environ. Manage.* **157** 287–296.

[8] Ogando F I B, Aguiar C L de, Viotto J V N, Heredia F J and Hernanz D 2019 *Food Res. Int.* **122** 643–652.

[9] Liu J, Xie J, Ren Z and Zhang W 2013 *Desalin. Water Treat.* **51** 3826–3831.

[10] Bedoui A, Saad M E K, Elaloui E, Rabaaoui N, Allagui M S and Moussaoui Y 2013 *J. Hazard. Mater.* **250–251** 447–453.

[11] Tadda M A, Ahsan A, Shitu A, ElSergany M, Arunkumar T, Jose B, Razzaque M A and Daud N N N 2016 *J. Adv. Civ. Eng. Pract. Res.* **2** 7–13.

[12] Jawad A H, Sabar S, Ishak M A M, Wilson L D, Ahmad Norrahma S S, Talari M K and Farhan A M 2017 *Chem. Eng. Commun.* **204** 1143–1156.

[13] Xu Z, Sun Z, Zhou Y, Chen W, Zhang T, Huang Y and Zhang D 2019 *Colloids Surfaces A Physicochem. Eng. Asp.* **582** 123934.

[14] Lamine S M, Ridha C, Mahfoud H M, Mouad C, Lotfi B and Al-Dujaili A H 2014 *Energy Procedia* **50** 393–400.

[15] Astuti W, Sulistyaningsih T, Kusumastuti E, Thomas G Y R S and Kusnadi R Y 2019 *Bioresour. Technol.* **287** 121426.

[16] Gebauer J, El-Siddig K and Ebert G 2002 *Gartenbauwissenschaft* **67** 155–160.

[17] Nedjai R, Kabbashi N A, Alkhatib M F R, and Md Alam Z 2021 J. Environ. Treat. Tech. 9 686–697. https://doi.org/10.47277/JETT/9(3)697.

[18] Vunain E, Kenneth D and Timothy B 2017 *Appl. Water Sci.* **7** 4301–4319.

[19] Vunain E and Biswick T 2018 *Sep. Sci. Technol.* **54** 27–41.

[20] Nedjai R, Alkhatib M F R, Md Alam Z and Kabbashi N A 2021 *IIUM Eng. J.* ***22*** *31–49.* https://doi.org/10.31436/iiumej.v22i2.1682.

[21] Razmi R, Ardjmand M, Ramavandi B and Heydarinasab A 2019 *Water Environ. J.* **33** 499–507.

[22] Mondal N K and Roy S 2016 *Clean Technol. Environ. Policy* **18** 429–47.

[23] Mohammad Q, Alghouti M, Mubarak N M and Sahu J N 2015 *J. Ind. Eng. Chem.* ***35*** *63-74.*

[24] Leong K-Y, See S, Lim J-W, Bashir M J K, Ng C-A and Tham L 2017 *Appl. Water Sci.* **7** 2009–20.

Enhancement of Indigenous Fungal Cellulase Production by Gamma Rays

F L N Hidayati[1], Sardjono[1], Giyatmi[2], and M N Cahyanto[1]*
[1]Department of Food and Agricultural Technology, Faculty of Agricultural Technology, Gadjah Mada University, Jl Flora, Bulaksumur, Yogyakarta 55281, Indonesia
[2]Polytechnic Institute of Nuclear Technology, National Nuclear Energy Agency of Indonesia, Jl Babarsari POB 6101, Yogyakarta 55281, Indonesia

*Corresponding author: mn_cahyanto@ugm.ac.id

Abstract. The increasing demand for cellulases causes the need for a high cellulase-producing microbe. Mutagenesis is an efficient way to produce a high-titer cellulase-producing strain. Mutagenesis using gamma rays irradiation has the advantage that it can cause a double strand break of DNA. Repair of double-strand break tends to has an error-prone repair that leads to the alteration of DNA sequence. The aim of this study was to screen high cellulase-producing indigenous fungal mutants produced by mutagenesis. *Trichoderma sp.* PK1J2 was subjected to gamma irradiation at 300 Gy. The mutants produced were screened using a plate medium containing cellulose as a sole carbon source. After staining with congo red, colonies with wider clear zones were grown in a liquid medium for four days, and the cellulase activities were analyzed. Mutant M8 produces endoglucanase, FPase, and β-glucosidase at 0.46 U/ml, 0.18 U/ml, and 1.10 U/ml, respectively, which were 90%, 50%, and 30% higher than those of the parental strain.

1. Introduction

Cellulase is a biocatalyst indispensable in various commercial sectors, such as the pulp and paper industry, textile industry, detergent industry, pharmaceutical industry, animal feed industry, and food processing industry. Cellulase is also used to extract bioactive compounds from plants, remove bacterial biofilm, and biorefinery production. In the food industry, cellulases are utilized in various processes such as olive oil extraction, juice clarification, concentrating purees, and reducing the viscosity of nectars. Cellulase is also used to improve the quality of bakery products, pigment extraction from plants, and modification of fruit sensory properties, including flavor, aroma, and texture of fruits [1].

Many organisms can produce cellulases, including bacteria, actinobacteria, fungi, plants, and animals. Among these organisms, filamentous fungi are widely used for cellulase production, of which the genera *Trichoderma, Penicillium*, and *Aspergillus* are known as models for industrial-scale cellulase production [2]. One alternative that can increase cellulase production is the genetic modification of fungal strains. Fungal strains can be modified by mutagenesis into more robust ones in producing higher titers with better cellulase quality. There are two approaches to mutagenesis, namely physical by UV or gamma rays and chemical mutagenesis by N-methyl-N-nitro-N-nitrosoguanidine (MNNG), Ethidium bromide (EtBr), ethyl methyl sulphonate (EMS) or nitrous acid (HNO2) [3], [4].

Gamma-ray irradiation is a low energy transfer radiation (LET radiation) with a LET value of 1 kEVμm-1 or lower produced by CO^{60} or Cs^{137} [5]. Gamma-ray radiation causes death or mutation in fungi due to DNA damage, either directly or indirectly. Gamma irradiation can directly break the single-strand and double-strand of DNA, remove nitrogen bases, and modify nitrogen bases. Indirectly, gamma-ray irradiation breaks the DNA by forming free radicals [6], [7]. The use of gamma-radiation has the advantage that it can cause double-strand breaks, besides single-strand breaks and deletion and modification of bases. In contrast to other types of DNA damage, which generally modify one strand of the double helix and can be repaired using the other strand as a template, double-strand breaks lead to the loss of integrity of both complementary strands. Error-prone repair of double-strand break causes more opportunities for inaccurate repair and leads to the transformation of the DNA sequence, resulting in alteration of genetic information [6], [8].

Gamma irradiation has been shown to cause changes in nitrogenous bases in genes in the form of transformations, transitions, or changes in the reading of the base sequence [9]. Gamma rays have been reported to enhance the production of endoxylanase, α, and β-galactosidase in *Aspergillus niger* [10], [11]. This study aimed to investigate the improvement of cellulase activity in indigenous fungi by mutation using gamma rays.

2. Material and Method

2.1 Microorganism

Five local isolates used in this were culture collections within Gadjah Mada University. *Trichoderma sp.* PK1J2, *Trichoderma sp.* MLT3J2, *Aspergillus sp.* FIG1, and *Penicillium sp.* G2J2 were obtained from the Laboratory of Biotechnology, Faculty of Agricultural Technology, Gadjah Mada University, while *Aspergillus sp.* FNCC 6151 was provided by Food and Nutrition Culture Collection, Gadjah Mada University. These microorganisms were maintained on Potato Dextrose Agar (PDA) and stored at 4 °C.

2.2 Screening indigenous fungi

Screening of the indigenous fungi was conducted by produced the cellulase of the five indigenous fungi and analyzed their cellulase activity. One milliliter of spore suspension (10^7 spores/mL) of the fungi was transferred to 25 mL of medium contains Mandel's mineral salt solution and 1% CMC as a substrate. The fermentation was carried out at 35 °C with a shaking speed of 100 rpm. After the fermentation, samples were withdrawn every day. The sample was filtered with filter paper and centrifuged at 12000 rpm and 4 °C for 10 min. The supernatant was analyzed for FPase activities using the Ghose method [12].

2.3 Mutagenesis with gamma rays treatment

The fungi were inoculated in a PDA medium and incubated at 30°C for 7 days. After the incubation, the spores were harvested using 0.05% tween 80. Ten milliliters of spore suspension (1×10^6 spores ml^{-1}) were exposed by Cobalt-60 gamma rays (at 100, 200,300, 400, and 500 Gy). The mutagenesis by gamma rays was held at Polytechnic Institute of Nuclear Technology, National Nuclear Energy Agency of Indonesia, Yogyakarta, Indonesia.

2.4 Analysis of fungal viability

Serial dilutions were made from each treated spore suspensions, then 0.1 ml of each dilution was poured onto DRBC (Dichloran Rose Bengal Chloramphenicol) medium using the spread plate method. After 3 days of incubation at 30°C, the fungal viability percentage was calculated by the plate count method.

2.5 Clear zone assay

Plate screening medium contains 10 g/L Carboxymethyl Cellulose (CMC) as a carbon source, 17.5g/L agar, 4 g/L sorboses, 0.1% Triton-X, Mandels mineral salt solution, and 1% glucose as a repressor. The fungi were inoculated in each plate screening and incubated at 28 ^{0}C for 3 days, then incubated at 50 ^{0}C for 18 h [13]. The plates were flooded with 0.5% congo red for 15 minutes and destained with 1 M NaCl solution twice [14]. The clear zone ratio was calculated by dividing the clear zone diameter by the colony diameter.

2.6 Enzyme production

One milliliter of spore suspension (10^7 spores/mL) transferred to 25 mL of medium contains Mandel's mineral salt solution and 1% CMC as a substrate. The fermentation was carried out at 35 $^{\circ}$C with a shaking speed of 100 rpm. After the fermentation, samples were withdrawn, filtered with filter paper, centrifuged at 12000 rpm and 4 $^{\circ}$C for 10 min, and the supernatant was analyzed for endoglucanase, β-glucosidase, and FPase activities.

2.7 Enzyme assay

Endoglucanase activity assay was carried out using an endo-cellulase CellG5 assay kit from Megazyme according to manufacture instruction.

The β-glucosidase activity was analyzed by adding 0.2 mL of enzyme sample to 1.0 mL p-nitrophenyl-β-D-glucopyranoside substrate and 1.8 ml of citrate buffer (0.1 M, pH 4.8). The solution was incubated at 50 ^{0}C for 30 min, then the solutions were added with 4.0 mL of glycine buffer (0.4 M, pH 10.8), and read the absorbance of the solutions at 430 nm [15].

The FPase activity was analyzed according to Ghose [12] by adding 0.5 mL of enzyme sample in 1.0 ml citrate buffer (0.05 M, pH 4.8) with one Whatman No.1 filter paper strip 1x6 cm (50 mg). The reaction mixture was incubated at 50 ^{0}C for 60 min, then added the solutions with 3.0 ml DNS. The reaction was stopped by boiled the solution for 5 min. After boiling, the solution was transferred to a cold water bath and added 20 mL distilled water. Wait for 20 min to settle the pulp. The color formed of the solutions was measured at 540 nm.

3. Result and Discussion

3.1 Screening indigenous fungi

Penicillium, Trichoderma, and *Aspergillus* are high titer cellulase-producing fungi. The results of cellulases activity assay produced by *Penicillium sp.* G2J2, *Trichoderma sp.* PK1J2, *Aspergillus sp.* FIG1, *Trichoderma sp.* MLT3J2, and *Aspergillus sp.* FNCC 6151 are presented in Figure 1.

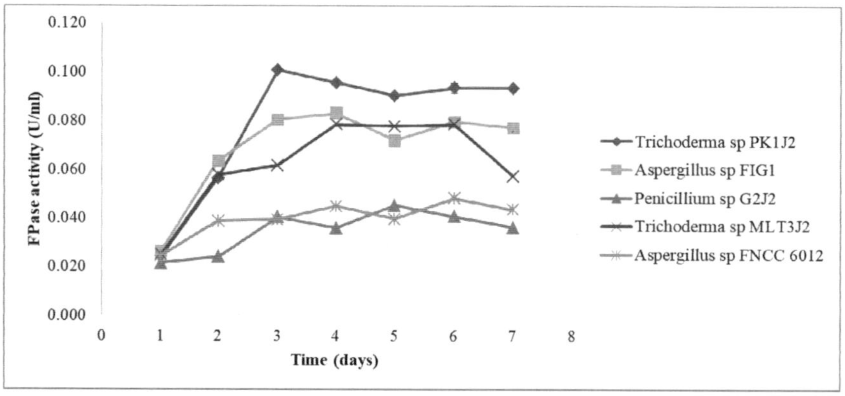

Figure 1. FPase activity of indigenous fungi

IOP Conf. Series: Materials Science and Engineering 1192 (2021) 012004 doi:10.1088/1757-899X/1192/1/012004

The result showed that *Trichoderma sp.* PK1J2 produces the highest FPase activity on the third-day incubation of 0.101 U/mL. *Aspergillus sp.* FIG1, *Aspergillus sp.* FNCC 6012, and *Penicillium sp.* G2J2 have the highest FPase activity on the fourth, fifth, and six-day incubation of 0.083 U/mL, 0.048 U/mL, and 0.045 U/mL, respectively, while *Trichoderma sp.* MLT3J2 produces the same amount of FPase on days four, five, and six of incubation with 0.078 U/mL. In this study, fungi with the highest cellulase activity would be selected for mutagenesis. *Trichoderma sp.* PK1J2 has the highest FPase activity from the 3rd until the 7th day compared with the four fungi, so *Trichoderma sp.* PK1J2 was selected for mutagenesis using gamma rays irradiation.

3.2 Analysis of fungal viability in different doses of gamma rays

Trichoderma sp PK1J2 was exposed to different doses of gamma rays. Each spore suspension was exposed to 100, 200, 300, 400, and 500 Gy of gamma rays. In this study, an irradiation rate of 3885 Gy / hour was used, which requires 1.54 minutes of exposure time to produce an irradiation dose of 100 Gy. The effect of various gamma radiation doses on fungal viability is presented in Figure 2.

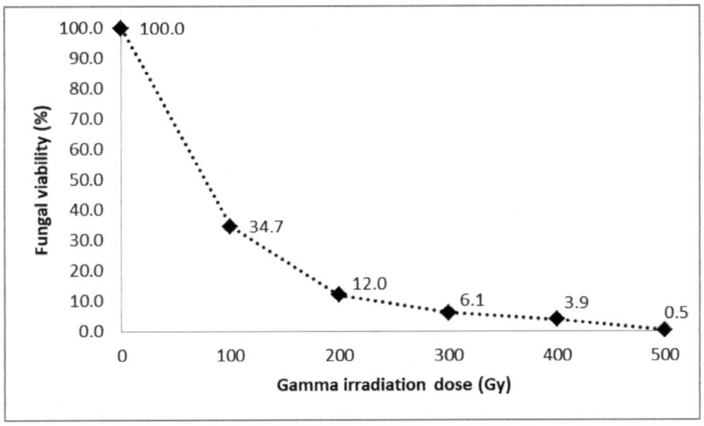

Figure 2. Fungal viability in different gamma irradiation doses

After exposure to gamma-ray irradiation on *Trichoderma sp.* PK1J2, it was found that there was decreasing fungal viability along with the increasing dose of gamma-ray irradiation. Higher radiation doses cause more DNA damage and increase the potential for fungal death. The spores viability was decreased 65% after exposure to 100 Gy gamma rays, and only 0.5% of spores survived in 500 Gy. The decreasing of spore viability after gamma-ray irradiation was reported elsewhere. About a half of *T. reesei*'s spore died under the 500 Gy gamma rays [16], and 90% lethality of *A. niger*'s spore on 1400 Gy [11]. Fungal viability was used to estimate the D_{10} value. D_{10} value represented the gamma irradiation dose that reduced the fungal viability of 1 log cycle or 90%. Gamma rays tended to generate multiple mutations in one gene at a dose higher than the D_{10} value [9], [17]. In this study, the dose of 300 Gy was selected for further analysis.

3.3 Screening mutant

Mutations of *Trichoderma sp.* PK1J2 was performed using gamma-ray irradiation at a dose of 300 Gy. After the mutation, the mutants were grown on plate screening media, and the clear zone formation was assayed using congo red. Clear zone considers as a qualitatively total cellulase. The wider clear zone formation represented the higher cellulase activity.

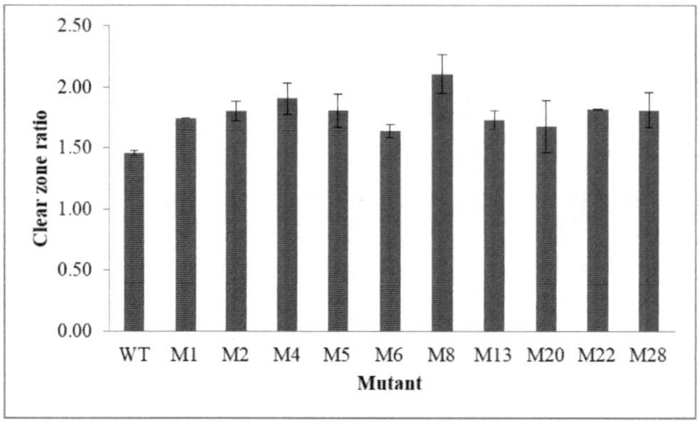

Figure 3. Clear zone ratio of mutants

Based on the clear zone assay, some mutants formed a wider clear zone than the parental strain. Figure 3 shows the top ten mutants with the highest clear zone ratio, and they were M1, M2, M4, M5, M6, M8, M13, M20, M22, and M28. The mutants formed a clear zone with a ratio ranged from 1.64 to 2.11, while the parental strain produced a clear zone ratio of 1.44. Mutant M8 has the widest clear zone compared to other mutants. For further analysis, mutants with wide clear zone formation inoculated in submerge fermentation media to produce cellulase and analyzed the enzyme activity.

3.4 Enzyme analysis of mutant

Enzyme activity assay was carried out on mutants that produce a clear zone that wider than parental strain. Ten mutants were analyzed for endoglucanase activity. The results of the endoglucanase activity of the mutants examined are presented in Figure 4.

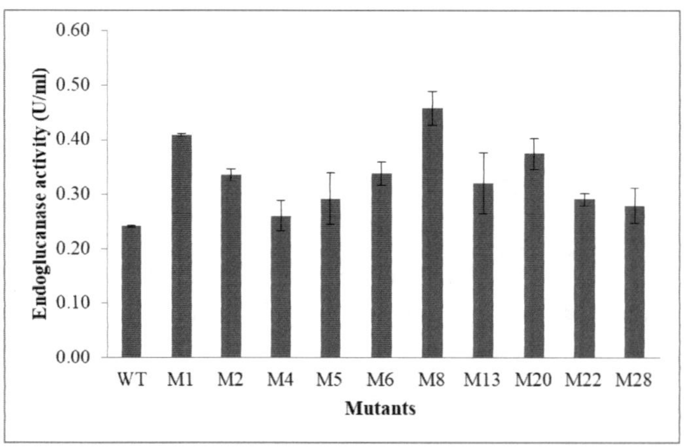

Figure 4. Endoglucanase activity of mutants

As seen in Figure 4, the ten mutants produced endoglucanase higher than the parental strain. Endoglucanase activity of the ten mutants after 4 days ranged from 0.24 U/mL to 0.46 U/mL, whereas the parental strain had endoglucanase activity of 0.24 U/mL. Mutant M4, M5, M13, M22, and M8 produced endoglucanase slightly higher than the parental strain. However, mutant M1, M2, M6, M8, and M20 produced endoglucanase considerably higher than the parental strain. Mutant M8 has the highest endoglucanase activity among the other mutants. This result is similar to the clear zone assay. Mutant M8 forms the widest clear zone compare with the other mutants.

Mutant M8 produces endoglucanase activity of 0.46 U/mL, which was 1.90 times higher than the parental strain. The increasing of cellulase production by gamma-ray irradiation was reported on *Chaetomium cellulyticum, Penicillium cyclopium,* and *Trichoderma reesei* 2414, which increased the CMCase activity by 1.60, 1.40, And 1.23 time, respectively [16], [18], [19]. The other study also reported that *T. reesei* produced CMCase activity by 1.50 times higher than its parent after exposed to gamma rays at 750 Gy [20].

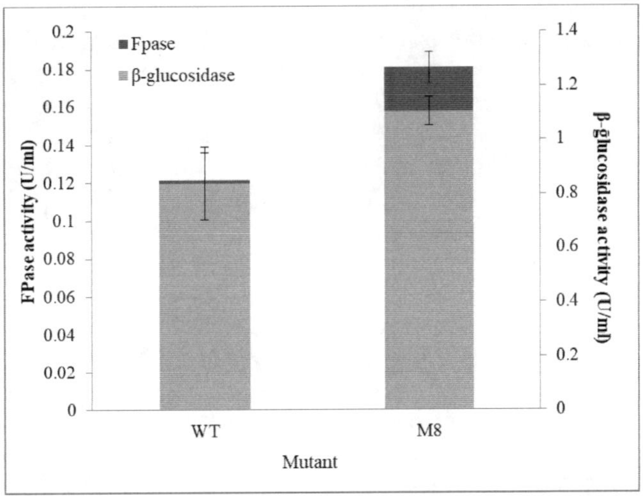

Figure 5. FPase and β-glucosidase activity of mutant

FPase and β-glucosidase activity of mutant M8 were analyzed. As seen in Figure 5, mutant M8 produces FPase and β-glucosidase of 0.18 U/mL and 1.1 U/mL, respectively, which was higher than the parental strain. FPase activity of mutant M8 was 50% higher than the parental strain. The other study was reported that mutant *T. reesei* CCT 2414 and *T. viride* had about 30% and 38% FPase activity higher than its parent after exposure by gamma irradiation [16], [21].

Gamma rays irradiation could increase the cellulase activity produced by fungi, probably due to the gamma rays alter the DNA sequence of fungi at protein secretion gene, transcriptional regulation gene, or gene of sugar transport and metabolism, either directly or indirectly. The research mentioned that the mutant with higher cellulase than the parental strain derived from random mutagenesis has a mutation in RNA metabolism, protein secretion/vacuolar targeting, and transcriptional regulation gene. The other mutant reported has a mutation in the gene of sugar transport and general metabolism [22]. Mutagenesis *Trichoderma sp.* PK1J2 using gamma rays irradiation produces mutant with higher endoglucanase, β-glucosidase, and FPase activity than the parental strain.

4. Conclusion

Mutagenesis of *Trichoderma sp.* PK1J2 using gamma rays irradiation at 300 Gy produced a mutant with cellulase activity higher than the parental strain. Mutant M8 had endoglucanase, FPase, and β-glucosidase activity at 0.46 U/ml, 0.18 U/ml, and 1.10 U/ml, respectively, which were 90%, 50%, and 30% higher than the parental strain.

5. Acknowledgments

This work was funded by PMDSU Batch III research grant No.: 2968/UN1.DITLIT/DIT-LIT/LT/2019.

6. Reference

[1] Sharma A, Tewari R, Rana S, Soni R, and Soni S 2016 *Appl. Biochem. Biotechnol.* **179** 1346
[2] Passos D, Jr N, and Castro A 2018 *Curr. Opin. Green Sustain. Chem.* **14** 60
[3] Yoon L, Ang T, Ngoh G, and Chua A 2014 *Biomass and Bioenergy* **67** 319
[4] Javed M, Ikram-Ul-Haq, and Mariyam I 2011 *Pakistan J. Bot.* **43** 669
[5] Hada M and Georgakilas A 2008 **49** 203
[6] Hennertg W, Grunbergsl S, and Haseltinell W 1982 *J. Biol. Chem.* **257** 11750
[7] Kreutzer D and Essigmann J 1998 *Proc. Natl. Acad. Sci. U. S. A.* **95** 3578
[8] Rodgers K and Mcvey M 2016 *J. Cell. Physiol.* **231** 15
[9] Toyoshima Y, Takahashi A, Tanaka H, Watabene J, Mogi Y, Yamazaki T, Hamada R, Iwashita K, Satoh K, and Narumi I 2012 *Mutat. Res. / Fundam. Mol. Mech. Mutagen.* **740** 43
[10] Awan M, Tabbasam N, Ayub N, Babar M, Rana S, and Rajoka M 2011 *Mol. Biol. Rep.* **38** 1367
[11] Ottenheim C, Werner K, Zimmermann W, and Chuan J 2015 *Biochem. Eng. J.* **94** 9
[12] Ghose T 1987 *Pure Appl. Chem.* **59** 256
[13] Chand P, Aruna A, Maqsood A, and Rao L 2005 *J. Appl. Microbiol.* **98** 318
[14] Qaisar S, Zohra R, Aman A, and Qader S 2014 *Carbohydr. Polym.* **104** 199
[15] Zhang Y, Hong J, and Ye X 2009 *Biofuels: Method and Protocols, Method in molecular biology* ed Mielenz J (Berlin: Humana Press) chapter 14 p 213
[16] Darabzadeh N, Hamidi-Esfahani Z, and Hejazi P 2018 *Appl. food Biotechnol.* **5** 11
[17] Ottenheim C, Werner A, Zimmermann W, and Wu J 2015 *Biochem. Eng. J.* **94** 9
[18] Fawzi E and Hamdy H 2011 *bangladesh J. Bot.* **40** 139
[19] Das D, Chakraborty D, Chakraborty A, and Santra S 2021 *Int. J. Bioprocess Biotechnol. Adv.* **7** 337
[20] Mulyana N, Larasati T, Nurbayti S, and A'yuni Q 2020 *J. Phys. Conf. Ser.* **1436** 1
[21] Bagheri K, Shahbazi S, Askari H, Mojerlou S, and Amirlou F 2018 *Nov. Biol. Reper.* **4** 329
[22] Kubicek C 2013 *J. Biotechnol.* **163** 133

Progress in the detection and quantification of collagens: a review

A H M Gameil[1], F Yusof[1,2], A S Azmi[1] and N I Mohamad Puad[1]

[1]Department of Biotechnology Engineering, Faculty of Engineering, International Islamic University Malaysia, P.O. Box 50728 Kuala Lumpur, Malaysia

[2]Corresponding author's email: yfaridah@iium.edu.my

Abstract Collagens are an important and ubiquitous family of proteins. They have many functions in the human body and similarly have found numerous, potent applications in various industries including the manufacture of biomaterials. The ever-increasing demand for collagen has made necessary the exploration of alternative sources such as bacterial collagen-like proteins which have a triple-helical domain of Gly-X-Y amino acid repeats. Detection and quantification of native collagens have been well-established. However, collagen-like proteins differ in their composition and do not have the unique abundance of hydroxyproline and hydroxylysine found in vertebrate collagens. Thus, this poses a problem in the detection and quantification of collagen-like proteins. This paper evaluates reports on the detection and quantification of collagens and collagen-like proteins. A systematic search of the PubMed database was conducted in May 2021, to which five additional papers were added. The 310 unique search results were then subjected to a screening and elimination process, at the end of which 22 papers were included in the study. The findings were summarized and presented in a table that highlights progress in this field. While novel methods have been developed for the detection and quantitation of collagens in general, mainly using enzyme digestion, hybridization, and fluorescence, there is a need for a rapid, one-step method that selectively and sensitively detects and quantitates collagen and collagen-like protein samples with ease.

1. Introduction

Collagens represent a diverse and ubiquitous family of proteins in the animal kingdom, forming the skin, cartilage, bones, and organs of vertebrates as well as the exoskeleton of invertebrates such as crustaceans and nematodes. The collagen family serves various functions such as tensile strength and elasticity, protection, tissue architecture, cell signaling, and some collagens carry out complex functions related to cell metabolism [1-3]. As we begin to understand and appreciate these functions, collagen finds more utility and a newer purpose in enhancing our quality of living. Collagen supplements serve various pharmaceutical and cosmeceutical functions, and collagen has innumerable uses in the surgical and wound care aspects of various medical departments. Collagen scaffolds and novel biomaterials are also being developed to regenerate tissues [4, 5]. The scientific field also experienced developments in order to draw level with these developments and enable progress in these fields. Testing and assays must be more robust to cater to small-scale experiments as well as large-scale productions. Concurrently, a relatively recent development in the area is the discovery of bacterial and viral collagens, known collectively as collagen-like proteins [6, 7]. These collagen-like proteins have collagen-like, triple-helical domains of glycine-X-Y repeats that show similar thermal and structural stabilities to animal collagens [4-7]. This development is very promising for the biomaterial, medical and biomedical applications of collagen, among others, in the effort to provide a sustainable supply of collagens as well as in making "designer collagens" for various applications [4]. However, new questions emerged to be dealt with, for instance, scientists need to easily ensure and confirm that the protein that has been produced is indeed the recombinant or collagen-like protein of interest. For instance, mammalian collagen is composed of large amounts of hydroxyproline and hydroxylysine, whereas prokaryotic collagens do not have these unique amino acids – what techniques, if any, are available to detect and quantify these different types of collagens? This study sought to answer these questions and to determine the gap(s) in the field.

2. Methodology

This review was carried out by using a systematic search guided by the Preferred Reporting Items for Systematic Reviews and Meta-analyses (PRISMA) guidelines [8]. The keywords were determined using the PICO (Population, Interventions, Comparisons, and Outcomes of study) system, as shown in Table 1. The keywords and synonyms presented were considered and utilized to formulate the search string. The information source used was the PubMed database (National Center for Biotechnology Information, https://www.ncbi.nlm.nih.gov/pubmed/advanced). The systematic search was performed in the database as shown in Table 1, in May 2021.

Table 1. Keyword summary used to develop the search terms.

Characteristics	Keywords	Synonyms
Population	Collagen*	Collagen-like proteins, bacterial, prokaryotic, collagen mimetic peptide, triple-helix
Intervention	method	technique, assay, fluorescen*, colorimetric, hybridization, enzymatic, derivatization, absorbance, spectrophotometric
Comparison	native and recombinant mammalian collagens	human, bovine, mammalian, type
Outcome	detection, quantification	detect*, quantif*, determination
Search string for PubMed database	((collagen*[Title]) OR ("collagen-like protein"[Title]) OR ("bacterial collagen"[Title])) AND ((method[Title/Abstract]) OR (assay[Title/Abstract]) OR (technique[Title/Abstract])) AND ((detect*[Title]) OR (quantif*[Title]) OR (hybridiz*[Title/Abstract]) OR (fluorescen*[Title/Abstract])) NOT (role[Title]) Filters: Journal Article, in the last 10 years	

These papers were then imported into the online Parsifal tool, using TexMed tool and Zotero software (Version 5.0.88). All titles and abstracts retrieved from the comprehensive search were screened and evaluated based on the inclusion and exclusion criteria. Only English-language, journal articles are included in this study, and review articles, conference papers, and book sections were excluded. Publications on the detection and/or quantification method of any type of collagen were included, except those considered as not assays. The data extracted from each study were its basis, principal reagent type, advantages and limitations, detection limit, and the name and type of collagen studied.

3. Results

The systematic search returned a total of 305 unique citation results via the PubMed database. An additional five articles were identified via citation-checking techniques. Imported studies were made up of all the 310 records. Of these, 310 abstracts were screened by title and/or abstract screening and 287 articles were excluded as they did not fit the inclusion and/or exclusion criteria. *In vivo* and *ex vivo* staining and imaging of collagenous tissue or fibers and other histochemical staining techniques in the study of numerous diseases formed the major portion of the excluded literature found on collagen detection and quantitation. The remaining 23 records [9-31] went through full-text screening. The full text for one article could not be obtained [25]. A total of 22 papers were identified and included in the data extraction process. The findings of each study are summarized in Table 2.

Table 2. Summary of findings.

Ref	Method Keywords	Collagen type	Application	Advantages	Limitations
9	Thermal nanoparticles, silver nanoparticles	Porcine, human type I, rat tail collagen	Denatured mammalian collagen	Rapid (30 minutes) Reproducible Highly sensitive Inexpensive	Not Available
10	enzyme-linked immunosorbent assay (ELISA), antibody, matrix-metalloproteinase (MMP)	human type III	Digested fragment of human type III collagen	Specific (high affinity)	For each specific biomarker, an antibody is required.
11	Hybridization, FRET (fluorescence resonance energy transfer) graphene oxide	Collagen peptide	Collagen triple-helical domains	Rapid Sensitive Multiple samples specific	Not Available
12	Alkaline hydrolysis	Mammalian collagen (porcine)	Hydroxyproline-containing collagen	Rapid Sensitive	Complex
13	Alkaline hydrolysis, Woessner method	Hoki skin collagen	Hydroxyproline-containing collagen	Accurate Reproducible High throughput	Requires strict adherence to protocol. Complex
14	chromatography, mass spectrometry	Animal collagen, type I collagen	Thermally denatured, trypsin digested animal collagens	Semi-non-destructive Sensitive Small sample quantity Reduced steps	Specific biomarker required
15	Hybridization, photo-triggered caged CMP, fluorescence, near infrared tags, (carboxyfluorescein and IR680)	Collagen mimetic peptides	Any triple-helical collagen peptide	Versatile	Not Available
16	Hybridization, fluorescence, carboxyfluorescein bound collagen hybridizing peptides	Collagen mimetic peptides	Denatured mammalian collagen peptides	Rapid	Not Available
17	Polyclonal antibodies, gold nanoparticles	Rat tissue	Any triple-helical collagen domain	Rapid sensitive	Not Available
18	chromatography and mass spectrometry	Raw and cooked pork	Hydroxyproline-containing collagens	Rapid sensitive multiple samples	Costly Complex

IOP Conf. Series: Materials Science and Engineering 1192 (2021) 012005 doi:10.1088/1757-899X/1192/1/012005

19	Colorimetric, Sirius red	Rat tail collagen, porcine skin gelatin	Hydroxyproline-containing collagens	Low-cost Simple Rapid Versatile Small sample volume	Interference from Bovine Serum Albumin and elastins.
20	Hybridization, colorimetric, fluorescence, dye-labeled collagen-like peptide	Collagen mimetic peptides	Any triple-helical collagen peptides	Sensitive	Not Available
21	Hybridization, fluorescence, graphene oxide, FAM	Collagen mimetic peptides	Hydroxyproline-containing collagen peptides	Specific. High affinity.	Not Available
22	Hybridization, fluorescence, graphene oxide, FAM (5-Carboxyfluorescein)	Collagen mimetic peptides	Hydroxyproline-containing collagen peptides	Specific Selective Sensitive	Not Available
23	Hybridization	Collagen mimetic peptides	Hydroxyproline-containing collagen peptides	Specific	Not Available
24	Isotope-labeling, LCMS, chromatography, mass spectrometry,	Collagen peptides	Hydroxyproline-containing collagens	Accurate Stable Multiple samples	Complex Costly
26	hybridization, fluorescence, fluorescein, cyclic collagen mimetic peptides	Collagen mimetic peptides	Hydroxyproline-containing collagens or peptides	Selective Sensitive	Not Available
27	chromatography and mass spectrometry, fluorescein, LC–ESI–TOF MS), histidine	Chicken type II	Hydroxyproline-containing animal collagen type II	Accurate	Costly Complex
28	closed-tube LAMP method	Animal collagens	Animal collagen DNA	Rapid detection	DNA-based
29	Enzymatic digestion, fluorescence, 3,4-dihydroxyphenylacetic acid	Bovine collagen, human Type I, others	Bacterial collagenase digested collagen peptides	Specific to collagens Sensitive Multiple samples Simple	Not Available
30	Fluorescence 4-chlorobenzene-1,2-diol	Proline and glycine containing peptides	Collagen peptides	Selective Sensitive Rapid Simple	Not Available
31	Acid hydrolysis, modified HPLC, chromatography and mass spectrometry	Animal collagens	Hydroxyproline-containing collagens	reliable	Costly Complex

4. Discussion

Several assay methods including colorimetric [9, 19, 20], chromatographic [14, 18, 24, 27, 31], antibody-based [10, 17], nanoparticles-based [9, 17] and radio-isotope labelling [24] approaches have been offered for the quantification of collagens. First, the picrosirius red method is mentioned in one included paper [19]. This colorimetric assay of collagens, using the Sirius red dye in picric acid solution, provides a simpler and more rapid quantification of collagens when compared to older techniques. However, drawbacks such as low selectivity of the colorimetric method have been a problem, in addition to its limitation for collagen-like proteins which do not have hydroxyproline. Hence, several modifications of the assay have been proposed.

Another approach to collagen determination involves the use of enzymatic or chemical degradation of collagens followed by high-performance liquid chromatography (HPLC) and spectrophotometry or mass spectrometry (MS). Detection of native eukaryotic collagens has often utilized the unique abundance of hydroxyproline residues that is characteristic of collagen proteins. Aside from collagen, only a handful of other mammalian proteins have hydroxyproline-containing collagen-like (triple-helical) domains, notably the elastins and argonaute-2. Thus, the detection of the presence of hydroxyproline is used in the detection of collagen, where hydroxyproline is derivatized to enable its colorimetric detection. Hydroxyproline, released by acid or alkaline hydrolysis of collagens, can be quantified by full amino acid analysis using chromatography followed by post-column detection. However, the most widely used method is a colorimetric assay based on the reaction of oxidized hydroxyproline with Ehrlich's reagent (p-dimethylaminobenzaldehyde). For example, hydroxyproline can be oxidized to pyrrole-2-carboxylic acid using hydrogen peroxide as an oxidizing agent in the presence of alkaline copper sulfate. The excess oxidizing agent is then removed by heating, and the oxidation product is reacted with Ehrlich's reagent in the presence of dilute sulfuric acid [13].

Finally, the quantification of collagens by enzyme-labeled immunosorbent assay (ELISA) using specific antibodies to each type of collagen or a specific collagen degradation product is another popular approach especially for the detection of collagen peptide biomarkers of diseases in biological samples. Even though it is quite specific, it is an expensive, complex, time-consuming approach. One included article [15] utilized a unique, novel approach that would enable quantitation with some modifications, as well as simultaneous imaging of collagen.

Newer approaches for the detection of collagens make use of another unique aspect of collagen proteins, which is its triple-helical structure. Inspired by the hybridization of denatured DNA strands with a probe, several articles have been published where denatured collagens have been detected by using triple-helical hybridization with single-stranded, fluorescence-tagged, and cyclic collagen-mimetic peptides. As mentioned previously, one included article involves a photo-triggered hybridizations step that can then be used to image the collagen *in vivo* and *ex vivo* as well [15]. During the data extraction, a method keywords column was created to analyze the data, and the most common method keyword that appears is fluorescence, appearing together with hybridization seven times [11, 15-16, 20-22, 26]. Hybridization methods offer higher affinity towards collagen peptides than antibodies, and thus can be useful as less expensive methods for biomarker studies.

Finally, a relatively recent assay proposed that the collagen can be derivatized via amine-labeling of its glycine residues by using chromophores or fluorophores, since the collagen triple helix is made up of consistent repeats of glycine residues. For this to be used, the collagen must first be digested to expose its free N-terminal glycine, by using bacterial collagenases such as *Clostridium histolyticum*. After the derivatization, the dyes fluoresce upon reaction with the amine residues in a linear manner of increasing fluorescence with protein concentration that can be compared against other collagen standards of known concentrations [29, 30]. This unique approach solves the crucial problem of detecting recombinant and prokaryotic collagen-like proteins which lack hydroxyproline and otherwise cannot be detected or quantified by the picrosirius method. In addition, it offers a sensitive alternative to assay collagen peptides. However, the method has its own limitations, chiefly, the need for a spectrofluorometer and the sample preparation time of a few hours. As the collagen triple helix is resistant to protease digestion by pepsin and trypsin, one way to enhance the selectivity of this approach would be to first digest a sample with trypsin, hence eliminating other non-collagenous proteins.

5. Conclusion

Collagens represent a vast family of useful biomaterials whose uses span several industries including the pharmaceutical and food industries. Collagen peptides are also important in the biomedical diagnostics field for understanding and diagnosing diseases using products of collagen degradation, or collagen biomarkers. In addition, collagen-like domains of proteins were found to be common in bacteria including pathogenic ones. This means that there will be more reasons to understand and utilize these unique triple-helical proteins. Hence, there is a need for novel, one-step methods to rapidly and selectively detect and quantitate the various types of collagens and collagen-like samples with ease. Most of the recent existing approaches, however, address the detection and quantification of hydroxyproline-containing biological samples. Here, the picrosirius red assay remains ideal for most applications of mammalian collagens where speed, low cost, and sensitivity are required. While relatively recent fluorometric techniques currently exist, the preparation steps are still laborious and assume the availability of the fluorophore, assay reagents, and costly spectrofluorometric instrumentation. The same can be said of liquid chromatography mass spectrometry, and thus, there exists room for an improved collagen detection and quantitation technique. As for the case of collagen-like proteins, only two studies offer a lucrative approach [29, 30]. Furthermore, this review provides a preliminary step towards systematic mapping of collagen detection and quantitation literature. By modifying the systematic search to be more inclusive, a more comprehensive search of databases can be used to map the literature and identify gaps in those areas of collagen detection and quantitation that were not relevant to this paper, such as in imaging of collagenous tissue samples.

Acknowledgments
This study is funded by the Ministry of Higher Education, Malaysia via the Transdisciplinary Research Grant Scheme (TRGS/1/2018/UIAM/01/1/3). The authors report no conflict of interest.

References
[1] Silvipriya K, Kumar K, Bhat A, Kumar B, John A and Lakshmanan P 2015 *J. Appl. Pharm. Sci.* **5** pp 123-7
[2] Sandhu S V, Gupta S, Bansal H, Singla K and Yadav N S 2012 *J. Orofac. Res.* **2** pp 153-9
[3] Gelse K, Poschl E and Aigner T 2003 *Adv. Drug. Deliv. Rev.* **55** pp 1531-46
[4] An B, Kaplan D L and Brodsky B 2014 *Front. Chem.* **2** 40
[5] Ramshaw J A, Werkmeister J A and Dumsday G J 2014 *Bioengineered* **5** pp 227-33
[6] Xu C, Yu Z, Inouye M, Brodsky B and Mirochnitchenko O 2010 *Biomacromolecules* **11** pp 348-56
[7] Yu Z, An B, Ramshaw J A and Brodsky B 2014 **186** pp 451-61
[8] Moher D, Liberati A, Tetzlaff J, Altman D G and PRISMA Group 2009 *PLoS Med.* **6**
[9] Ahumada M, McLaughlin S, Pacioni N L and Alarcon E I 2016 *Anal. Bioanal. Chem.* **408 pp** 1993-6
[10] Barascuk N, Vassiliadis E, Larsen L, Wang J, Zheng Q, Xing R, Cao Y, Crespo C, Lapret I, Sabatini M *et al.* 2011 *Clin Biochem.* **44** pp 900-6
[11] Cai X, Yang Q, Ding J, Ye W, Li X and Xiao J 2016 *J. Mater. Chem.* B. **4** pp 7009-13
[12] da Silva C M, Spinelli E and Rodrigues S V 2015 *Food Chem.* **173** pp 619-23
[13] Hofman K, Hall B, Cleaver H and Marshall S 2011 *Anal Biochem.* **417** pp 289-91
[14] Kumazawa Y, Hattori S and Taga Y 2019 *Anal. Chem.* **91** pp 1796-800
[15] Li Y, Foss C A, Pomper M G and Yu S M 2014 *J Vis Exp.*
[16] Lin A H, Zitnay J L, Li Y, Yu S M, and Weiss J A 2019 *J. Orthop. Res.* **37** pp 431-8
[17] Masiri J, Benoit L, Barrios-Lopez B, Thienes C, Meshgi M, Agapov A, Dobritsa A, Nadala C and Samadpour M 2016 *Meat Sci.* **121** pp 397-402
[18] Qiu B, Wei F, Sun X, Wang X, Duan B, Shi C, Zhang J, Zhang J, Qiu W and Mu W 2014 *Mol. Med. Rep.* **10** pp 1157-63
[19] Rodriguez-Rodriguez P, Arribas S M, de Pablo A L, Gonzalez M C, Abderrahim F and Condezo-Hoyos L 2013 *Anal. Bioanal. Chem.* **405** pp 6863-71
[20] Sun X, Fan J, Li X, Zhang S, Liu X and Xiao J 2016 *Chem. Commun. (Camb).* **52** 3107-10
[21] Sun X, Fan J, Ye W, Zhang H, Cong Y and Xiao J 2016 A highly specific graphene platform for sensing collagen triple helix. *J. Mater. Chem.* B. **4** pp 1064-9
[22] Sun X, Qiao Y, Li W, Sui Y, Ruan Y and Xiao J 2020 *J. Mater. Chem.* B. **8** pp 6027-33
[23] Sun X, Yao L, Fu C, Luo L, Wang J and Xiao J 2019 *J. Mater. Chem.* B. **7** pp 7676-82

[24] Taga Y, Kusubata M, Ogawa-Goto K and Hattori S 2014 *J. Agric. Food Chem.* **62** pp 12096-102
[25] Takita K K, Fujii K K, Ishii K and Koide T 2019 *Org. Biomol. Chem.* **17** pp 7380-7
[26] Takita K K, Fujii K K, Kadonosono T, Masuda R and Koide T 2018 *Chem. Biochem.* **19** pp 1613-7
[27] Vatansever B, Senal M O, Akgoz M and Goren A C 2015 *Anal. Bioanal. Chem.* **407** pp 1981-7
[28] Wang Y, Yi L, Pan X, Zhang J and Duan R 2018 *Food Chem.* **256** pp 40-4
[29] Yasmin H, Kabashima T, Rahman M S, Shibata T and Kai M 2014 *Sci. Rep.* **4** 4950
[30] Yasmin H, Rahman M S, Shibata T, Kabashima T and Kai M 2015 *Chemical Papers* **69**
[31] Yuswan M H, NH A J, Mohamad H, Keso S, Mohamad N A, Tengku Md Yusoff T S, Ismail N F, Abdul Manaf Y N, Mohd Hashim A, Mohd Desa M N, *et al.* 2021 *Food Chem.* **337** 127762

IOP Conf. Series: Materials Science and Engineering 1192 (2021) 012006 doi:10.1088/1757-899X/1192/1/012006

Continuous assessment of sweat lactic acid secretion using microfluidic sweat lactic acid monitoring system

Y Goto[1], Y Suzuki[1], K Morisawa[2], A Hosoyama[2], Y Taira[2] and H Kudo[1,3]

[1]Electrical Engineering Program, Graduate School of Science and Technology, Meiji University, 1-1-1 Higashimita, Tama-ku, Kawasaki, Kanagawa 214-8537, Japan
[2]St. Marianna University School of Medicine, 2-16-1 Sugao, Miyamae-ku, Kawasaki, Kanagawa 216-8511, Japan

[3]Corresponding Author's E-mail: hkudo@meiji.ac.jp

Abstract. A microfluidic sweat lactic acid (LA) monitoring system for non-invasive and continuous assessment of sweat LA was constructed and tested. Our system transports whole secretions from the skin to an electrochemical LA biosensor using a continuous flow of phosphate buffered saline at the skin's surface. The LA biosensor was fabricated by modifying screen printed carbon electrodes with osmium wired horseradish peroxidase and lactate oxidase. For continuous monitoring purposes, the microfluidic LA biosensor was connected with a flow cell made of polydimethylsiloxane (PDMS). The sampling device was fabricated by laminating a medical plaster on a PDMS flow-channel. Before the clinical tests, the characteristics of the LA biosensor were evaluated. Sufficient sensitivity (0.4232 nA/μM) and selectivity to LA (> 50fold against glucose, glutamic acid, phenylalanine, ammonia and ethanol) were confirmed. Besides, the medical plaster-based attachment contributed to the reduced leakage between the skin and the sampling device. Under the approval of the ethical committees, LA monitoring of patients with cardiovascular disease at the intensive care unit was carried out. As a result, temporal changes of sweat LA were successfully monitored.

1. Introduction

In clinical sites, vital signs such as heart rate and pulse are used in the treatments of severely dysfunctional patients. Temporal changes of those vital signs result in performing appropriate treatment because they reflect changes in physical condition. Kenzaka *et al.* reported that increased respiratory rate and shock index were correlated with the severity of illness in patients with sepsis [1]. Vital signs are mainly based on physical information for which measurement techniques are mature. On the other hand, chemical information, such as lactic acid and glucose, is also useful in assessing pathological conditions. Lactic acid (LA) is mainly produced in anaerobic metabolism and is released into the blood. Therefore, blood lactic acid (BLA) reflects oxygen metabolism in tissues [2–5]. Cocchi *et al.* reported that BLA correlated with in-hospital mortality [6]. For example, LA is used in the evaluation of sepsis, heart failure, liver disease, and lactic acidosis associated therewith [7]. Lactic acidosis is a condition in which BLA is 5 mM or more and the blood is acidic, and the mortality rate can be as high as 50% unless appropriate treatment is performed early [8–10]. Currently, BLA measurement requires regular blood sampling. Repeated blood sampling is a heavy burden for patients as well as it causes medical staff with extra work. Also, continuous measurement is difficult. Therefore, non-invasive and continuous LA measurement techniques that can be easily used even in clinical sites are required.

We focused on sweat, which is one of the biomarker-rich body fluids. Sweat is secreted from the sweat glands mainly for thermoregulation. There are 1.6 to 4 million sweat glands throughout the body, and their density varies depending on the body part [11, 12]. LA is also contained in sweat. Sweat LA (SLA) is produced by the metabolism of the sweat glands themselves, and the concentration of SLA is about ten times higher than that of BLA [13, 14]. Many of the previous literature represent that SLA mainly depends on the sweat rate, it also fluctuated for some other reasons such as the intensity of exercise [15–19]. However, the detailed kinetics of SLA secretion is not clarified. This can be associated with the fact that an appropriate technique for monitoring sweat contents has not been developed. In some previous studies, SLA was measured by collecting a large amount of sweat; thus it was likely that changes in SLA cannot be monitored under low sweating conditions. Hence, a simplified technique of sweat content monitoring, that is capable of use in clinical sites or in daily activities where it is difficult to collect a large amount of sweat, is strongly requested.

In this study, a non-invasive and continuous monitoring system for SLA was developed. The novelty of this system is that a small amount of lactic acid secreted from the skin's surface immediately dissolved in the career flow, and transported to a biosensor, which is specific to LA. We applied this system to patients under intensive care and verified the utility of SLA for evaluating oxygen metabolism in tissues. In this paper, we report on the design and characteristics of the microfluidic LA biosensor and the results of clinical trials.

2. Materials and methods

2.1. *Construction of the sweat lactic acid monitoring system*

The microfluidic SLA monitoring system (Figure 1) consists of an electrochemical LA biosensor, a sampling device with a flow channel and a tube pump (APSN23535RC, A&M, Japan). The sensor unit is connected to a potentiostat (ALS-701EX, CH Instruments Inc., USA). The LA biosensor measures LA as a change of the level of hydrogen peroxide generated by the enzymatic reaction of lactate oxidase (LOD). The production of hydrogen peroxide is converted into a redox current. The sampling device is used for collecting sweat contents. The system utilizes phosphate buffered saline (PBS) as a carrier flow. Whole secretions from the skin are dissolved in PBS and transported to the LA biosensor. Then, LA is selectively measured by the LA biosensor. PBS after the measurement is discharged as a waste liquid.

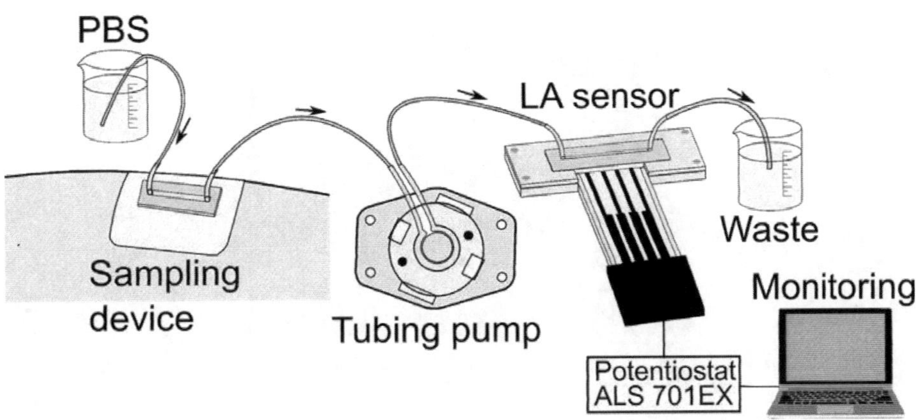

Figure 1. Structure of the microfluidic SLA monitoring system.

2.1.1. *The microfluidic LA biosensor*

The LA biosensor was formed by screen printing using carbon graphite paste. Figure 2 shows the structure of the LA biosensor. It has two working electrodes (WE), a counter electrode (CE) and a silver/silver chloride reference electrode (RE). The electrodes are covered with a polyvinyl chloride adhesive sheet having a microchannel in order to keep the sensing area constant. Two WEs is coated with an Os-HRP redox polymer (002096, Bioanalytical Systems, USA).

In addition, LOD (T-47, Asahikasei Pharma Co., Japan) is immobilized on WE1. The microfluidic LA biosensor was fabricated by combining the LA biosensor with a flow cell made of polydimethylsiloxane (PDMS) with a microchannel (1 mm × 1 mm × 15 mm). The flow cell was prepared by pouring PDMS (SYLGARD184 Silicone Elastomer Kit, Dow Corning Corp., USA) into the flow channel mold made of polymethyl methacrylate (PMMA). In order to improve the durability against external impact, the biosensor was also installed in a PMMA casing as presented in Figure 3.

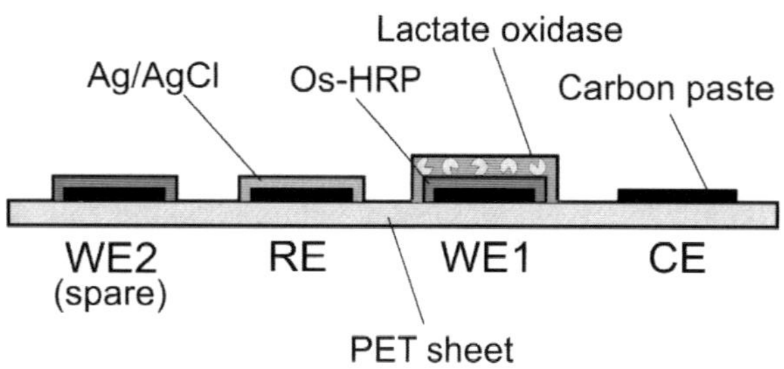

Figure 2. Structure of the LA biosensor.

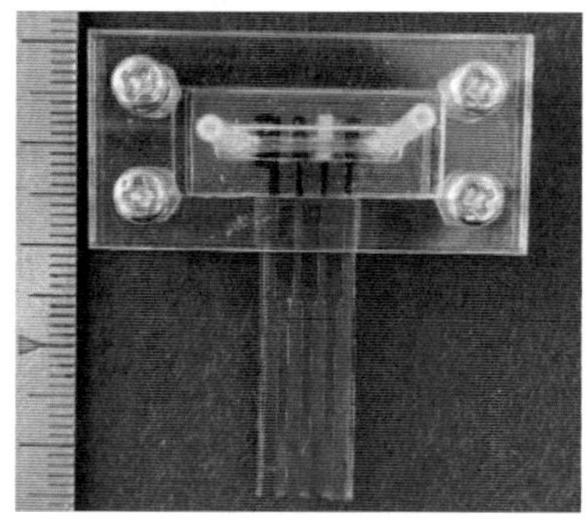

Figure 3. Appearance of the microfluidic LA biosensor.

2.1.2. *The sampling device*

The sampling device was fabricated by combining a PDMS flow cell and a medical plaster (NSD3-2, NIPRO Co., Japan) with an elastic adhesive. The medical plaster was used to improve adhesion to the skin and prevent leakage of PBS during the measurement. The appearance of the sampling device is shown in Figure 4.

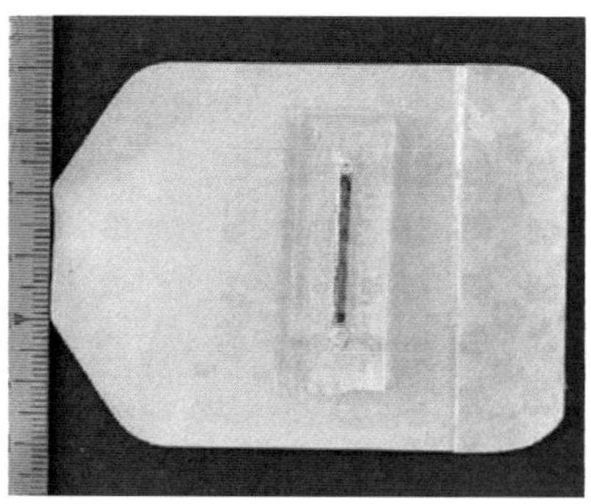

Figure 4. Appearance of the sampling device.

2.2. *Characterization of the microfluidic LA biosensor*

As shown in Figure 5, the experimental system for evaluating characteristics of the microfluidic LA biosensor was constructed by using a tube pump (APSN23535RC, A&M, Japan) and silicon tubes. The flow rate of the carrier flow was set to 50 μL/min. Cyclic voltammetry (CV) was performed under the presence of various concentrations of LA (0, 50, 100, 500, 1000 μM) in PBS. The sweep rate was 50 mV/s, and the sweep range was −0.1 to 0.6 V vs. Ag/AgCl. The driving potential of the biosensor was determined based on the result of CV. Subsequent experiments were conducted under that potential. Constant potential amperometry (CPA) was performed to investigate the quantitative characteristic of the LA biosensor. The concentration of introduced LA solutions were 0, 10, 25, 50, 100, 200, 500, 1000 and 2000 μM. Then, the selectivity to LA was evaluated by introducing each solution (500 μM) of LA, glucose, glutamic acid, phenylalanine, ammonia, and ethanol into the system.

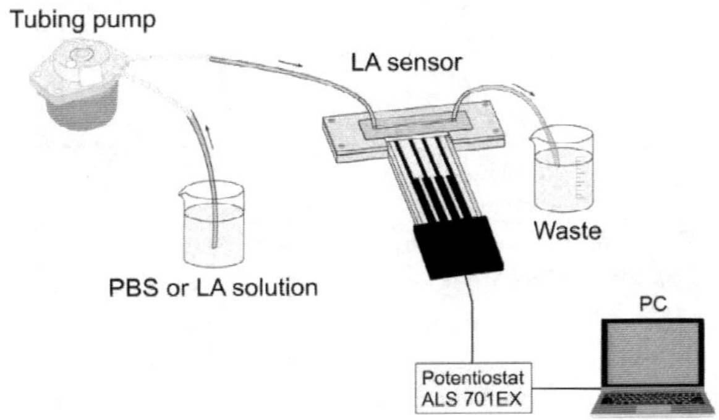

Figure 5. Experimental system for evaluating characteristics of the microfluidic LA biosensor.

2.3. SLA measurement at the intensive care unit

This system was applied to SLA monitoring of patients with cardiovascular disease (approval of ethical committees: No. 3401 St. Marianna Univ. and No. 17-534 Meiji Univ.). The experiment was performed at the intensive care unit (ICU) of St. Marianna University School of Medicine Hospital. The sampling device was attached on the forearm of the patient. At the same time, sweat rate was also measured. Sweat rate was measured by using a micro sweat meter (TPL3520, Techno Next, Japan) near the sampling device. In addition, the part to which the sampling device was attached was sterilized with an alcohol sheet before the measurement.

3. Results and discussion

3.1. Sensitivity and selectivity of the LA biosensor

Figure 6 shows the cyclic voltammograms obtained by introducing LA solutions with different concentrations to the LA biosensor. As shown in the figure, reduction currents (potential sweep in the negative direction) were almost similar within the potentials from 0.6 to 0.4 V vs. Ag/AgCl in any concentrations. In contrast, the output currents around 0 V vs. Ag/AgCl depended on the LA concentrations because of the redox reaction of Os-HRP. Therefore, the driving potential of the LA biosensor was determined to 0 V vs. Ag/AgCl. The operating potential was sufficiently low for reducing the influences of possible reductants in the body (e.g. uric acid, ascorbic acid, etc.). Figure 7 shows the calibration curve of the LA biosensor. The sensitivity of the LA sensor was 0.4232 nA/μM, and the calibration range was 25–500 μM. Although SLA is about 5–60 mM in normal cases, which exceeds the linear range, this calibration range is sufficient for the measurement. This can be associated with the mechanism of our system, which dissolves whole sweat into the carrier flow. SLA is diluted about 100-fold in the carrier flow. Table 1 shows the result of evaluating the selectivity of the LA biosensor. The responses to glucose, glutamic acid, phenylalanine, ammonia and ethanol were 2.04%, 1.78%, 1.72%, 1.80% and 1.69%, respectively, comparing with that of LA. That is, the selectivity was about 50-fold even for glucose with the lowest selectivity. Therefore, the biosensor possessed high selectivity to LA. From the above, it was confirmed that the LA biosensor was suitable for SLA measurement.

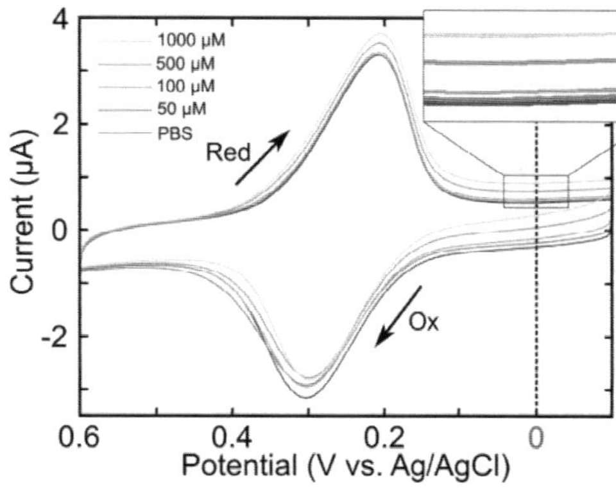

Figure 6. Cyclic voltammograms for standard LA solutions.

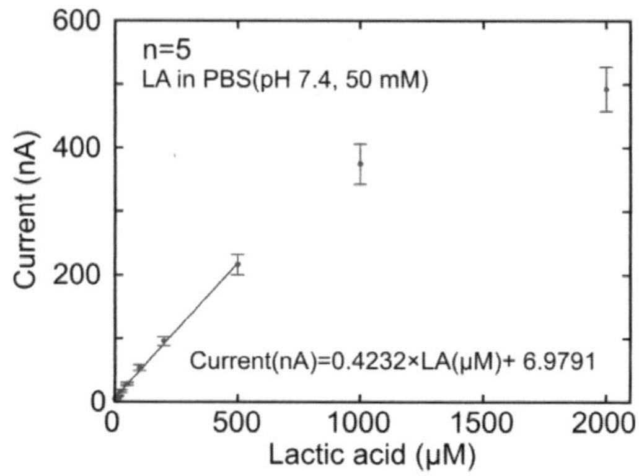

Figure 7. Calibration curve of the microfluidic LA biosensor.

Table 1. Selectivity of the microfluidic LA biosensor.

	Lactic acid	Glucose	Glutamic acid	Phenylalanine	Ammonia	Ethanol
Output (%) (*n*=5)	100	2.04	1.78	1.72	1.80	1.69
Standard deviation	0	0.25	0.31	0.37	0.41	0.23

3.2. Real-time SLA monitoring at the ICU

SLA monitoring was also performed in the ICU. From the ethical point of view, the time of clinical test was limited to two hours at maximum. The system setting could be completed in about 2–3 minutes, which was appropriate for the use in regular operation. The medical plaster-based sampling device was responsible for reducing the time for attachment. Since the sampling device was tightly adhered to the skin, no peeling of the sampling device and no liquid leakage occurred throughout the experiment. Previously, the PDMS sampling device was attached directly to the skin and secured with an armband, which was time-consuming and resulted in frequent leakage. A typical result obtained from clinical tests is shown in Figure 8. Regardless of the sweat rate, it is possible to estimate the total secretion of LA because the flow rate and the area of the sweat sampling part are constant. The concentration of LA was calculated by dividing the secretion of LA by the sweat rate. As can be seen in the figure, LA secreted on the surface of the skin was monitored successfully. No noise was generated by the patient's movements or the medical staffs' work. These suggest that our system is useful for SLA evaluation. LA secreted from the skin may not only be produced at a constant rate but also depend on sweating. We will continue experiments using this system and investigate the dependence of SLA on sweating and the correlation between SLA and BLA. For clinical applications, we will also consider the appropriate sampling area and location.

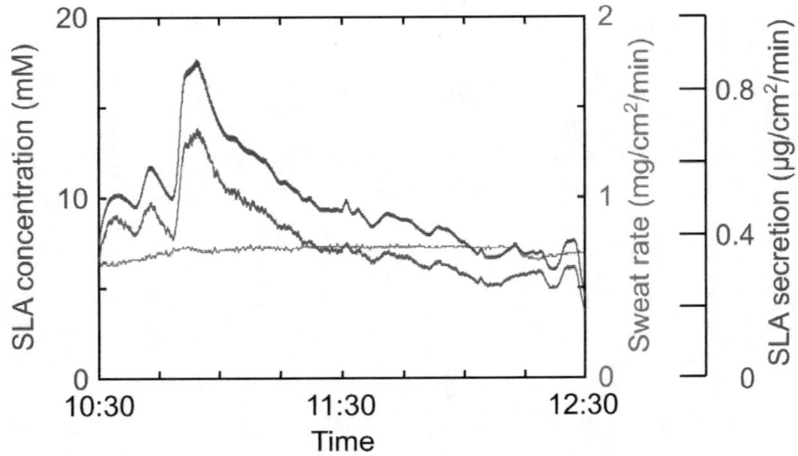

Figure 8. Typical result of real-time SLA monitoring at the ICU.

4. Conclusions

We described the construction of the system for monitoring SLA to evaluate oxygen metabolism in tissues. The LA biosensor using LOD enabled specific detection of LA and had a sufficient response. This confirmed that the biosensor had the necessary performance for monitoring SLA on the skin surface, so it was decided to apply it to patients at the ICU. By using a medical plaster for the sampling device, it was able to adhere to the skin and measure without leakage. The SLA monitoring system was constructed by combining the LA biosensor and sampling device. As a result of conducting clinical tests using this system, we succeeded in real-time monitoring of SLA secreted from the surface of the skin. Although there are still many problems to be solved, practical application in clinical settings is expected in the future.

Acknowledgments

This study was partly supported by JSPS Grant-in-Aid for Scientific Research (C-19K04432).

References

[1] Kenzaka T, Okayama M, Kuroki S, Fukui M, Yahata S, Hayashi H, Kitao A, Sugiyama D, Kajii E and Hashimoto M 2012 *Intern. Med.* **51** 871–76
[2] Bakker J, Nijsten M WN and Jancen T C 2013 *Ann. Intensive Care* **3** 12
[3] Allen S E and Holm J L 2008 *J. Vet. Emerg. Crit. Care* **18** 123–32
[4] Doğan E A, Ünal A, Ünal A and Erdoğan Ç 2017 *Epilepsy Behav.* **75** 13–17
[5] Brooks G A 2018 *Cell Metab.* **27** 757–85
[6] Cocchi M N, Miller J, Hunziker S, Carney E, Salciccioli J, Farris S, Joyce N, Zimetbaum P, Howell M D and Donnino M W 2011 *Minerva Anestesiol.* **77** 1063–71
[7] Andersen L W, Mackenhauer J, Roberts J C, Berg K M, Cocchi M N and Donnino M W 2013 *Mayo Clin. Proc.* **88** 1127–40
[8] Kamel K S, Oh M S and Halperin M L 2020 *Kidney Int.* **97** 75–88
[9] Xu Q, HowlettClyne S, Fuezery A and Cembrowski G S 2017 *Clin. Biochem.* **50** 1164–67
[10] Luft F C 2001 *J. Am. Soc. Nephrol.* **12** S15–19
[11] Hwang K and Baik S H 1997 *Acta. Anat.* **158** 112–20
[12] Smith C J and Havenith G 2012 *Med. Sci. Sports Exerc.* **44** 2350–61
[13] Wolfe S, Cage G, Epstein M, Tice L, Miller H and Gordon R S, Jr. 1970 *J. Clin. Investig.* **49** 1880–84
[14] Sonner Z, Wilder E, Heikenfeld J, Kasting G, Beyette F, Swaile D, Sherman F, Joyce J, Hagen J, Kelley-Loughnane N and Naik R 2015 *Biomicrofluidics* **9** 031301
[15] Buono M J, Lee N V and Miller P W 2010 *J. Physiol. Sci.* **60** 103–07
[16] Bijman J and Quinton P M 1987 *Pediatr. Res.* **21** 79–82

IOP Conf. Series: Materials Science and Engineering 1192 (2021) 012006 doi:10.1088/1757-899X/1192/1/012006

[17] Kondoh Y, Kawase M and Ohmori S 1992 *Eur. J. Appl. Physiol.* **65** 88–93
[18] Fellmann N, Grizard G and Coudert J 1983 *J. Appl. Physiol.* **54** 355–60
[19] Meyer F, Laitano O, Bar-Or O, McDougall D and Heigenhauser G J F 2007 *Braz. J. Med. Biol. Res.* **40** 135–43

IOP Conf. Series: Materials Science and Engineering 1192 (2021) 012007 doi:10.1088/1757-899X/1192/1/012007

Comparison Study Between Encapsulation of *Acalypha indica* Linn Extracts with Chitosan-PCL and Chitosan-OA

M A Johari[1] , A S Azmi[1], J Jamaluddin[2], R Hasham[3], C F Chee[4] and F Ali[1,5]

[1] Department of Biotechnology Engineering, Kulliyyah of Engineering, International Islamic University Malaysia, P.O Box 10, 50728 Kuala Lumpur, Malaysia.
[2] Department of Chemical Engineering, School of Chemical and Energy Engineering, Universiti Teknologi Malaysia, 81310 Johor Bahru, Johor, Malaysia
[3]Department of Bioprocess and Polymer Engineering, School of Chemical and Energy Engineering, Faculty of Engineering, Universiti Teknologi Malaysia, 81310 Johor Bahru, Johor
[4] Institute for Advanced Studies, University of Malaya, Nanotechnology & Catalysis Research Centre (NANOCAT), Level 3 Block A, 50603 Kuala Lumpur, Federal Territory of Kuala Lumpur

[5]Corresponding Author's E-mail: fathilah@iium.edu.my

Abstract. Drug delivery is one of the major applications in the biodegradable polymer science. Chitosan is a non-toxic and naturally biodegradable polymer. It is soluble in acidic aqueous media and insoluble in higher pH media. Chitosan has been modified to improve its properties such as stability and the modified derivatives have been widely used in many applications especially for drug delivery. There are several kinds of chitosan modification. This study investigated the modification of chitosan with polyester-types of polymer and the fatty acid-based polymer. The aim of this study is to compare the formation encapsulation of chitosan-oleic acid conjugate (Ch-OA) and chitosan- polycaprolactone (PCL) copolymer for the encapsulation of *Acalypha indica* active compounds for drug delivery by using emulsion-solvent evaporation technique. The crude extracts have been extracted and the phytochemicals inside the crude extracts are less stable in nature. These active compounds need to be encapsulated to stabilize them and delivered well into the body system. As for conclusion, chitosan:PCL is able to form better encapsulation using this method.

1. Introduction

The idea of controlling drug delivery has been proven to improve the delivery of the therapeutic substances by increasing their target to specific cells, tissue or organs [1]. Selecting a good drug carrier is the most promising so that the side effects can be minimized and able to sustain the drug releasing. Drug delivery is one of the major applications of the biodegradable polymers. This is due to the main role of the biodegradable polymers as a controlling and regulating several biochemical and biophysical functions of living cells. The applications of the biodegradable polymers for drug delivery is very promising because of good biocompatibility thus may safe for body system [2]. Chitosan is one of the excellent candidates for polymers drug carrier. It is produced from the natural resources as well as biodegradable and biocompatible plus it comprises the most abundant polymer occurring in nature. Also, chitosan has good mucoadhesion property [3].

However, chitosan has its drawback which is less soluble in neutral or alkaline pH (Bruschi and Luciano, 2015). Thus it needs several modifications of the chitosan backbone (Bhavsar et al., 2017). There are two methods of copolymerization which are blending and grafting method. Blending technique is a quite simple method as this method involves the mechanically mixing of two different polymers whereas grafting involves the chemical bonding of functional groups between the different polymers. The chitosan-modification with another polymer (natural or synthetic polymer) is to boost hydrophilic property, hence increasing its solubility.

Acalypha indica is a herbal plant and less get attention and occured as a weed. Nevertheless, there is an interesting benefit to this plant. It was reported that *Acalypha indica* has been used as a traditional medication by native people for hundreds of years ago. It was found that the presence of quebrachitol compound that found inside leaves can act as active inhibition of anti-cancer activity against small cell lung cancer [6]. The encapsulation of the extract by the modification of chitosan copolymer will regulate the secretion of *Acalypha indica* crude extract for the targetted cancer cells. The upcoming outlook on this research can go further to produce a dietary product for controlling the cholesterol level to targetted consumers.

To produce a good and suitable microencapsulated of *Acalypha indica* extracts, freeze drying technique will be selected. This method is one of the quite common methods in the production of microcapsules designed for drug delivery systems and has been applied in pharmaceutical industries [7]. The end product will be in the form of the dry powder thus making the extracts more stable and efficient for transporting to the targetted part in the body system. Less efficient extracts encapsulated may lead to wastage and also the accumulation of phytochemicals in the liver.

2. Materials and Methods

2.1. Chemical and Materials

Chitosan was obtained from Sigma Aldrich, poly-(ε-caprolactone) (PCL), *Acalypha indica* (AI) extracts were obtained from Universiti Teknologi Malaysia (UTM), ethanol absolute denatured (99. 8% v/v) brand HmbG® Chemicals, polyvinyl alcohol (PVA) (MW 31,000- 50,000 Da), dichloromethane (DCM), and acetic acid glacial were purchased from Sigma Aldrich.

2.2. Encapsulation of Acalypha indica Linn *extracts with chitosan-PCL*

PCL was emulsified in an organic phase of DCM (0. 5 % w/v), PVA was dissolved in deionized water (2 % w/v), and chitosan was dissolved in aqueous 0. 5 % v/v acetic acid (0. 2 % w/v). Stirring were done for 24 hours under vigorous stirring without applying heat except PVA heated under 60°C. The preparation process of polymer solutions was done in a fume hood to ensure the environmental conditions under controlled.

Next, the preparation of microparticles with and without *Acalypha indica* extracts were performed by emulsion-solvent evaporation method according to El. Hadi et al. (2019) with slight modification. 5 mL emulsions of PCL were transferred into an empty 50 mL-sized tube followed by 5 mL of *Acalypha indica* extracts, 5 mL of chitosan and 5 mL of PVA. The solutions were homogenized under ultrasonic homogenizer (Fisher Scientific FB705,700 watts) with amplitude 90 % for 5 minutes to form a water-in-oil (W/O) emulsion. After that, the solutions were stirred moderately on the magnetic stirrer for 24 hours under room temperature in a controlled environment. During stirring, the particles start to ageing and forming microencapsulation. After 24 hours, the solutions was centrifuged under 13,000 rpm for 1 hour using a mini centrifuge (Eppendorf, speed x 1,000). After centrifuge, two layers of supernatant and precipitate were formed. The particles were separated from the solution. The next step was the washing step. The solvents were removed and replaced by deionized water. Then the precipitates and the deionized water were mixed by vortex. Again, the solution was centrifuged under 13,000 rpm for 1 hour. After that, the centrifugation steps were repeated several times until the solutions turn the pH to neutral using a blue litmus paper as an indicator. Then the samples are ready for lyophilization. The blank microparticles encapsulation were prepared using similar procedure without the addition of *Acalypha indica* extracts.

2.3. *Encapsulation of chitosan-OA conjugate*

2.3.1. Synthesis of N-oleoyl-chitosan (Ch-OA). N-oleoyl chitosan was prepared by a method described by Yan et al with slight modification [8]. Briefly, 100 mg of low molecular weight chitosan (448869, Sigma-Aldrich, USA) was dissolved in 10 ml of 0.01 M hydrochloric acid solution through magnetically stirring at 28 ± 1°C. 50 mg of oleic acid (0.18 mmol), 69 mg of N-(3-dimethylaminopropyl)-N′-ethylcarbodiimide hydrochloride, EDC (0.36 mmol, 2 equiv, Merck, USA) and 41 mg of N-hydroxysuccimide, NHS (0.36 mmol, 2 equiv, Merck, USA) were dissolved in 5 ml of dimethyl sulfoxide (Fisher Scientific, UK). Both chitosan and oleic acid solutions were allowed to stir for 2 h at 28°C. Then, these solutions were mixed and stirred for 48 h at 28°C. The formed CS-OA conjugate was dialyzed (Standard regenerated cellulose membrane, Spectra/Por® 1, molecular weight cut off 6-8 kD, Spectrum Lab, USA) with dimethyl sulfoxide for 24 h and subsequently with deionized water for another 48 h at 28°C to remove the excess EDC, NHS, and oleic acid. It was freeze-dried (-40 °C, 0.1 mbar, Alpha 1-2 Ldplus, Martin Christ, Germany) and the chitosan-OA conjugate in dry form was kept in desiccator at 28 ± 1°C until use.

2.3.2. Encapsulation of Acalypha indica Linn extracts with chitosan-OA. The chitosan-OA conjugate was dissolved in the 0. 5% acetic acid. The stirring of chitosan-OA took 24 hours to fully dissolve. Then, 0. 1 % w/v of Acalypha indica extracts were mixed with 0. 4 % w/v chitosan-OA conjugate and vortex. 2 % w/v of PVA was added into the solution in dropwise. Then, the next steps were continued exactly as for chitosan- PCL.

2.4. *Surface morphology study*

The surface morphology studies was carried out using a scanning electron microscope (SEM) model JEOL JSM-IT 100 InTouchScope™. The samples were deposited on a Nisshin EM conductive carbon tape and coated. Then the coated samples were placed on a brass holder under the vacuum holder. The voltage were accelerated under 5 kV.

2.5. *Encapsulation efficiency (EE%) identification.*

2.5.1. Preparation of standard curve. The standard sample of known concentrations were prepared and inspected in the range 200-1100 nm wavelength to identify the wavelength with a maximum peak of absorbance indicates the most dominant active compounds. Ethanol absolute as the blank solution. After the wavelength of the maximum absorbance obtained, five standard samples of known concentrations (0.02, 0.05, 0.1, 0.15 and 0.2 % w/v) were tested according to the previous wavelength. The linear relationship and the UV absorbance (y) were created according to the standard curve of spectrophotometry and the equation obtained and further used to find the unknown concentration of the extracts being encapsulated in the microcapsules.

2.5.2. Releasing the Acalypha indica extract from the microcapsules using ethanol. The microcapsules with Acalypha indica extracts were dispersed in the 99.98% ethanol and centrifuged using a mini centrifuge (Eppendorf, speed x 1,000) under 13,000 rpm and 1 hour to release the Acalypha indica extracts from the microcapsule suspension into the ethanol. After centrifuge, the supernatant was taken and transfer into a cuvette for testing. The UV absorbance (y) of the free Acalypha indica extracts were tested by UV-vis spectrophotometer. To obtain the concentration (x) of encapsulated extract, the Y value were substituted into the linear equation from the standard curve.

2.5.3. Determination of encapsulation efficiency (EE%). Finally, the value of EE% were calculated using Equation 1 and the data were tabulated.

$$EE\% = \frac{Amount\ of\ extract\ entrapped\ in\ the\ microcapsules\ (g)}{Total\ extract\ added\ (\%w/v)} x100\% \quad (1)$$

IOP Conf. Series: Materials Science and Engineering 1192 (2021) 012007 doi:10.1088/1757-899X/1192/1/012007

3. Results and Discussions

Encapsulation is the most desired method to stabilize the active compounds and prolong the shelf life. After undergoes emulsion-solvent evaporation process, the samples were freeze dried to prevent polymer degradation [9]. Ultrasonic homogenization device was selected to mix the solutions forming O/W emulsion as the intermediate form before hardening forming microparticles [10]. Table 1 shows the percentage of encapsulation efficiency of chitosan-PCL higher than chitosan-OA. Morphological studies by scanning electron microscope (SEM) supports the encapsulation efficiency data to show that chitosan-PCL manage to form better encapsulation in spherical shape (Figure 2 and 3).

Table 1. The encapsulation efficiency (EE%) of chitosan-PCL and chitosan-OA.

Chitosan modification	Encapsulation efficiency (EE%)
Chitosan-PCL	59.83
Chitosan-OA	20.00

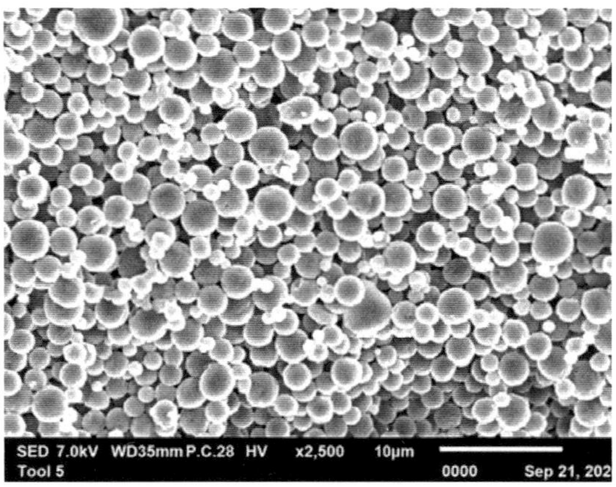

Figure 1. Chitosan/PCL microcapsules without *Acalypha indica* extract.

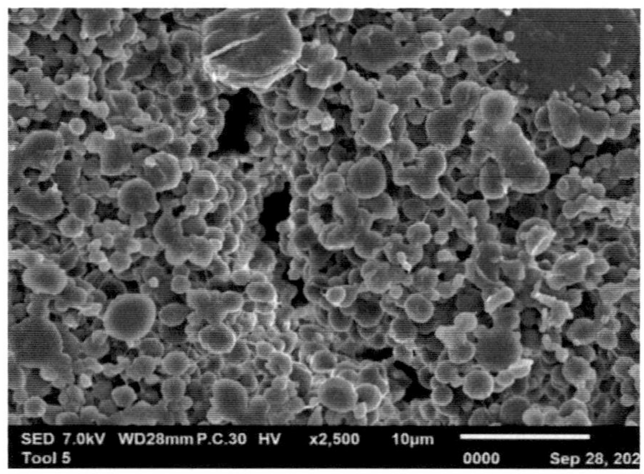

Figure 2. Chitosan/PCL with *Acalypha indica* extract encapsulations.

Figure 3. SEM image for the chitosan-OA loaded *Acalypha indica* microcapsules.

Figures 1 and 2 show that the encapsulations occurred in the chitosan-PCL microparticles without and with *Acalypha indica* extracts by 5 minutes ultrasonic homogenization. The particles formed in smooth and spherical shape and uniformly size. This is due to the vibration of ultrasonic probe mixed the PCL, chitosan, and PVA reduce the sizes, homogenize and easy to self- assemble forming tiny droplets of microparticulates [11]. When *Acalypha indica* extracts were added, the microcapsules formation slighly changed and increased in size as shown in Figure 2. This morphological surface characteristics supported by Lu et.al [12]. Conversely, the shape of chitosan-OA as shown in Figure 3 does not form particle shape and due to that, the performance in encapsulations less efficient than chitosan-PCL encapsulations.

4. Conclusions

In conclusion, chitosan-PCL managed to form microencapsulations in spherical shape by solvent-evaporation method and the chitosan-PCL encapsulation gave higher encapsulation than chitosan-OA. SEM image showed the spherical shape of encapsulation which is one of the important characteristics of encapsulations.

Acknowledgements

This work was financially supported by the Collaborative Research Initiative Grant Scheme (C-RIGS) 2019 (CRIGS-19-004-0010) from International Islamic University Malaysia.

References

[1] Islam N (ed). 2014 *Nanotechnology Recent Trends, Emerging Issues and Future Directions*. New York: Nova Science Publishers, Inc

[2] Roy S M and Sahoo S K 2016 Controlled Drug Delivery: Polymeric Biomaterials for. *Encycl Biomed Polym Polym Biomater* 2135–2146

[3] Szymanska E and Winnicka K 2015 Stability of chitosan- A challenge for pharmaceutical and biomedical applications. *Mar Drugs* 1819–1846

[4] Bruschi M L 2015 *Strategies to Modify the Drug Release from Pharmaceutical Systems*. Elsevier Inc.

[5] Bhavsar C, Momin M, Gharat S, et al. 2017 *Functionalized and graft copolymers of chitosan and its pharmaceutical applications*. Taylor & Francis

[6] Zahidin N S, Saidin S, Zulkifli R M, et al. 2017 A review of Acalypha indica L. (Euphorbiaceae) as traditional medicinal plant and its therapeutic potential *J Ethnopharmacol* **207** 146–173

[7] Santos D, Maurício A C, Sencadas V, et al. 2018 Spray Drying: An Overview. *Biomater - Phys Chem - New Ed.* 72247

[8] Yan Y L, Xi G C, Cheng S L, et al. 2007 Effect of the molecular mass and degree of substitution of oleoylchitosan on the structure, rheological properties, and formation of nanoparticles. *J Agric Food Chem* **55** 4842–4847

[9] Vauthier C and Bouchemal K 2009 Methods for the Preparation and Manufacture of Polymeric Nanoparticles. *Pharmaceutical Research* **26** 1025–1058

[10] Awad TS, Moharram HA, Shaltout OE, et al. 2012 Applications of ultrasound in analysis, processing and quality control of food: A review. *Food Res Int* **48** 410–427

[11] Taha A, Ahmed E, Ismaiel A, et al. 2020 Ultrasonic emulsification: An overview on the preparation of different emulsifiers-stabilized emulsions. *Trends Food Sci Technol* **105** 363–377

[12] Lu B, Lv X and Le Y 2019 Chitosan-modified PLGA nanoparticles for control-released drug delivery *Polymers (Basel)* **11**

Point-of-use upflow sand filter for rural water treatment using natural local sand: Understanding and predicting pressure drop

F Abd Lahin*, R Sarbatly and C Chel-Ken
Membrane Technology Research Group, Material and Mineral Research Unit, Faculty of Engineering, Universiti Malaysia Sabah, Jalan UMS, 88400 Kota Kinabalu, Sabah, MALAYSIA

Email: farhana.abdlahin@ums.edu.my

Abstract. A simple, small scale upflow sand filter was fabricated using a locally obtained sands at three different rivers in Sabah, Malaysia: Liwagu River (SL), Tamparuli River (ST), and Kaingaran River (SK). The grain size, porosity, bulk density, particle density and sphericity of the sands were characterized to associate with the corresponding pressure drop across the sand bed. The highest pressure drop per unit length for SK, PT, and SL are 15.85 kPa m^{-1} at 0.747 m s^{-1} v_s, 10.18 kPa m^{-1} at 0.352 m s^{-1} v_s, and 9.24 kPa m^{-1} at 0.747 m s^{-1} v_s, respectively. The pressure drop per unit length at different filter bed depth were plotted, and compared against three theoretical models of Ergun, Kozeny-Carman, and Fair and Hatch. By analyzing the experimental-theoretical comparison using RMSE and Chi-Test, prediction of pressure drop in an upflow sand filter is able to be predicted using the Kozeny-Carman equation preceding filter bed fluidization and subsequently Fair and Hatch's equation after bed is fluidized.

1. Introduction

To this day, provision of clean water in isolated areas is still an issue in most developing countries. Without treating water for consumption, these communities are highly susceptible to waterborne diseases such as dysentery, typhoid, polio, cholera and diarrhea [1]. In Sabah, Malaysia alone, absence of treated water in rural areas has caused numbers of severe cases of cholera and diarrhea [2]. To provide clean water in rural areas that is outside of a main water supply network, point-of-use treatment system is usually preferred. For such purpose, sand filters constructed using local sands have been widely used due to its sustainability, low cost and high efficiency in removing pollutants. However, a conventional sand filter with downflow configuration is often associated with fouling, owing to the presence of Schmutzdecke layer, formed at the top most of the sand bed. This layer is a collection of alluvial mud, microorganism, organic waste and algae sediments during settlement and straining [3].

In view of a simple treatment method for isolated areas, an upflow configured sand filter has been viewed to be advantageous [4], especially as a point-of-use water treatment. Upflow sand filter has been applied in numerous water purification processes, such as in the removal of bacteria, algae, phosphorus, nitrogen, heavy metals, turbidity, total suspended solids (TSS), and chemical oxygen demand (COD) [5]. The absence of the Schmutzdecke layer indicates that filter clogging is less likely to occur and filter cleaning will be less frequent. Backwash can be conducted rapidly, require a shorter time, and filter downtime is thus minimized [6]. Similar to the conventional down-flow sand filter, the upflow sand filter separation mechanism operates based on screening, filtration, sorption, and ion exchange [7]. As opposed to the down-flow sand filter, removal mechanism of its upflow counterparts, mainly occurs at depth where particles are attached to the sand grain or to previously retained particles. Such filtration mechanism is commonly associated with rapid filtration as the pressure driven filter causes contaminants to be removed deep in the filter bed.

Pressure drop is one of the most important aspects in designing a sand filter. It is resulted from the obstruction and drag produced by the sand media and the internal auxiliary of sand filters, such as the diffuser plate and nozzle underdrain [8]. Excessive pressure drop significantly reduces the water flux, requiring higher pumping power on sand filters, which reduces its overall efficiency. Conventionally, pressure drop monitoring can be conducted through manual observation of operating pressure at the start and the end of a filter bed. On the other hand, prediction using modelling is equally possible. Ergun and Kozeny-carman equations are often used to predict the pressure drop in a packed bed (equation 1 and 2), where $\Delta P/\Delta L$ = pressure loss per unit length (Pa/m), μ = viscosity of water (Pa s), d_{eq} = particle equivalent diameter (m), φ = sphericity (dimensionless), ρ = fluid density (kg/m^3), ε = media porosity, v_s = filtration rate (m/s). However, in order to accurately estimate the pressure drop throughout the filter bed, physical parameters of the sand media such as the porosity, sphericity, and equivalent diameter must be determined [9].

$$\frac{\Delta P}{\Delta L}=150\frac{\mu v_s}{{d_{eq}}^2\varphi^2}\frac{(1\text{-}\varepsilon)^2}{\varepsilon^3}+1.75\frac{\rho {v_s}^2}{d_{eq}\varphi}\frac{1\text{-}\varepsilon}{\varepsilon} \quad (1)$$

$$\frac{\Delta P}{\Delta L}=180\frac{\mu v_s}{{d_{eq}}^2\varphi^2}\frac{(1\text{-}\varepsilon)^2}{\varepsilon^3} \quad (2)$$

Previous literatures had designed the upflow sand filter in a towering vessel with a height of more than 30 feet [6], which is not practical for a small-scale application in a rural setting. To scale down the sand filter, Li and Pashley [10] constructed a miniaturized upflow sand filter using glass particles, industrial-produced quartz particles and activated charcoal. However, to enable the construction of simple, efficient sand filters in rural regions, local sands are preferred. In this study, locally obtained sands are used to fabricate small scale upflow sand filters. The pressure drop across the sand bed caused by varying grain size and filter bed depth is closely monitored, where the obtained pressure drop data is compared against the theoretical models. The findings of this study will be used for future design of an upflow sand filtration unit that is able to be self-assembled and operated in rural areas.

2. Material and Methods

2.1. Sand media

Sand sampling was conducted at three different rivers located in rural areas in Sabah, Malaysia: Liwagu River (SL), Tamparuli River (ST), and Kaingaran River (SK). All three locations were upstream from any residential areas and ensured to be less frequented by human or animals. Sand selection was carried out based on a few simple guides, i.e.: (i) Coarseness of the sand should be felt, (ii) the individual particle of sand can be observed, (iii) while squeezing the sand and letting go, the sand particles should smoothly pour out of the hand. Washing of the sand was done by swirling the sand and water in a container, before decanting the dirty water from the container. The process was repeated multiple times.

2.2. Physical characteristics

The bulk density, particle density, porosity, sphericity, and equivalent diameter were determined for each sand sample. The sand sphericity, φ was assessed visually under the microscope with 20× magnification and compared to the Krumbein Roundness Sphericity Chart [11]. Sphericity is described as the ratio of the surface area of an equivalent volume sphere to the surface area of the sand grain. Sphericity increases with the roundness of the particle. $\varphi = 1$ defines perfectly the spherical grains and

$\varphi < 1$ defines the angular particles [12]. The equivalent diameter, d_{eq} was determined based on the method presented by Bové et al. [8]. d_{eq} of sand grain is the diameter of a sphere with a volume equal to the average volume of the media grains [9].

2.3. Laboratory scale upflow sand filter

The experiment was carried out with a laboratory scaled sand filter fabricated from a clear acrylic tube with 0.154 m inner diameter and 1.5 m height. A 4 mm thick inner plate was affixed 150 mm from the bottom of the filter as support for sand media, and 2-mm holes of 6 mm apart were drilled across the plate to act as support as well as a diffuser. A variation of sand media was placed ranging from 0.4 to 0.8 m above the plate supported by two gravel layers to ensure no sand media travel to the lower layer of the filter. Manometer outlet was placed across the sand filter at intervals of 100 mm.

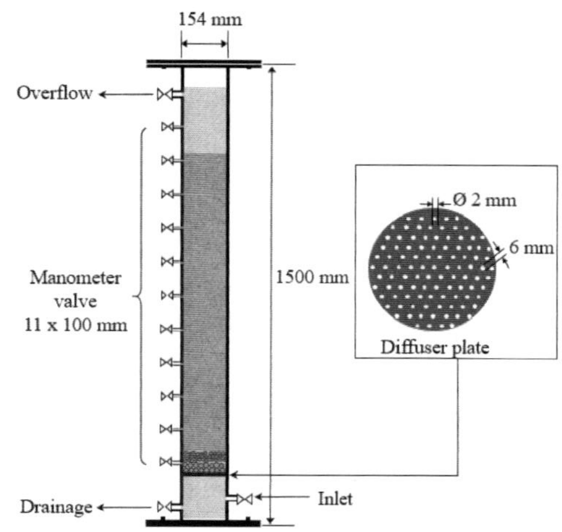

Figure 1. Upflow sand filter configuration

At the beginning of each trial set, the sand filter was operated at filtration velocity 10×10^{-4} m.s^{-1} for 10 minutes, allowing the sand bed to expand and ensuring a standardized initial compaction of the sand media. This step would ensure the experimental replicability and reduce the effect of different initial porosities for each trial set. After performing this initial procedure, the sand filter was allowed to reach a steady-state condition and data acquisition for pressure variables was initiated. Tap water was used throughout the experiment at a velocity ranging from 1.34×10^{-4} to 12.99×10^{-4} m.s^{-1}. The pressure head was recorded when the water flow and manometer reading stabilized.

3. **Results and discussion**

3.1. Physical characterization of filtration granules

The particle density of each sand sample was relatively in the same range, considering the equivalent diameter varied from 1.5 to 2.5 mm (Table 1). In general, smaller grain size will have greater porosity, however, the shape and angularity of the grain will have similar to greater effect on porosity. The angularity of the grain, which also determined by its sphericity affected the porosity of sand as pointed out by Rutledge and Gagnon [13]: porosity increases with more angularity of the sand. Bové et al. [8] have reported the same finding. Overall, the analysis shows that the acquired sands are suitable to be utilized as a media for upflow filtration as 0.4 porosity has been reported to provide deep filtration for removal of bacteria and viruses [14].

Table 1. Sand and gravel support physical parameters

Sample	d_{eq} (mm)	Porosity	ρ_b (kg/m^3)	ρ_r (kg/m^3)	φ
SL	2.5 ± 0.0002	0.42 ± 0.013	1477 ± 19.76	2536 ± 31.93	0.8
PT	1.7 ± 0.0001	0.40 ± 0.016	1507 ± 34.27	2512 ± 17.55	0.8
SK	1.5 ± 0.0010	0.43 ± 0.025	1461 ± 62.17	2575 ± 28.99	0.7

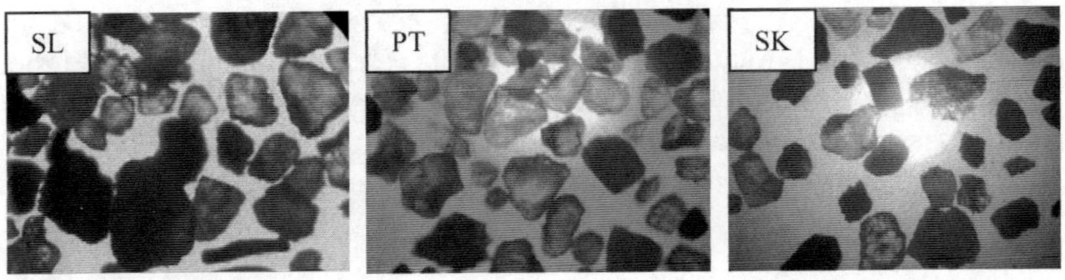

Figure 2. Sand samples under 20× magnification

3.2. Pressure drop across bed

An increase in v_s results in an increase of pressure drop. The highest pressure drop per unit length for each grain size was recorded as the following: SK (15.85 kPa m^{-1} at 0.747 m s^{-1} v_s) > PT (10.18 kPa m^{-1} at 0.352 m s^{-1} v_s) > SL (9.24 kPa m^{-1} at 0.747 m s^{-1} v_s). The result indicated that the smaller grain size resulted in a greater pressure drop. According to Ergun equation (equation 1), the pressure drop is also influenced by the collective effect of the sphericity and porosity of the grains. The highest pressure drop per unit length was observed for sand sample SK at 15.8 kPa m^{-1} before the pressure drop stabilized during fluidization. Other researchers have observed a similar trend while applying different types of granular media [8, 15]. This study reveals that, although selection of smaller grain size may potentially have greater pollutant removal rate, it is not necessarily desirable in order to avoid a greater pressure drop in the filter bed.

It was interesting to witness that after a linear increase of pressure drop, at one point the pressure drop decreased before stabilizing with a further increase in v_s. The decline in pressure drop occurred after the fluidization of the filter bed. For each sand sample, this phenomenon occurred at a different velocity: SL at 0.747 m s^{-1} v_s, SK at 0.747 m s^{-1} v_s, and PT at 0.352 m s^{-1} v_s. All experiments showed that the pressure drop remained relatively constant after the filter bed was fluidized. Increases in the v_s expanded the filter bed, yet pressure drop remained comparatively steady. The fluidized pressure drop was recorded at 7.79 kPa m^{-1}, 8.48 kPa m^{-1}, and 14.1 kPa m^{-1} for SL, PT, and SK, respectively (figure 3). This pressure drop should remain constant until 90% porosity is achieved [16]. The pressure drop during fluidization can be calculated by the equation developed by Fair and Hatch (1933) [17], which relates the pressure drop across a fully fluidized bed to the bed height and the porosity of the sand media. Same pressure drop behavioral tendencies have been explained by previous researchers [16, 18, 19]. It is also important to note that the equation is independent of the size of granular media.

$$\rho h \Delta h = g(\rho_p - \rho)(1 - \varepsilon)l \tag{3}$$

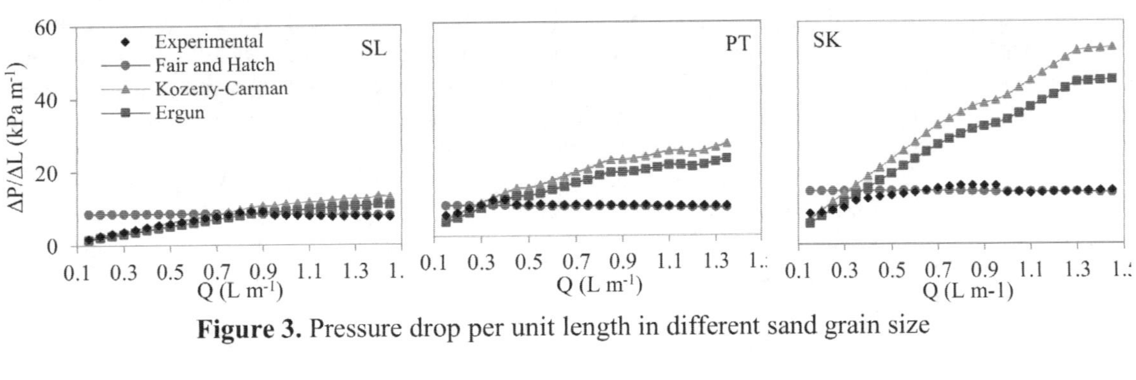

Figure 3. Pressure drop per unit length in different sand grain size

The result indicates that higher pressure drop is developed by deeper filter bed (figure 4). The maximum pressure drop occurred at a similar v_s (0.835 m s^{-1} v_s) except for 40 cm at 0.747 m s^{-1} v_s and 80 cm at 0.878 m s^{-1} v_s. The respective maximum recorded pressure drops in each setup were: 30 cm at 7.96 kPa m^{-1}, 40 cm at 13.86 kPa m^{-1}, 50 cm at 18.03 kPa m^{-1}, 60 cm at 22.94 kPa m^{-1}, 70 cm at 28.68 kPa m^{-1}, and 80 cm at 32.54 kPa m^{-1}. Succeeding the maximum pressure drop, the filter bed was fluidized and the pressure drop stabilized. Previous studies have also recorded that deeper filter beds significantly increases the operating pressure drop [20]. This is primarily caused by the higher energy required for the water getting through the barriers provided by the depth of the media. Hence, demanding a higher energy consumption for a certain water yield. Additionally, gravity may play a small role as the current configuration is in upflow direction. For that reason, a higher filter bed may not be necessarily beneficial for removal of pollutants as similar removal efficiency can be achieved through a shallow bed depth [21].

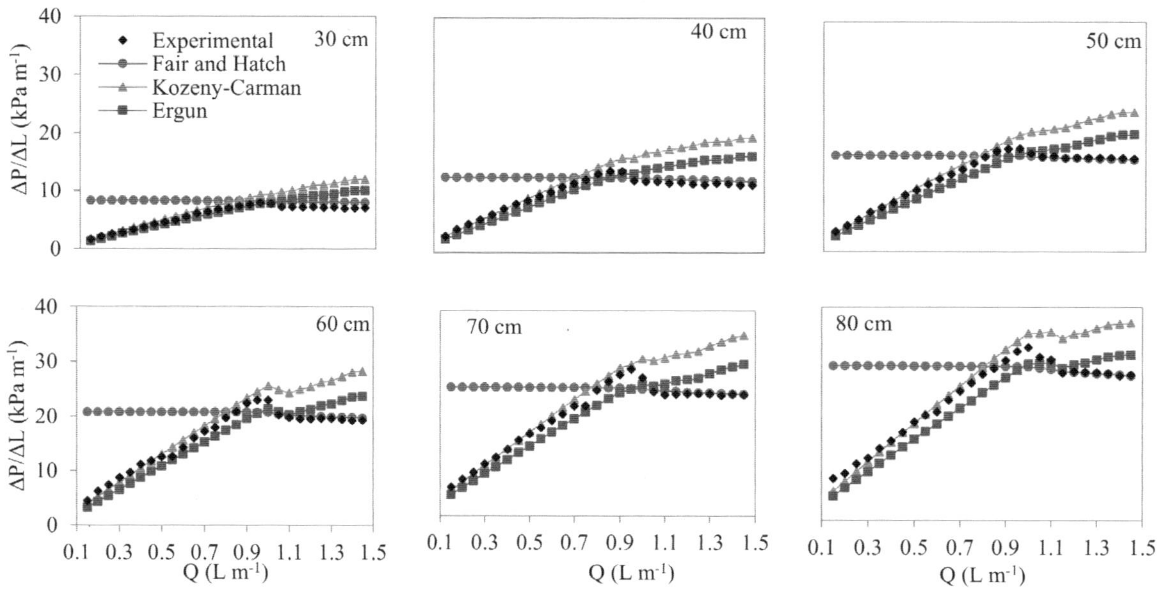

Figure 4. Pressure drop per unit length in difference filter bed depth

3.3. Pressure Drop Prediction Model

Evaluation of significant difference between the experimental and empirical results were performed using Chi-test and Root Mean Square Error (RMSE). Experimental results (figure 3 and figure 4) showed two stages of pressure drop behavior, namely before and after fluidization occurred and therefore the analytical comparison was conducted based on these two stages (figure 5). Before fluidization, Chi-test showed no significant difference (p-value>0.05) between the experimental result and the empirical value predicted by Kozeny-Carman and Ergun equation in varied filter bed height. RMSE shows the Kozeny-Carman equation yielded to have the lowest value, indicating higher prediction performance. After fluidization, the condition was reversed, as Fair and Hatch's values showed no significant difference, while Kozeny-Carman and Ergun equations showed p-values smaller or equal to 0.05. RMSE value further concurred with the analysis as Fair and Hatch's equation displayed the smaller values, revealing a better fit to the experimental values in all experimental variations.

This analysis shows that prediction of pressure drop in an upflow sand filter is possible with consideration of its minimum fluidization velocity. The pressure drop behavior can be predicted using Kozeny-Carman equation up until fluidization is about to take place and Fair and Hatch's equation, subsequent to fluidization. Prediction models are an imperative tool for variable modification, especially sand filter flowrate. Reduction in straining efficiency during filtration due to a fluidized bed can be avoided by pre-determination and working under the minimum fluidization velocity. Although, the precision of prediction is highly reliant on on the accuracy of the values of parameters, such as sphericity, sand media porosity, and sand equivalent diameter. Nevertheless, the obtained outcomes give a comprehensive insight on the filtering process of upflow filters with local sands, which gives promising potential to the realization to construct simple, low-cost sand filters in various rural regions.

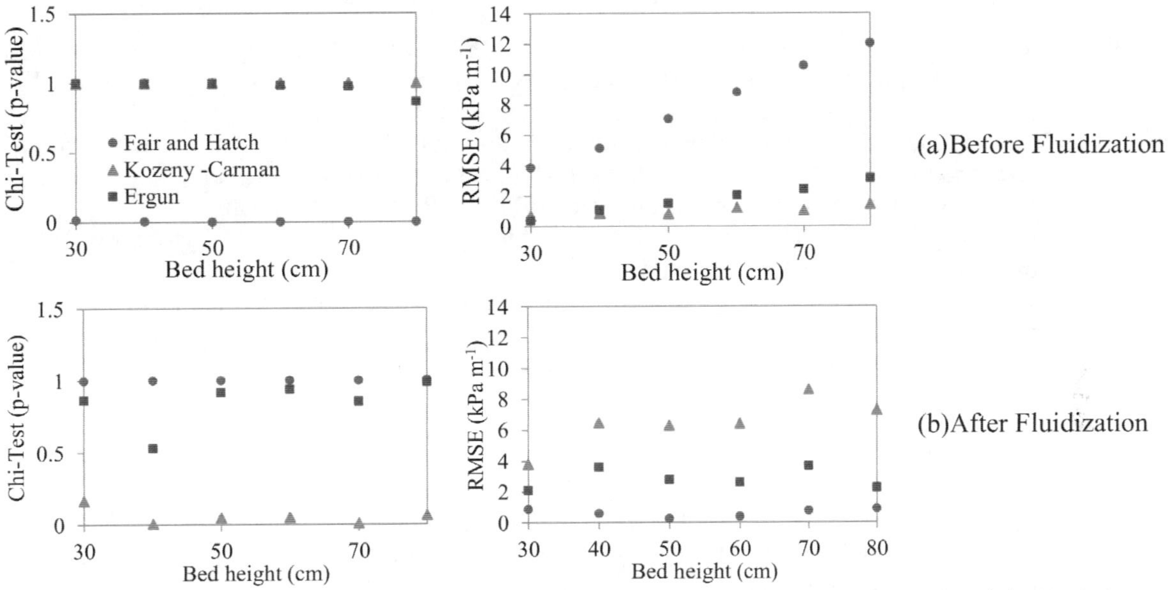

Figure 5. Experimental Result Analysis against Empirical Model Value with Bed Height Variation

4. Conclusion

A small scale upflow sand filter was constructed using natural sands obtained from three local rivers around Sabah, Malaysia. Physical characterization was conducted to determine the grain size, porosity, bulk density, particle density and sphericity of the sand, where the grain size were SL (2.5 mm) > PT (1.7 mm) > SK (1.5 mm). The sand porosities (0.4 to 0.43) are suitable for removal of pollutants, especially bacteria in a deep filtration process. Pressure drop analysis showed that a smaller grain size contributes to a higher pressure drop. Maximum pressure drop recorded were: SK (15.85 kPa m^{-1} at 0.747 m s^{-1} v_s) > PT (10.18 kPa m^{-1} at 0.352 m s^{-1} v_s) > SL (9.24 kPa m^{-1} at 0.747 m s^{-1} v_s). The effect of the sphericity of the sand on the pressure drop was not observed since the variance of sphericity index was not obvious. Smaller sand grain, although could increase the removal efficiency, it also increases the pressure drop and pumping energy requirement. The pressure drop per unit length in the upflow filter bed was greatly affected by the depth of the filter bed. A deeper filter bed showed a higher pressure drop across the filter: 30 cm at 7.96 kPa m^{-1}, 40 cm at 13.86 kPa m^{-1}, 50 cm at 18.03 kPa m^{-1}, 60 cm at 22.94 kPa m^{-1}, 70 cm at 28.68 kPa m^{-1}, and 80 cm at 32.54 kPa m^{-1}. Although, the maximum pressure drop occurred at around similar v_s (0.747 – 0.878 m s^{-1} v_s). A deeper filter bed may not be ideal for optimum operation of the filter in terms of energy consumption as smaller depth could present better or equivalent efficiency in pollutant removal. Therefore, the optimization of filter bed height can only be done with the additional knowledge of pollutant removal efficiency with respect to filter height. Comparison of experimental results against prediction model showed that the pressure drop in the upflow sand filter could be predicted using the Kozeny-Carman equation preceding filter bed fluidization. After fluidization occurs, pressure drop lowers and then stabilizes. The fluidized filter bed pressure drop can then be predicted using the Fair and Hatch's equation. The obtained knowledge of prediction models provides a preliminary understanding on upflow sand filters with locally obtained sands. This shows that further development of small scaled, point-of-use upflow sand filter that can be assembled locally are highly promising in order to deliver solutions for rural areas water provision.

5. Acknowledgment
This research has been financially supported by Universiti Malaysia Sabah through UMSGreat Research Grant (GUG0247) and the Ministry of Science, Technology and Innovation through PRGS Grant (PRGS0011).

6. Reference
[1] World Health Organization WEDC 2013 Emergency treatment of drinking-water at the point of use *Tech. notes Drink. water, Sanit. Hyg. emergencies* 3–6
[2] Zin T, SabaiAung T, Saupin S, Myint T, KhinSN D, Soe Aung M 2015 Influencing Factors for Cholera and Diarrhoea: Water Sanitation and Hygiene in Impoverished Rural Villages of Beluran District, Sabah Malaysia. Malaysian *J. Public Heal. Med.* **15** 30–40
[3] Barret, JM, Bryck J, Collins MR, Janonis BA, Logsdon GS 1991 Manual of Design for Slow Sand Filtration. *AWWA Res. Found. AWWA* 288
[4] Sarbatly R, Lahin FA, Ken CC 2020 The outlook of rural water supply in developing country: Review on Sabah, Malaysia *J Borneo Sci.* **41** 19–43
[5] Heikal G, Wagdy, R, Eldidamony G 2017 Bacteriophage Removal Using Upflow Biosand filter: A Laboratory Study. *Am. Sci. Res. J. Eng. Technol. Sci.* **1** 118–125
[6] Smith CK 1979 *Tangentially fed upflow sand filter method and apparatus*. 845341
[7] Pratap MR 2005 *Upflow Filter for Rapid and Effective Treatment of Stormwater at Critical Source Areas*. Pennsylvania State University
[8] Bové J, Arbat G, Duran-Ros M, Pujol T, Velayos J, Ramírez de Cartagena F, Puig-Bargués J 2015 Pressure drop across sand and recycled glass media used in micro irrigation filters. *Biosyst. Eng.* **137** 55–63
[9] Soyer E, Akgiray O 2009 A new simple equation for the prediction of filter expansion during backwashing *J. Water Supply Res. Technol. - AQUA* **58** 336–345
[10] Li X, Pashley RM 2016 A study on the characteristics of upflow matrix filter materials for the treatment of domestic sewage water *J. Water Process Eng.* **9** 179–187
[11] Krumbein WC 1941 Measurement and geological significance of shape and roundness of sedimentary particles *J. Sediment. Petrol.* **11** 64–72
[12] McCabe WL, Schmith JC, Harriot P 2001 *Unit operations of chemical engineering* (New York: McGraw-Hill) p 133
[13] Rutledge SO, Gagnon GA 2002 Comparing crushed recycled glass to silica sand for dual media filtration *J. Environ. Eng. Sci.* **1** 349–358
[14] Aronino R, Dlugy C, Arkhangelsky E, Shandalov S, Oron G, Brenner A, Gitis, V 2009 Removal of viruses from surface water and secondary effluents by sand filtration *Water Res.* **43** 87–96
[15] Mesquita M, Testezlaf R, Ramirez JCS 2012 The effect of media bed characteristics and internal auxiliary elements on sand filter head loss *Agric. Water Manag.* **115** 178–185
[16] Summerfelt ST 2006 Design and management of conventional fluidized-sand biofilters *Aquac. Eng.* **34** 275–302
[17] Fair GM, Hatch LP 1933 Fundamental factors governing the streamline flow of water through sand *Am. Water Work. Assoc.* **25** 1551–1565
[18] Brouckaert BM 2004 *Hydrodynamic detachment of deposited particles in fluidized bed filter backwashing* (Georgia Institute of Technology)
[19] Cocco R, Karri SBR, Knowlton T 2014 Introduction to fluidization *Chem. Eng. Prog.* **110** 21–29
[20] Escudero D, Heindel TJ, 2011 Bed height and material density effects on fluidized bed hydrodynamics *Chem. Eng. Sci.* **66** 3648–3655
[21] Solé-Torres C, Puig-Bargués J, Duran-Ros M, Arbat G, Pujol J, Ramírez de Cartagena F 2019 Effect of underdrain design, media height and filtration velocity on the performance of micro irrigation sand filters using reclaimed effluents *Biosyst. Eng.* **187** 292–304

IOP Conf. Series: Materials Science and Engineering 1192 (2021) 012009 doi:10.1088/1757-899X/1192/1/012009

Rapid Screening of Microfibrillated Cellulose Tructure Through FTIR and Principal Component Analsysis

N F N M Nor[1], N F M Azmin[1*], A L Asnawi[2*], D N Jimat[1], S Abdullah[3]

[1]Department of Biotechnology Engineering, International Islamic University Malaysia, Jalan Gombak, 53100 Kuala Lumpur, Malaysia
[2]Department of Electrical Engineering, International Islamic University Malaysia, Jalan Gombak, 53100 Kuala Lumpur, Malaysia
[3]Faculty of Chemical and Process Engineering Technology, Universiti Malaysia Pahang, Lebuhraya Tun Razak, 26300 Kuantan, Pahang, Malaysia

*Corresponding author's email: norfadhillah@iium.edu.my; aniliza@iium.edu.my

Abstract. Analysis of FTIR spectra combined with multivariate statistical analysis technique specifically Principal Component Analysis was used for rapid screening of microfibrillated cellulose (MFC) structure. The current methods used to extract the MFC are by using the chemical and physical approaches. To date, most researchers focused on bench (lab) scale experiment to identify the structure of MFC. Lack of mathematical models focusing on this goal has motivated this project. Principal component analysis is applied to identify the chemical composition of the MFC. The dataset comprises FTIR spectra of 12 samples that comes from MFC with different particles sizes, 200 µm, 250 µm and 800 µm. The result shows that the wavelength region which represents the MFC structure is in the range of 2950 cm^{-1} to 2978 cm^{-1} for particle size of 200 micrometer since it has larger surface area for penetration of fungal into the biomass due to lower diffusion of air, water and metabolite intermediates of which cellulose can be easily hydrolyzed due to increase in pore size of substance through greater removal of hemicellulose and lignin. The overall result indicates that the combination of FTIR analysis and PCA is a useful technique for rapid screening of MFC structure.

1. Introduction

Microfibrillated cellulose (MFC) can be referred as a larger unit of approximately more than thirty individual cellulose molecules brought together by biomass [1]. It is obtained through fibrillation of lignocellulose materials [2]. The current methods for screening of MFC structure are slow, laborious, and use harsh reagents which are impractical to use in screening process. This necessitates the search for alternative methods and one of the under exploration is an approached called chemometrics. Chemometric and Intelligent Laboratory Systems, the leading journal in the field of chemometric defines chemometric as the chemical discipline that uses mathematical, statistical, and other methods employing formal logic design or select optimal measurement procedures experiments, and to provide maximum relevant chemical information by analysing chemical data [3].

FTIR spectroscopy was selected in this study due to its simplicity, rapid and non-destructive measurements of chemical and physical compositions, high sensitivity and selectivity as well as real

IOP Conf. Series: Materials Science and Engineering 1192 (2021) 012009 doi:10.1088/1757-899X/1192/1/012009

time detection on the effect of fungal pre-treatment and ultrasonication for different particle sizes of the MFC structure. As shown by Amir [4] in their study to identify wheat varieties, the FTIR analysis enables them to quickly visualize the chemical composition and estimate the quality parameters of different types of wheat.

However, additional data processing technique is required due to the complexity in unmasking the spectra. Application of the multivariate statistical technique for the analysis of the FTIR data was necessary in order to perform feature extraction process, proper evaluation, and identification of the obtained spectra. Process that involves analysing the biomass structure and monitoring changes that occur from physical, chemical or biologicaltreatments currently required the non-destructive analytical tool which generated hundreds of spectra within short period of time [5]. Basically, pre-treatment process aimed to remove lignin and hemicellulose, reduce crystallinity of cellulose as well as increase the porosityof the lignocellulosic materials [6,7]. This study aimed to identify the effect of particle sizes on the screening process since large particle size will limit the penetration of fungi into biomass due to lower diffusion of air, water and metabolite intermediates.

Principal Components Analysis (PCA) involves the establishment of the grouping structures of various lignocellulosic biomasses by transformation of the correlated variables in the multivariate data into the uncorrelated variables which is principal components (PCs).The main goals in PCA can be illustrated as the process of extracting the important information from data tables in order to reduce the size of data by providing the important information that would simplified the set of data based on the analysis of the observations and variables structure. In PCA, important data is express as principal components in which it represents a set of orthogonal variables. Thus, the combination of FTIR and multivariate statistical analysis are well suited for identification and differentiation purposes even in very large data sets.

2. Materials and Methods

2.1. Raw Data

FTIR analysis was carried out from run 1 until run 12 with different particles size and inoculum concentrations. All samples were triplicate in SSF making the total of 12 samples with two independent variables particularly granular raw particle size (RPS) at 200, 525, and 850 μm and initial inoculum size (IS) with the compositions of 2.0, 3.5, and 5.0% (w/v) of CPH.

2.2. Data Pre-processing

The result obtained from FTIR analysis was pre-processed by organizing the spectral data to obtain the appropriate data matrix subjected to the PCA algorithm. FTIR dataset consists of 12 samples. The dataset was transposed and as a result, the rows represent the wavenumbers whilst the columns represent the samples of the MFC. Direct data matrix is not commonly practice in characterization of MFC structure since the components was analysed according to the variation of the absorbance frequency in each band while in transposed matrix, factors are computed based on the intensity of absorbance relative to the whole spectrum. Hence the identification of MFC structure was performed by applying PCA on the transposed matrix.

2.3. Principal Component Analysis

PCA was executed using MATLAB software using the following steps.

Step 1: Normalization of the dataset

Each row in the dataset was normalised using the zscore function in MATLAB's library.

Step 2: Obtaining the eigenvalues and eigenvectors

Eigenvalues and eigenvectors were obtained from the covariance matrix. Variabilities in the data were explained by the magnitude of eigenvalues. Scree plot was used to explain the data variability. The x-

IOP Conf. Series: Materials Science and Engineering 1192 (2021) 012009 doi:10.1088/1757-899X/1192/1/012009

axis in scree plot represents the index of the number of eigenvalues whereas y-axis denoted as magnitude of the eigenvalues. Thus, values in vector Y must not include any non-negative function and NaNs.

Step 3: Generating plots from the principal components analysis
A scree plot, score plots and loading plot were generated subsequent to the PCA to enables extraction of information.

3. Results

3.1. Scree Plot
The scree plot enables the determination of number of principal components to be kept or discarded because it indicates the relative importance of each principal component. The number of principal components retained should represents at least 70 to 80 percent of the variance in the data. In this study, only the first two principal components. PC1 and PC2, were selected while the rest of the principle components can be discarded as shown in Figure 1. PC1 explained 92% variability of the data and PC2 explained 2% of data variability. Total percentage of variance explained by PC1 and PC2 is 94% in which within the acceptable range for a validated result.

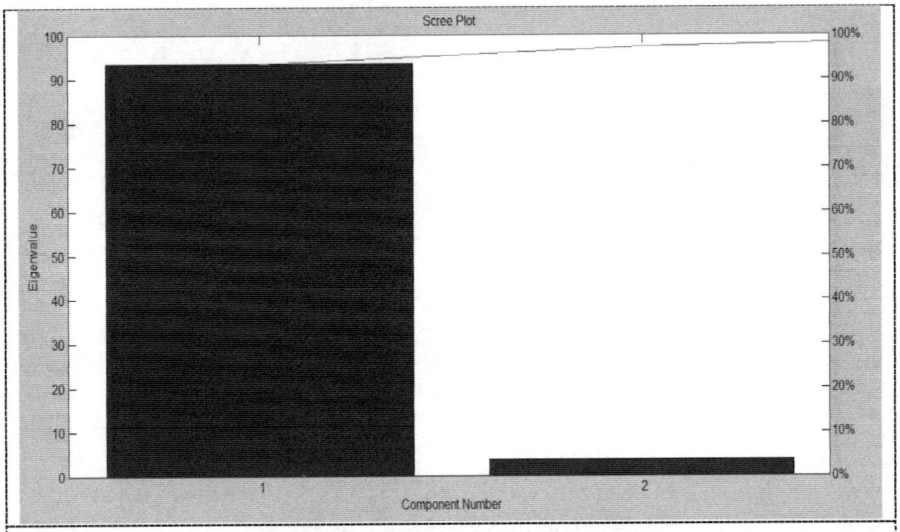

Figure 1. Scree plot of the data.

3.2. Score Plot
Score plot indicates the relationship between the samples by providing the summaries of the observations. For PC1, the samples can be grouped into two groups which are the positive region and negative region. Sample 1,3,4,5 and 9 can be group into one group which is positive region in PC1 as they are lie on the positive region of PC1 while sample 2, 6, 7, 8, 10, 11 and 12 lie on the negative side of PC1. So, they can be grouped into one group. Sample 1, 3 and 4 have similarity in their particle size which is 200 μm. Sample 6, 7 and 8 represents the sample with particles size of 525 μm. Particle size of 850 μm can be represented by the sample 10, 11 and 12. The control samples denoted as sample 1, 5 and 9. Based on the observation and inference, a conclusion can be made for PC1 which it is distinguishing the particles size of 200 μm with particles sizes of 525 μm and 850 μm with the exception

of sample 2 which is lie on negative side of PC1 due to less amount of fungus penetrate into the biomass in which it provide inefficient degradation of hemicellulose and lignin.

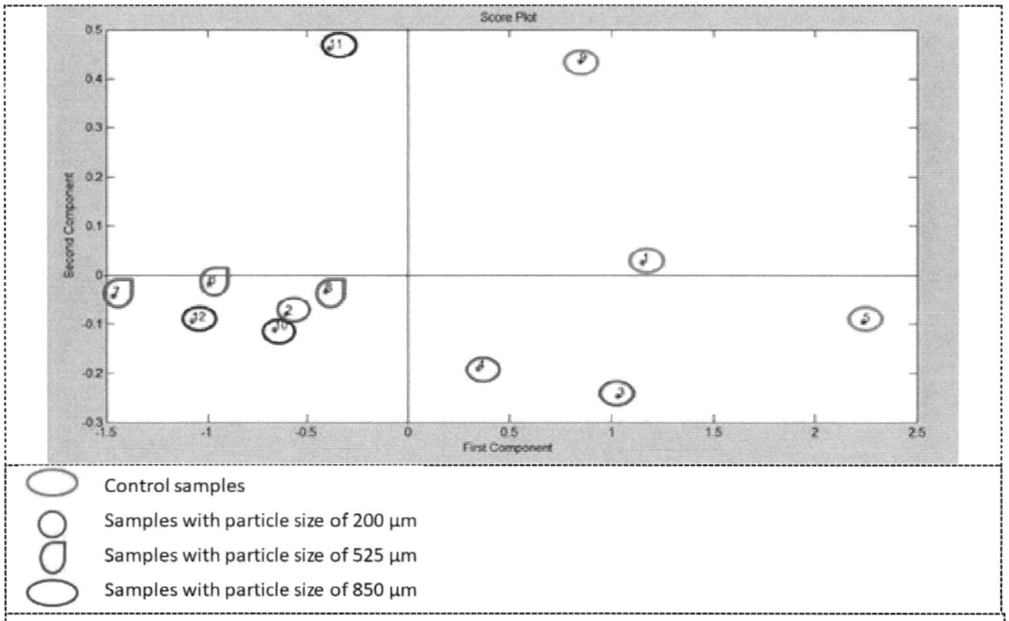

Figure 2. Score plot of PC1 versus PC2.

3.3. Loading Plot

This loading plot provides relationship between all variables. In loading plot, each variable is represented as a vector while direction and length of the vector indicate how variable contributes to the principal components in the plot. Figure 3 represents the bi-plot of which it consists of both score and loading of a selected component.

Based on this loading plot, it can be observed that MFC structure was presence within the wavelength ranging from 2950 cm-1 to 2978 cm^{-1}. It can be clearly seen that the particle size of 200 µm indicate the presence of MFC structure in which it is important variable to PC1 and the sample of run 3 and 4 are closest to this region of MFC structure since these samples have highest correlation with the wavelength representing the region of MFC structure. FTIR data currently focused on the optimization of MFC from cocoa pod husk (CPH) substrate in two distinct processes. Substrate was subjected to fungus-pre-treatment at the first stage through a solid-state fermentation by *Aspergillus nig*er to promote enzymatic splitting of cellulose. The fungus pre-treated substrate was then mechanically processed by the application of high-intensity ultrasonication. The optimum process of the fungus pre-treatment can be obtained from samples 3 and 4 of 200 µm since in these samples; fungus facilitated the splitting and reducing size of individual fiber into the microfibrils size. It can be concludes that particle size of 200 µm have larger surface area for penetration of fungus into the biomass due to lower diffusion of air, water and metabolite intermediates in which cellulose can be easily hydrolysed due to increase in pore size of substrate through greatly removal of hemicellulose and lignin. Thus, fungal pre-treatment provides a significant effect towards the MFC particle size produced by providing the required enzyme of cellulase to assist the ultrasonic process to split and reduce the size of individual fiber into the microfibrils size.

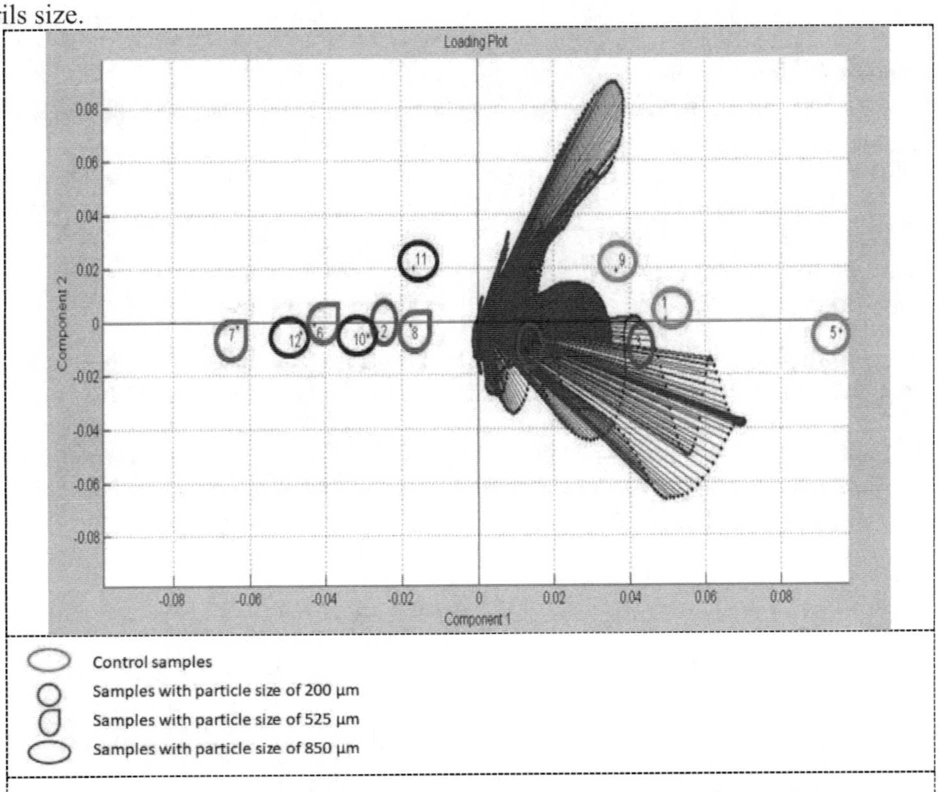

Figure 3. Bi-plot, a combination of score plot and loading plot.

4. Discussion

Based on the scree plot obtained, only the first two principal components, PC1 and PC2 were selected while the rest of the principal components were discarded. Remaining subsequent principal components explained percentages less than 5. PC1 explained 92%of the data variance and whereas PC2 explained 2%. Total percentage of variance explained by PC1 and PC2 is 94% which within the acceptable range for a validated result. The reliability of the analysis depends on the variance explained and the number of principal component (PCs) used.

Rapid screening of the MFC in cocoa pod husk biomass is based on score plot and divided according to the sample types in order to provide further understanding on the relations between them. The division is based on the number of runs with differentparticles size through combinations of fungal pre-treatment with the high-intensity ultrasonification. Samples were divided into control runs and actual runs. A control run indicates the sample undergoes ultrasonification without fungal pre-treatment. Basically, there are 3 control runs for 3 different particle sizes. There are 9 actual runs in which it undergoes fungal pre-treatment and ultrasonification. These runs are categorized equally according to 3 different particle sizes. Ultrasonification reduces the size of biomass by altering structure of hemicellulose and lignin. Efficiency of the ultrasonic pre-treatment depends on the size of particles. Samples obtained through combination of fungal pre-treatment and ultrasonification will remove the major portion of the hemicellulose and lignin.

For PC1, the samples can be grouped into two groups which are the positive regionand negative region. Sample 1,3,4,5 and 9 can be grouped into one group which is positive region in PC1 as they lie on the positive region of PC1 while sample 2, 6, 7, 8, 10, 11 and 12 lie on the negative side of PC1. So, they can be grouped into one group. Sample 1, 3 and 4 have similarity in their particle size which is 200 μm. A conclusion can be made for PC1 which it is distinguishing the particles size of 200 μm with particles sizes of 525 μm and 800 μm with the exception of sample 2 which is lie on negative side of PC1 due to less amount of fungal penetrate into the biomass in which it provides inefficient degradation of hemicellulose and lignin.

Similar to PC1, PC2 also can be grouped into positive region and the negative region. Only sample 1,9 and 11 are in the positive region of PC2. Sample 9 and 11are located closed to each other because they are underlying within similar particles sizes which are 850 μm. Sample 1 is located far away from sample 9 and 11 because they are different types of particles size. Based on the observation that have been made, PC2 differentiates the particles size of 850 μm with the rest of the other particles' sizes with the exception of sample 10 and 12 in which it lies between negative region of PC2. This might due to the increasing in particle size result in smaller surface area to volume ratio in which cellulose structure is more accessible for degradation by fungal.

Based on the loading plot, it can be observed that MFC structure was presence within the wavelength ranging from 2950 cm^{-1} to 2978 cm^{-1}. It can be clearly seen that the particlesize of 200 μm indicate the presence of MFC structure in which it is important variable to PC1 and the sample of run 3 and 4 are closest to this region of MFC structure since these samples have highest correlation with the wavelength representing the region of MFC structure. It can be concluded that particle size of 200 μm have larger surface area for penetration of fungal into the biomass due to lower diffusion of air, water and metabolite intermediates in which cellulose can be easily hydrolyzed due to increase in pore size of substrate through greatly removal of hemicellulose and lignin.

5. Conclusion

FTIR analysis is an infeasible approach in screening of MFC since large number of spectral involves and hardly distinguishable by visual inspection. Integration between FTIR analyses with PCA analysis will establish the underlying groups' structure of lignocellulose biomass in order to improve the screening process. Algorithm will record the peak present in each spectrumincluding the presence of their functional groups as well as corresponding peak area. Thus,PCA analysis provides a relationship between the effects of particle sizes on the screening process as the optimum fungal pretreatment occurred in samples 3 and 4 for particle sizes of 200 μm in the wavelength region ranging between 2950 cm^{-1} to 2978 cm^{-1} since it reduces the size of individual fiber into the microfibrils size.

Acknowledgments

This paper was part of works conducted under the IIUM-UMP-UITM Sustainable Research Collaboration 2020 grant (SRCG20-020-0020). The authors would also like to acknowledge all supports given by the IIUM Research Management Centre through the grant.

References

[1] Y. Habibi, L.A. Lucia and O.J. Rojas. (2010). Cellulose nanocrystals: Chemistry, self-assembly, and application. *Chemica Reviews*, 110 (6), pp. 3479-3500.

[2] D. N. Jimat, H. M. Salleh, and H. Anuar. (2016). Effect of Fungal Pretreatment by Aspergillus Niger Assisted by Ultrasonication on Microfibrillated Cellulose Obtained from Cocoa Pod Husk. *Akademia Baru*, 21(1), pp. 1–13.

[3] Tauler, R. *Chemometric and Intelligent Laboratory Systems*. https://www.journals.elsevier.com/chemometrics-and-intelligent-laboratory-systems

[4] R. M. Amir, F. M. Anjum, M. I. Khan, M. R. Khan, I. Pasha and M. Nadeem. (2013). Application of Fourier transform infrared (FTIR) spectroscopy for the identification of wheat varieties. *Journal Food Science Technology*, 50(5), pp. 1018-1023. [Online]. Available: http://doi: 10.1007/s13197-011-0424-y

[5] K. Banas, A. Banas, H. O. Moser, M. Bahou, M. Cholewa, P. Yang and S. K. Lim. (2010). Multivariate statistical methods in the forensic investigation of the post-blast residues measured by Fourier transform infrared spectroscopy. *Diamond Light Source Proceedings*, 1, pp. 7–10.

[Online]. Available: http://doi.org/10.1017/S2044820110000900.
[6] J. S. Lupoi, S. Singh, B. A. Simmons and R. J. Henry. (2014). Assessment of Lignocellulosic Biomass Using Analytical Spectroscopy: an Evolution to High-Throughput Techniques. *BioEnergy Resources*, pp. 1–23. [Online]. Available: http://doi.org/10.1007/s12155-013-9352-.
[7] A. K. Kumar and S. Sharma. (2017). Recent updates on different methods of pretreatment of lignocellulosic feedstocks: a review. *Bioresources and Bioprocessing*, 4. [Online]. Available: http://doi.org/10.1186/s40643-017-0137-9.

Performance Test of Solar – Powered Ice Maker Machine: Case Study in Bantul, Yogyakarta

P Wullandari[1] and B B Sedayu[1]

[1]Indonesian Research Institute for Fisheries Post-harvest Mechanization, Ministry of Marine and Fisheries, West Imogiri Street km 11.5, Jetis, Bantul, Special Region of Yogyakarta, Indonesia 55781

E-mail: utides@gmail.com

Abstract. Research on performance test of a solar-powered ice maker machine has been conducted in Bantul, Yogyakarta. This study aimed to observe the correlation between intensity of sunlight to the power battery capacity rates generated from solar panels in regard with the performance of ice maker machine. The testing was conducted during various weather conditions i.e. sunny, cloudy and light rain. The type of ice maker observed was a flake ice maker machine with specifications of the production capacity of 105 - 120 kgs/day, producing flake ice with dimensions of 2 x 3 x 3 mm^3. The energy of the machine was generated by nine solar panels with maximum power of 200 Wp (watt peak) per panel. A set of three panels was arranged in series, it was then coupled to other sets in parallel. The results showed that the power battery capacity was in corresponded to the sunlight intensity during sunny weather with the correlation: $y = 0.009x - 26.08$, while during cloudy dan raining conditions, the power capacity rates of the battery showed a declining with the correlation: $y = 0.008x - 23.92$ and $y = 0.007x + 69.41$, respectively. The ice production capacity during sunny, cloudy and light rainy weathers were 4.34 kg ice/hour; 4.63 kg ice / hour and 4.17 kg ice / hour respectively. Input power from solar panels depends on the intensity of sunlight. The ice produced by ice makers in cloudy weather conditions is much greater than the ice produced during sunny or rainy weather conditions.

Keywords: ice maker, solar power, flake ice, various weather, performance

1. Introduction

Nowadays, we still rely heavily on fossil energy such as coal, natural gas and oil, which accounted for 78.3% of global energy consumption in 2014. Fossil energy is mainly used in the form of direct combustion to generate electricity and heat energy, which can cause environmental problems, such as air pollution and intensive CO_2 emissions. Renewable energy with the advantages of unlimited resource and cleanliness can help us solve the problems from fossil energy. Renewable energy includes solar, wind, geothermal energy and so on, where solar energy is recognized as an important energy because of its equal distribution and cleanliness characteristics. Solar energy reaches the earth's surface with radiation, and can be utilized in two ways, namely: 1) with a solar collector to produce heat energy, 2) by using photo voltaics to produce electrical energy. Statistics show that the total global solar thermal capacity and solar PV capacity reached 435 GW and 227 GW in 2015 respectively [1]

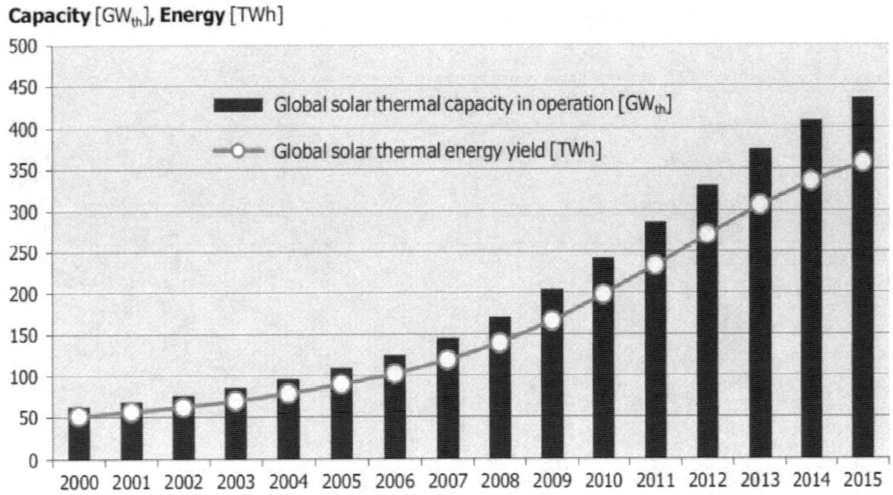

Figure 1. Global solar thermal capacity in use and annual energy output from 2000 to 2015.
(Source : [2])

Based on irradiation data collected from 18 locations in Indonesia, solar radiation in Indonesia can be classified as follows : for the western and eastern regions indonesia with the distribution of illumination in the Western Region of Indonesia (KBI) around 4.5 kWh/m^2/day with a monthly variation of about 10%; and in the Eastern Region Indonesia (KTI) is about 5.1 kWh/m^2/day with a monthly variation of about 9%. Thus, Indonesia's average solar exposure potential is around 4.8 kWh/m^2/day with a monthly variation of about 9%. [3]

Solar energy can not only be used directly in heating but can also be used to generate cooling power. When compared with heating systems with solar energy, cooling systems require more solar irradiation energy. Solar thermal energy is transferred to cooling power which operates on an adsorption / absorption principle using different absorption / absorption work pairs [1].

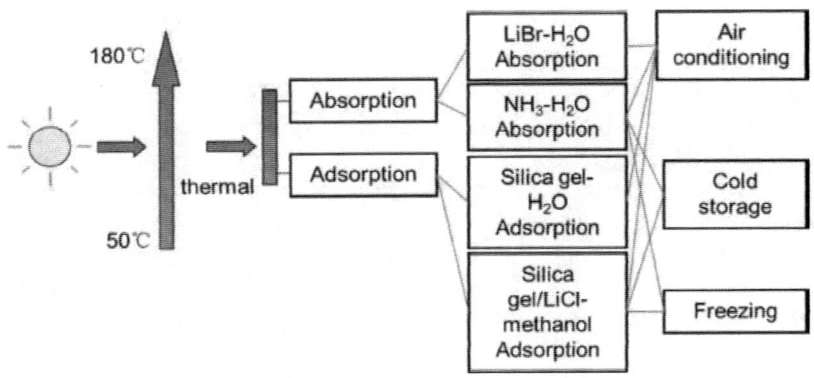

Figure 2. Schematic drawing of solar thermal cooling technology.
(Source : [1])

IOP Conf. Series: Materials Science and Engineering 1192 (2021) 012010 doi:10.1088/1757-899X/1192/1/012010

2. Materials and Methods

The hybrid ice machine (powered by electricity and solar energy) consists of three circuits, namely : (1) Ice maker, (2) Solar panels and (3) Box contains an inverter. Specification of ice maker machine are as follows : has a capacity of 105 - 120 kg ice / day, produces flake ice with dimensions of 2 x 3 x 3 mm [4].

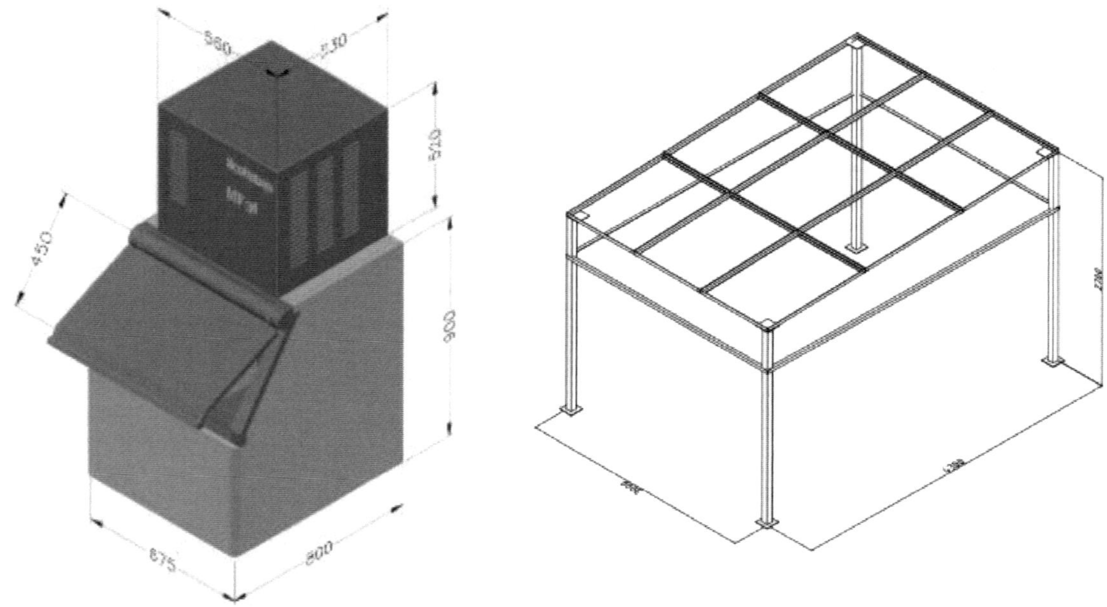

(a) (b)

Figure 3. (a) Ice maker machine (Source : [4]) and (b) Arrangement of two hundred Watt peak solar panels. (Source : [4])

There are nine solar panels with polycrystalline tipe used in this machine with a maximum power of 200 Wp (watt peak) per panel. The three panels are arranged in series, then the three panels in the series are arranged in parallel.

Measurement of sunlight intensity is done with a lux meter. The working principle of lux meter is as follows: the sensor captures light, the light energy that shines on the photo cell is transmitted by the photo cell into electric current energy, the results of the measurements taken will be displayed on the panel screen, the reading of the results displayed by the panel screen is a combination of the wavelength effects that are captured by the light sensor. Inside the light sensor that has been opened, a green color appears which is a transducer sensor. In the transducer sensor there is a photoresistor or LDR (Light Dependent Resistor), which has a function of the luxmeter working principle [5]. LDR or Light Dependent Resistor is a type of resistor that its resistance value is influenced by the light being received. Magnitude resistance value in LDR depends on the amount of light received by the LDR itself [6].

Sunlight is captured by the solar panel, then the solar panel generates a voltage, after that the voltage enters the control, then enters the battery with the load, the heavier the load takes electricity from the battery. The ice maker performance test with solar power sources is carried out by measuring the intensity of sunlight (lux) per unit time, recording power input and voltage from the solar panel, battery voltage and capacity, battery charging current, current discharge current from the battery and the weather conditions at the time of testing.

3. Results and Discussion

In photovoltaic (PV) solar systems, one of the main problems at present is that only 15 - 20% of the solar energy captured in PV panels can be converted into electricity, with the rest converted to heat [2]. The unit used is lumen / m^2. Lumen (lm) is the photometric equivalent of the watt, incorporating the "standard observer" eye response. One watt is equivalent to 683 lumens at a wavelength of 555 nm [3]. The solar panel frame assembly is carried out in the southern part of our office (Figure 5).

Figure 4. Knockdown of the solar panel frame.

Figure 5. Research's place.
Source : Anonym, 2021 [7]

3.1. *The relationship between the intensity of sunlight and power input from solar panel in a sunny weather*

The relationship between the intensity of sunlight and power input from solar panel in a sunny weather is presented in Figure 6. Figure 6 shows that the relationship between sunlight intensity and power input from solar panels is linear with the equation of $y = 0.009x - 26.08$. The higher the intensity of sunlight received by solar panels, the greater the power input obtained from solar panels. The minimum intensity of sunlight is 5,890 lumens / m^2, while the maximum sunlight intensity is 102,000 lumens / m^2. The minimum power input from the solar panels is 27 watts, while the maximum power input from the solar panels is 1,026 watts.

The minimum capacity of the battery is 42% (at a voltage of 22.3 volts) while the maximum capacity is 100% (at a voltage of 24.6 - 25.8 volts). The minimum battery voltage is 22.3 volts, while the maximum battery voltage is 25.8 volts. The total ice production for 8 hours was 34.71 kg of ice, or about 4.34 kg of ice / hour.

In clear weather conditions, the earth's surface receives about 1000 watts of solar energy per square meter. Less than 30% of the energy is reflected back into space, 47% is converted into heat, 23% is used for the entire working circulation above the earth's surface, a small part of 0.25% is accommodated by wind and waves and there is still a very small part (0.025%) is stored through the process of photosynthesis in plants [3].

Figure 6 shows the relationship between the intensity of sunlight and the power input from the solar panels in Bantul in a sunny weather condition is linear with the equation $y = 0.009x - 26.08$. Thus it can be concluded as follows:

1. The power input from the solar panel depends on the sunlight intensity,
2. The value of 0.009x is called the slope which determines the direction of the linear regression. The higher the intensity of sunlight, the higher the power input from the solar panels (indicating a positive relationship),
3. This slope also estimates the rate at which the power input from the solar panels increases on a daily basis,
4. The value - 26.08 is referred to as the intercept. The results of this study mean that at the value of X = 0, the value of the power input is - 26.08 watts.

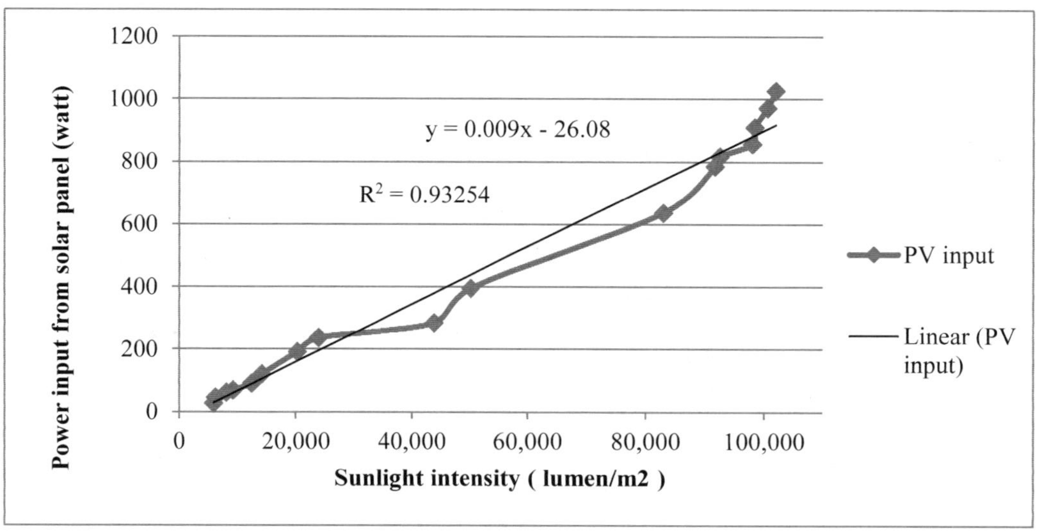

Figure 6. The relationship between the intensity of sunlight and power input from solar panel in a sunny weather.

The determination coefficient (R^2) is 0.933, so the correlation coefficient (R) is 0.966. This means the correlation between the intensity of sunlight with the power input of solar panel is 0.966. The meaning of the correlation value can be seen in Table 1.

Table 1. The explanation of correlation coefficient's value. [8]

Correlation coefficient's value	Explanation
0 – 0.199	Very low
0.2 – 0.399	Low
0.4 – 0.599	Enough
0.6 – 0.799	Strong
0.8 - 1	Very strong

From Table 1, it can be concluded that the correlation between the intensity of sunlight with the power input of solar panel is very strong.

3.2. The relationship between the intensity of sunlight and power input from solar panel in a cloudy weather

The relationship between the intensity of sunlight and power input from solar panel in a cloudy weather is presented in Figure 7. Figure 7 shows that the relationship between sunlight intensity and power input from solar panels is linear with the equation of $y = 0.008x - 23.92$, the higher the intensity of sunlight received by the solar panels, the greater the power input obtained from the solar panels. The minimum intensity of sunlight is 6,440 lumens / m^2, while the maximum sunlight intensity is 109,500 lumens / m^2. The minimum power input from the solar panels is 58 watts, while the maximum power input from the solar panels is 1,038 watts.

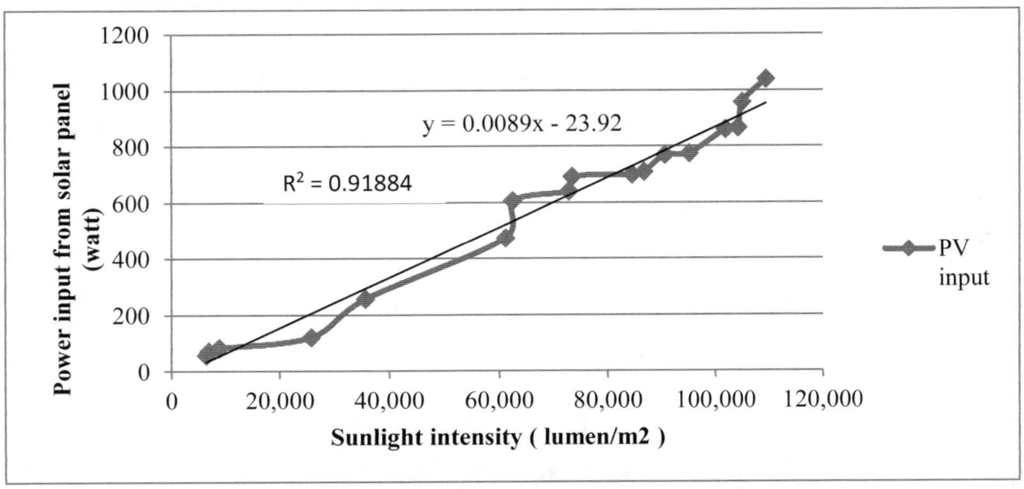

Figure 7. The relationship between the intensity of sunlight and power input from solar panel in a cloudy weather.

The minimum capacity of the battery is 60% (at 23 volts) while the maximum capacity is 100% (at 24.6 - 26.5 volts). The minimum battery voltage is 23 volts, while the maximum battery voltage is 26.5 volts. The total ice production for 8 hours was 37.03 kg of ice, or about 4.63 kg of ice / hour.

Figure 7 shows the relationship between sunlight intensity and the power input from solar panels in Bantul in cloudy weather conditions is linear with the equation $y = 0.008x - 23.92$. Thus it can be concluded as follows:

1. The input power from the solar panel depends on the intensity of the sunlight,

IOP Conf. Series: Materials Science and Engineering 1192 (2021) 012010 doi:10.1088/1757-899X/1192/1/012010

2. The value 0.008x is called the slope which determines the direction of the linear regression. The higher the intensity of sunlight, the higher the power input from the solar panels (indicating a positive relationship),
3. This slope also estimates the rate at which the power input from the solar panels increases on a daily basis,
4. Value - 23.92 is referred to as an intercept. The results of this study mean that at the value of X = 0, the value of the power input is -23.92 watts.

The determination coefficient (R^2) is 0.919, so the correlation coefficient (R) is 0.959. This means the correlation between the intensity of sunlight with the power input of solar panel is 0.959 (very strong) [8].

3.3. The relationship between the intensity of sunlight and power input from solar panel in a rainy weather

The relationship between the intensity of sunlight and power input from solar panel in a rainy weather is shown in Figure 8. Figure 8 shows that the relationship between sunlight intensity and power input from solar panels is linear with the equation y = 0.007x + 69.41, the higher the intensity of sunlight received by the solar panels, the greater the power input obtained from the solar panels. The minimum intensity of sunlight is 13,200 lumens / m^2, while the maximum intensity of sunlight is 84,000 lumens / m^2. The minimum power input from the solar panels is 97 watts, while the maximum power input from the solar panels is 739 watts.

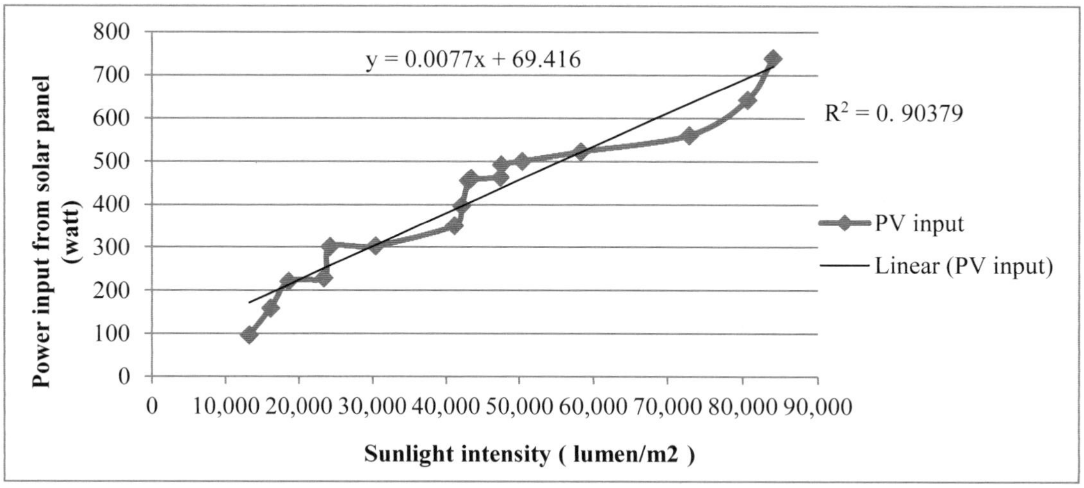

Figure 8. The relationship between the intensity of sunlight and power input from solar panel in a rainy weather.

The minimum capacity of the battery is 67% (at 23.3 volts) while the maximum capacity is 100% (at 24.6 - 25.3 volts). The minimum battery voltage is 23.3 volts, while the maximum battery voltage is 25.3 volts. The total ice production for 8 hours was 33.32 kg ice, or about 4.17 kg ice / hour. The initial water discharge is 0.18 liters / second.

Figure 8 shows the relationship between the intensity of sunlight and the power input from the solar panels in Bantul in rainy conditions is linear with the equation y = 0.008x - 23.92. Thus it can be concluded as follows :

1. Input power from solar panels depends on the intensity of sunlight,

2. The value 0.008x is called the slope which determines the direction of the linear regression. The higher the intensity of sunlight, the higher the power input from the solar panel (indicating a positive relationship),
3. This slope also estimates the rate at which the power input from the solar panels increases on a daily basis,

The determination coefficient (R^2) is 0.904, so the correlation coefficient (R) is 0.951. This means the correlation between the intensity of sunlight with the power input of solar panel is 0.951 (very strong) [8].

4. Conclusions

1. Input power from solar panels depends on the intensity of sunlight.
2. The ice produced by ice makers in cloudy weather conditions is much greater than the ice produced during sunny or rainy weather conditions
3. The higher the intensity of sunlight, the higher the power input from the solar panel (indicating a positive relationship),
4. The correlation between the intensity of sunlight with the power input of solar panel is very strong

5. References

[1] Anonym 2016 Renewables ; global status report http://www.ren21.net/wp-content/uploads/2016/10/REN21_GSR2016_FullReport_en_11.pdf , April 23, 2021
[2] Wang R Z, Xu Z Y, and Ge T S 2016 Introduction to solar heating and cooling systems 1. https://doi.org/10.1016/B978-0-08-100301-5.00001-1
[3] Widayana G 2012 Pemanfaatan energi surya. *Jurnal pendidikan teknologi dan kejuruan*. **9**. 10.23887/jptk-undiksha.v9i1.2876.
[4] Wullandari P, Hakim A R and Sarwono W 2019 Mesin pembuat es hibrid untuk mencukupi kebutuhan es di daerah 3T *Aplikasi teknologi pengelolaan perikanan tangkap* **Amafrad Press**, p. 59 – 189.
[5] Dewi P S 2011 Luxmeter https://www.slideshare.net/pramitasylvia/luxmeter-ppt, 13 Desember 2018
[6] Supatmi S 2021 Pengaruh sensor LDR terhadap pengontrolan lampu *Majalah Ilmiah UNIKOM* **8** 2
[7] Anonym 2021 Loka riset mekanisasi pengolahan hasil perikanan Retrieved from Loka Riset Mekanisasi Perikanan - Bantul - Google Maps
[8] Anggraeni M 2008 Kajian penggunaan *poly Aluminium Chloride* (PAC) dalam proses pemurnian nira aren dan lama pemurnian terhadap karakteristik nira aren (*Arenga pinnata Merr*). *Skripsi Fakultas Teknologi Industri Pertanian*, Universitas Padjadjaran.

IOP Conf. Series: Materials Science and Engineering 1192 (2021) 012011 doi:10.1088/1757-899X/1192/1/012011

Incorporation of Various Nanoclays in Semi-refined Carrageenan Film Composite

B B Sedayu[1,3], D Fransiska[2], P Wullandari[1], T D Novianto[1], W T Handoyo[1] and L Assadad[1]

[1]Indonesian Research Institute for Fisheries Postharvest Mechanization. Ministry of Marine Affairs and Fisheries. Jl.Imogiri Barat km.11,5. Bantul. Yogyakarta. Indonesia. 55781.

[2]Research and Development Center for Marine and Fisheries Product Processing and Biotechnology, Ministry of Marine Affairs and Fisheries. Jl.KS. Tubun, Petamburan VI. Jakarta. Indonesia. 10260

[3]Corresponding Author's Email: baktibsedayu@gmail.com, bakti.sedayu@kkp.go.id

Abstract. In current investigation, various commercial nanoclays with different level of hydrophilicity i.e., hydrophilic bentonite (HB), cloisite 10A, and cloisite® 30B were incorporated to a hydrophilic semi-refined carrageenan (SRC) films to observe their presence in film matrix in regards with the final film's properties. High shear mixing and sonication were applied to the film preparation to obtain high dispersion of nanoclays in the matrix. As expected, the hydrophilic clay showed better dispersion within the matrix as shown in Energy Dispersive X-Ray Spectroscopy (EDS) image. The more hydrophilic clay inclusion resulted in higher tensile strength, while the more hydrophobic resulted in higher stiffness of the films. The water vapour permeability was decreased in corresponded to the more hydrophobic clay incorporated, and the thermal properties of the films were overall enhanced by the nanoclays reinforcement. In general, nanoclays incorporation in SRC film improved the overall properties of the SRC film.

1. Introduction

Among many potential bio-based materials for replacement of conventional synthetic plastics, seaweed has recently attracted scientists for bioplastic development, as a promising candidate of biopolymer showing more advantages over other biomass particularly in term of its availability and sustainability [1]. Farmed seaweed from red algae species has been well known to be grown massively in tropical waters with consistent production increase every year. From this red seaweed species, we can harvest their main component polysaccharides namely carrageenan, and develop this material further into bioplastic [2]. There are two common types of carrageenan products available in market namely a refined-carrageenan (carrageenan) and a semi-refined carrageenan (SRC). For bioplastic production purposes, the later product is economically preferable due to its lower price. In comparison with carrageenan, SRC still contains residual cellulose since it takes less processing steps for the production.

Nonetheless, like many other bio-based plastic materials, carrageenan-based plastic has also exhibited some weaknesses such as high-water sensitivity and poor gas barrier properties [3] which are essential whenever they are used for food packaging applications. To obtain a fully functional bioplastic made of carrageenan, it requires some modification to enhance its properties such as blending with other biopolymer, incorporated with hydrophobic elements, or reinforce with nanoparticles [4-6].

The aforementioned techniques are common to be applied in biodegradable plastics production to obtain improved properties such as higher water resistant and greater mechanical or barrier properties. For example, starch that has been blended with hydrophobic biodegradable polymer like polylactic acid (PLA), polycaprolactone (PCL), and other polymers to improve its water sensitivity and tensile properties, and gas barrier properties [7]. Moreover, another more efficient method is also frequently applied in many biopolymers composite production i.e., nanoparticle reinforcement, which only incorporating a low level of nanofiller in the formulation, but it can improve the final biopolymer composite properties that equally loaded with 40−50 % of classical fillers [8, 9].

Among many commercially available nanomaterials used in biopolymer composite preparation, nanoclay has been well known as one of the most prevalent filler showing an ability to enhance the mechanical, barrier, and thermal properties of biopolymer film composites [10-12]. However, the successful reinforced polymer is dictated by interaction and interfacial chemistry between incorporated nanomaterials and polymer molecules, which results in intercalation or exfoliation morphologies within the polymer matrix [13]. Regarding this, compatibility between nanofillers and biopolymer matrix is a key factor to obtain property enhancement, as well as their amount in the biopolymer formulation [11].

In the present study, three different types of commercial nanoclays (montmorillonite nanoclays) were used in semi-refined carrageenan (SRC) film preparation, and their effects on the mechanical, barrier and thermal properties of the final film composites were investigated.

2. Materials and Methods

2.1. Materials

A food grade powdered semi-refined carrageenan (SRC) derived from red algae *Eucheuma cottonii* was purchased from W-Hydrocolloids, Inc. (the Philippines). Hydrophilic bentonite (montmorillonite clay), closite 10A (Clo.10A), and cloisite® 30B (Clo.30B) were purchased from Sigma Aldrich (Australia). Glycerol (92.08 g mol^{-1} MW) obtained from Sigma Aldrich was used as the plasticizer during film preparation.

2.2. Film preparation

For each film composite, a 5 percent mass of nanoclay relative to SRC (w/w) was firstly dispersed in 250 mL of Milli-Q and stirred for a night (±16 h) at 40 °C in a lid Pyrex bottle. It was then followed by high-speed shearing at 20,000 rpm for 10 min using a homogenizer CAT Unidrive X1000 (Germany) and sonication at 50 Hz and 80% amplitude for 10 min using a Hielscher ultrasonic processor UP400S (Germany). After allowing the solution reached ambient temperature, a 5 mg of SRC was added into the nanoclay solution and stirred for 10 minutes. The mixture was then heated at 90 °C for 30 minutes while stirring. During heating, a 40 % of glycerol relative to SRC (w/w) was added into the mixture. The mixture was then poured onto a 38 X 18 cm^2 acrylic rectangular tray and allowed to dry at ambient temperature for 36 hours.

2.3 Film's properties

Thickness. Three random locations on each film composite were measured the thickness using digital micrometer. The measurement was performed in triplicates.

Opacity. Prior to the measurement, each of film formulations was cut into an 80 × 12.5 mm^2 film strip, and at least three specimens were measured for each formulation. Film strip was mounted directly into the test cell of spectrophotometer (Biochrom Libra S12) and measured at λ = 550 nm following procedure by Gómez-Estaca, Giménez, Montero and Gómez-Guillén [14].

Water vapor permeability. Water vapor permeability (WVP) of the film specimens was measured following the previous method performed by Sedayu, Cran and Bigger [15].

Mechanical properties. Each of film formulations was cut into film strip of 120 × 20 mm^2 before testing by using Instron Universal Testing Machine Model 4301 (US). A 5 kN loadcell was used to measure the film samples with initial grid separation of 2.5 cm, and the testing was performed in accordance with the ASTM D 882-12 method [16]. The tensile strength (TS), E-modulus (EM), and elongation at break (EAB) values of the tested specimens were recorded and calculated automatically using an Instron BlueHill Universal software.

Fourier transform infrared (FTIR). Structural properties of the film specimens were observed using PerkinElmer Frontier FTIR spectrometer (USA) with attenuated total reflectance (ATR). The transmittances of infrared spectra were measured at ranges of 4000–600 cm^{-1} frequency by 16 scans.
Thermal properties. Thermographic analysis (TGA) was performed using Mettler Toledo Thermal Analyzer (Switzerland). Approximately 5 – 8 mg of composite film was placed into alumina sample pan and heated from 30 to 400 °C under nitrogen circumstance at a 10 °C/min heating rate to obtain TGA data with non-oxidative decomposition.

Statistical Analysis. The collected data were analyzed using One-way analysis of variance (ANOVA), and the significant different among the mean values were determined using a Duncan test at 5% significant level.

3. Results and Discussion

3.1. Physical and barrier properties
As shown in Table 1, the reinforcement of 5% (w/w) nanoclay in polymer matrix in general did not alter the thickness of the reinforced films, and this might be caused by very little amount of nanoclays included in the matrix, or too thin layer formation of the nanoclays across the polymer matrix, thus the changes were neglected. However, it can be seen that the opacity of composite films was increased after being reinforced with the nanoclays. The hydrophilic bentonite clay appeared to create more opaque film in compared with two other nanoclays with less hydrophilicity. Seems that level of dispersion of the nanoclay particles interfered the opacity of the film, the higher dispersion of the particles in the matrix the opaquer the film. Similar result was also observed by Zakuwan and Ahmad [17] after reinforcing carrageenan polymer with organically modified montmorillonite.

Table 1. Physical and barrier properties of the SRC formulations.

Film samples	Thickness/mm	Opacity	WVP/g mm cm^{-2} h^{-1} Pa^{-1} ($\times 10^{-8}$)
SRC	0.069 ± 0.001^{a}	10.92 ± 0.12^{a}	12.60 ± 0.15^{a}
SRC/HB	0.069 ± 0.001^{a}	12.42 ± 0.20^{b}	12.42 ± 0.09^{ab}
SRC/Clo.10A	0.069 ± 0.003^{a}	11.40 ± 0.29^{c}	12.36 ± 0.12^{b}
SRC/Clo.30B	0.069 ± 0.001^{a}	11.40 ± 0.41^{c}	12.22 ± 0.09^{b}

Values are given as mean with one standard deviation. Any two means in the same column followed by the same letter are not significantly different ($p > 0.05$) as determined by a Duncan's test.

The water vapor barrier of SRC film increased after nanoclay reinforcement. From the results above, hydrophilicity of the clays interferes the barrier properties of the films whereas the less hydrophilic clay inclusion resulted in greater decrease in WVP of the composite films. In principle, decreased WVP of composite film by nanoclay incorporation is attributed to tortuous pathway created by distribution of clay particles within the polymer matrix, which obstructs water vapor diffusion passing through the film [11]. In this investigation, the more hydrophobic clay visibly facilitates more impermeable layer bringing about more effective water vapor barrier than those of the hydrophilic clays. This can be explained by less hydrogen bonding, or physical/chemical interaction, between the hydrophobic nanoclay formation in polymer matrix and surrounding water molecules rendering a lower WVP than those of using hydrophilic nanoclay [18].

3.2. *Mechanical properties*

Mechanical properties of the composite films i.e., tensile strength (TS), E-modulus (EM), and elongation at break (EAB) are presented in Figure 1. According to the graphs below, incorporation of hydrophilic bentonite nanoclay increased the TS of SRC film by 3% ($p < 0.05$), however the SRC film's TS was oppositely decreased by 5 – 8 % after incorporation of the Cloisite 10A and Cloisite® 30B. These different behaviors might be interfered by compatibility of the clay particles within the SRC polymer structure, which is more compatible with the hydrophilic clay. Interaction such as chemical bonding between SRC polymer and clays particles determined the final properties of the film composite, and hydrophilic bentonite nanoclay in the current study appeared to have higher compatibility with SRC polymer. This phenomenon was also observed in our previous study by using energy dispersive X-ray spectroscopy (EDX) showing that the incorporation of hydrophilic nanoclays (hydrophilic bentonite nanoclay) resulting in greater dispersion across the carrageenan matrix in comparison with hydrophobic nanoclay (Cloisite® 30B) [19] (see Figure 2).

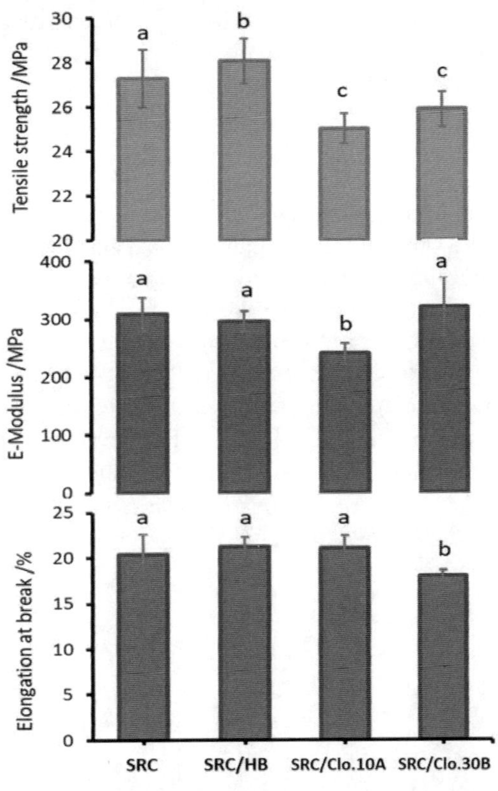

Figure 1. Mechanical properties of SRC reinforced films with hydrophilic bentonite (HB), Clo.10A, and Clo.30®B nanoclays. Values are plotted as the mean with a standard deviation error bar. Any labelled with the same letter are not significant different ($p < 0.05$) as determined by a Duncan test.

E-modulus (EM) value represents the stiffness of a composite film. As can be seen, there was little changes in the EM values of the SRC films due to nanoclay inclusion ($p > 0.05$), with the most prominent change was found in the Clo.10A inclusion (see **Fig.2**). While the nanoclay inclusion effects to stretchability (EAB) of SRC film, the increased EAB values were observed in the hydrophilic nanoclay inclusion. On the other hand, hydrophobic nanoclay (Clo.30B) oppositely decreased the film's stretchability. From these results, we can assume that the more hydrophobic nanoclay inclusion, the more rigid the film is. This result is also in agreement with previous study reported by Müller, Laurindo and Yamashita [3] showing the antiplasticization effect of nanoclay in biopolymer film.

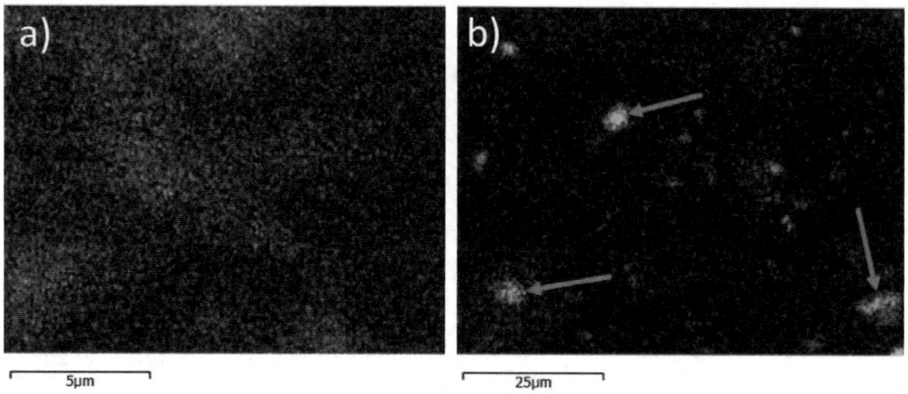

Figure 2. EDS images of SRC film reinforced with a) hydrophilic bentonite nanoclay; and b) Clo.30B. Silica particles (representing the nanoclay) agglomerations were showed by the red arrows (image b). *Source*. Sedayu, Cran and Bigger [19] with modification.

3.3. Thermal properties

The reinforcement of nanoclays in SRC composite film showed overall enhanced thermal stability of the films, this is reflected from the mass losses of samples during heating in TGA thermograms (see Figure 3a). Moreover, among the three different nanoclays used, the hydrophilic bentonite clay showed to increase the thermal stability of SRC film higher than those of other nanoclays by slightly shifting decomposition peak to higher temperature during the second stage of decomposition (see Figure 3b). From the dTGA curve (see Figure 3b), we can see a three major decomposition process of the SRC composite film during heating. The first stage decomposition occurred at temperature of ca. 90 °C that was attributed to evaporation of water and low-weight volatile compounds from the SRC composite film, it was followed by the second decomposition stage at temperature of ca. 205 °C corresponded to volatilization of plasticizer (glycerol) and middle-weight volatile compounds. At heating temperature above 260 °C, the third decomposition stage took place ascribed to the decomposition of main carrageenan polymer structure [6].

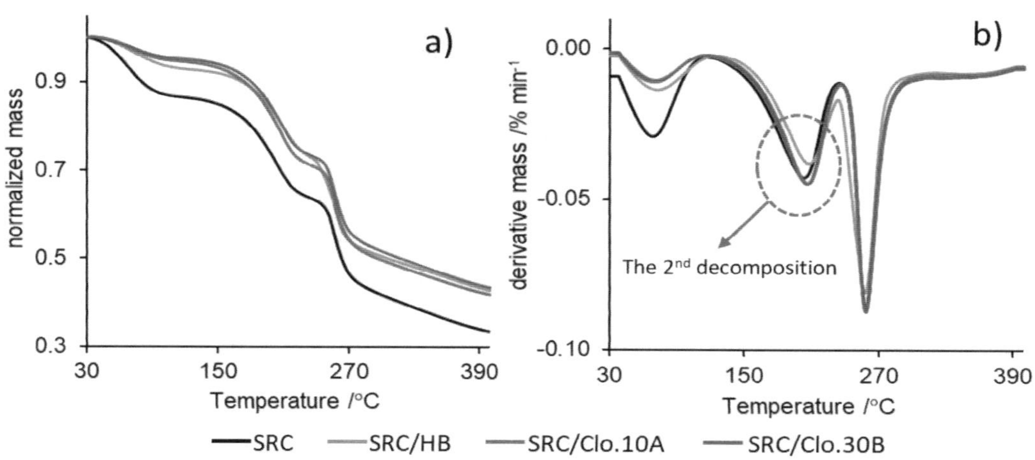

Figure 3. TGA and dTGA of the nanoclay reinforced SRC films.

3.4. Fourier Transform Infrared (FTIR)

The chemical bonds and functional groups of the composite films were analyzed by observing any distinguish stretching or peaks in the FTIR spectra compared to the control SRC film as presented in Figure 4. In general, all the spectra recorded from each film formulations exhibited very similar pattern, which shows a typical spectra of SRC film reported by Sedayu, Cran and Bigger [15]. This similarity can be caused by neglected small amount of nanoclay included in SRC film matrix, or it may also be an indicator of good miscibility or compatibility of nanoclay within the SRC polymer.

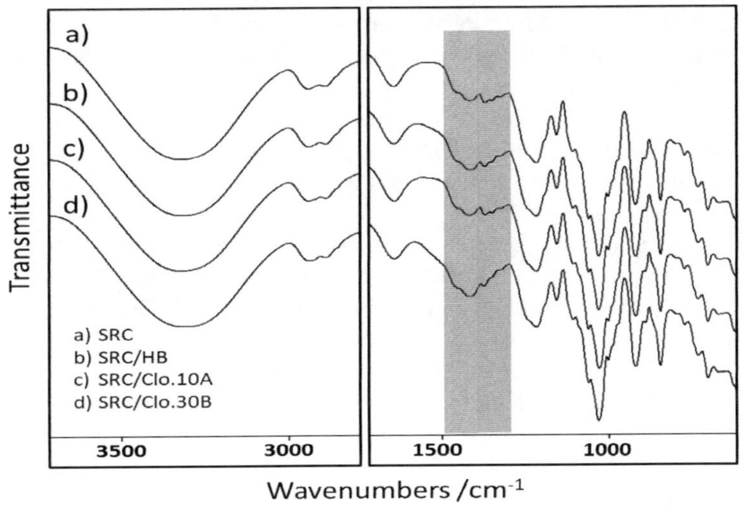

Figure 4. FTIR spectra of the nanoclay reinforced SRC films.

However, some minor changes are found in stretching bands at 1424 cm^{-1} corresponded to symmetric CH2 bending vibration [20], which was increased after the nanoclays reinforcement. Increasing of –CH2– alkyl groups may have correlation with the increasing of crystalline region in SRC polymer structure after nanoclay inclusion [21], and it is also in corresponded to the increased stiffness (EM) of the reinforced films (see **Fig. 1**).

4. Conclusion

Reinforcement of different types of nanoclays into SRC polymer matrix in general enhanced the barrier and thermal properties of the composite films. The SRC film was more compatible with hydrophilic nanoclay facilitating greater clay particles dispersion within the polymer matrix that subsequently resulting in higher tensile and thermal properties in compared with hydrophobic nanoclays. These results suggest a potential development of carrageenan-based plastic for wide applications, including for alternatif food packaging material.

References

[1] Rajendran N, Puppala S, Sneha Raj M, Ruth Angeeleena B and Rajam C 2012 Seaweeds can be a new source for bioplastics *Journal of Pharmacy Research* **5** 14769

[2] Sedayu B B, Cran M J and Bigger S W 2019 A Review of Property Enhancement Techniques for Carrageenan-based Films and Coatings *Carbohydrate Polymers* **216** 287-302

[3] Müller C M O, Laurindo J B and Yamashita F 2011 Effect of nanoclay incorporation method on mechanical and water vapor barrier properties of starch-based films *Industrial Crops and Products* **33** 605-10

[4] Harumarani S, Ma'ruf W F and Romadhon 2016 The Effect of Different Glycerol Concentration on Characteristic of Composite *Eucheuma cottonii* Semi Refined Carrageenan Edible Film and Beeswax *Jurnal Pengolahan & Bioteknologi Hasil Perikanan* **5** 101-5

[5] Rekemin N F N and Abedin N H Z 2017 Physicochemical characterization of kappacarrageenan (Euchema cottoni) based films incorporated with various plant oils *Carbohydrate Polymers* **157** 1479-87

[6] Sedayu B B, Cran M J and Bigger S W 2020 Reinforcement of Refined and SemiRefined Carrageenan Film with Nanocellulose *Polymers (Basel)* **12**

[7] O Fabunmi O, Tabil L, Panigrahi S and Chang P 2011 *Effects of Incorporating Polycaprolactone and Flax Fiber into Glycerol-Plasticized Pea Starch* vol 19

[8] Hejri Z, Ahmadpour A, Seifkordi A A and Zebarjad S M 2012 Role of Nano-Sized TiO2 on Mechanical and Thermal Behavior of Starch/Poly (vinyl alcohol) Blend Films *International Journal of Nanoscience and Nanotechnology* **8** 215-26

[9] Ray S S and Bousmina M 2005 Biodegradable polymers and their layered silicate nanocomposites: In greening the 21st century materials world *Progress in Materials Science* **50** 962-1079

[10] Aouada F A, Mattoso L H C and Longo E 2011 New strategies in the preparation of exfoliated thermoplastic starch–montmorillonite nanocomposites *Industrial Crops and Products* **34** 1502-8

[11] Rhim J W 2011 Effect of clay contents on mechanical and water vapor barrier properties of agar-based nanocomposite films *Carbohydrate Polymers* **86** 691-9

[12] Alboofetileh M, Rezaei M, Hosseini H and Abdollahi M 2013 Effect of montmorillonite clay and biopolymer concentration on the physical and mechanical properties of alginate nanocomposite films *Journal of Food Engineering* **117** 26-33

[13] Weiss J, Takhistov P and McClements D J 2006 Functional Materials in Food Nanotechnology *Journal of Food Science* **71** R107-R16

[14] Gómez-Estaca J, Giménez B, Montero P and Gómez-Guillén M C 2009 Incorporation of antioxidant borage extract into edible films based on sole skin gelatin or a commercial fish gelatin *Journal of Food Engineering* **92** 78-85

[15] Sedayu B B, Cran M J and Bigger S W 2018 Characterization of Semi-refined Carrageenan-Based Film for Primary Food Packaging Purposes *Journal of Polymers and the Environment* **26** 3754-61

[16] ASTM 2012 Standard test method for tensile properties of thin plastic sheeting. (West Conshohocken, PA: ASTM International)

[17] Zakuwan S Z and Ahmad I 2018 Synergistic Effect of Hybridized Cellulose Nanocrystals and Organically Modified Montmorillonite on kappa-Carrageenan Bionanocomposites *Nanomaterials (Basel)* **8**

[18] Mirzataheri M, Atai M and Mahdavian A R 2010 Physical and mechanical properties of nanocomposite barrier film containing encapsulated nanoclay *Journal of Applied Polymer Science* **118** 3284-91

[19] Sedayu B B, Cran M J and Bigger S W 2020 Improving the moisture barrier and mechanical properties of semi-refined carrageenan films *Journal of Applied Polymer Science* **137** 49238

[20] Schwarzova I, Stevulova N, Singovszka E and Terpakova E 2015 Surface Treated Natural Fibres as Filler in Biocomposites *IOP Conference Series: Materials Science and Engineering* **96** 012028

[21] Abdulkhani A, Hojati Marvast E, Ashori A, Hamzeh Y and Karimi A N 2013 Preparation of cellulose/polyvinyl alcohol biocomposite films using 1-n-butyl-3methylimidazolium chloride *International journal of biological macromolecules* **62** 379-86

A Review on Diversity of Anticancer Compounds Derived from Indonesian Marine Sponges

Tutik Murniasih*, Masteria Yunovilsa Putra, Asep Bayu, and Joko Tri Wibowo

Research Center for Biotechnology -Indonesian Institute of Sciences
Jl. Raya Bogor Km.46 Cibinong Bogor, West Java Indonesia

*e-mail address: tutikmurniasih71@gmail.com

Abstract. As a tropical archipelago country, Indonesia has a mega diversity of marine organisms, such as sponges. About 850 species of sponges were identified from the east part of Indonesia. The uniqueness of Indonesian marine sponges attracted many researchers to explore the sponge's potential in producing active substances. During 1995-2016, about 40 genera of Indonesian sponges were investigated for their potential in producing pharmacological activity such as antimicrobial, anticancer, antivirus, multidrug-resistant (MDR), etc. The data showed that 56.7% of 430 reported compounds were confirmed as new compounds. The research trend on Indonesian sponges was high during 2004-2013, but decreasing after 2014. However, researches in the term of active substances from marine sponges should find provide the basic skeleton of anti-cancer drug lead compounds. Chemical structure diversity plays an important role in the exploration of anticancer lead compounds. The purpose of this paper is to review the potential of anticancer diversity compounds derived from Indonesian sponges, to get comprehensive data for further investigation. As the conclusion of our review, the most anticancer compounds derived from Indonesian marine invertebrates are alkaloid groups (such as aaptamine, manzamine, and bromopyrrole derivatives), then terpenoid groups (such as diterpene, coelodiol, and coeloic acid, sesquiterpene aminoquinone, and also (+)-curcuphenol and (+)-curcudiol), and also from the other groups such as sterole, peptide, polyketide, amino acid derivatives, natural organic acid, and quinone. The most effective anticancer compounds were 5-benzoyldemethylaaptamine, isoaaptamine, 3-bromofascaplysin, hyrtioreticulins A, stylissamide X, sigmosceptrellin B, and diacarperoxide S.

Keywords: diversity, active compounds, anti-cancer, sponges, marine resources

1. Introductions

Cancer is the third highest cause of mortality in Indonesia after heart disease and stroke [44]. Cancer death cases in Indonesia increased from 1.3-1.8% during 2013- 2018 [1]. According to the data from the Indonesian Ministry of Health, the number of death cases caused by cancer in 2018 was 207,221 cases in Indonesia and 9.6 million cases worldwide. The costs incurred for handling cancer reached 3.41 trillion rupiahs/year. Indonesian government implemented several programs such as the socialization of healthy living, cancer early detection, improvement of health facilities, and availability of cancer drugs. Currently, the availability of imported drugs for cancer treatment is limited and expensive. Consequently, the drugs are not affordable for cancer patients from middle to lower class society. The development of cancer drugs, either herbal medicine or modern medicine for chemotherapy, aimed to solve this crucial issue. The investigation to develop cancer drugs in

Indonesia mostly focused on the terrestrial plant with a small part that focused on marine organisms. Nonetheless, marine organisms have been recognized as the producer of active compounds with high potential, especially for anticancer.
The collected publications during 1985 to 2012, were reported there were 16,617 new compounds derived from marine organisms. Almost 30% of the reported compounds were derived from the marine sponges [4]. A total of 4196 or 25.25% of the reported compounds were pharmacologically active and 56% of them were confirmed as anticancer [2]. The review about the potency of marine biotas as a source of potential active natural products was done by hanif et al [7]. Ye et al also reported that marine sponges were the vital source of anticancer diversity such as sterol, terpene, alkaloid, peptide, macrolide , polyketone, etc [45].
The purpose of this study was to find the potential anticancer compounds by reviewing the previous study concerning to the diversity of chemicals derived from Indonesian sponges. Studies on the chemical diversity of potential anticancer substances are needed to find the potential lead compounds. By increasing the chemical diversity, the chance to obtain the potential lead compound candidate becomes bigger. Screening of the chemical structures using the cheminformatics approach will be more effective by cutting several steps in the drug discovery strategy.. "This study is one of the interesting issues to be discussed to find a new strategy in anticancer drug discovery. The one approach was by the compilation of other studies and previous findings in this paper

2. Indonesian sponge as the bioactive compounds reservoir.

Sponge-derived bioactive compounds became interesting issue to be explore since the finding of several potential drug candidates. The spongouiridine group with more than 300 analog compounds was part of the results of antitumor drug exploration [4]. The study on Indonesian sponge biodiversity was firstly conducted in 1889 through Sibolga expedition, but most of the specimen was unpublished. The second expedition was Snellius II (1987) that reported 830 species of sponge identified from eastern Indonesia in combination with the specimen from the previous expedition. Most of them were *H. massa*, *Druinella p.*, *Haliclona violacea*, *Acanthella c.*, *H. fascigera*, *C. schulzei*, *Spirastrella decumbens*, *Cliona spec.* and *Ircinia irregu.* [6].
The investigation of Indonesian Sponge's natural products were started in early 1970. During 1970-2017, A total number of 498 novel molecules were reported produce by Indonesian sponges [7]. The reported sponges-derived natural products were mostly produced by three classes: Demospongiae (91.8%), Homoscleromorpha (6.6%), and Calcarea (1.6%). The highest genera reported were Achantostrongylophora (4.4%), Xestospongia (3.6%), Sinularia (3.4%), Apysinella, Theonella and Strepsichordaia (2.9% each), Plakortis, Petrosia, Spongia, Melophlus (2.6% each), Agelas, Rhabdastrella, and Lissoclinum (2.5% each) [7]. The profile of the Indonesia sponge could be seen in figure 1

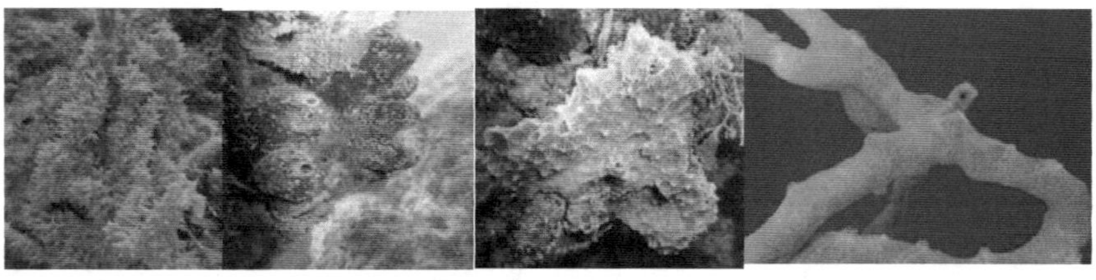

Figure 1. The profile of Indonesian sponge (*Callyspongia* sp, *Theonella* sp, *pinky sponge*, *red rope sponge*)

Sponges specimen were collected using scuba diving on the 5-10 m depth. The specimen was split for identification purposes and chemical separation. During transportation from the sampling location the sponge were keep fresh by using dry ice. After arriving to the laboratory fresh sample then transfer to

IOP Conf. Series: Materials Science and Engineering 1192 (2021) 012012 doi:10.1088/1757-899X/1192/1/012012

the -20°C freezer. The sponges' specimen was identified by the taxonomist and the extraction until profiling the active compounds was done for producing the publication.
The review of the Indonesian sponge publications during 1995-2016, get the data that among 244 papers, 430 reported compounds were derived from sponges, 244 of them confirmed as a novel metabolite. Most of the isolated compounds were active anticancer with various cell lines and mechanisms.
The Indonesian sponge's natural products publications were started in 1995, became increasing since 2000 until 2013. Start from 2014 right now become decreasing, it's caused by changing trends of investigation from unculturable to culturable sources like bacteria. Thus, the investigation of new anticancer drug compounds derived from sponge's associated bacteria is still in progress.

3. Diversity of anticancer compounds derived from Indonesian Sponge.

The diversity of anticancer compounds derived from Indonesian sponges will be discussed in this paper as the result of our review to the previous publication wich focus on the chemical class of steroid, terpenoid, alkaloid, peptide, and the mixture basic skeleton groups.

3.1 Steroid & Terpenoid

A-nor sterol with rare D-ring unsaturated, 3 -(hydroxymethyl)-A-nor-5 -cholest-14-en-16-one was reported derived from sponge *Axinella carteri* Dendy, 1889 (Demospongiae: Halichondrida: Axinellidae), that collected in Derawan Island, Indonesia. The bioactivity of a fraction where this compound was isolated showing moderate inhibiting tumor cell lines (P 388, A 549, and HT 29) with IC^{50} about 5 mg mL^{-1} [8]. Another sterol reported produced by Indonesian sponge *Dasychalina* sp. was haplosamate A, the sterol containing a sulfate and a methyl phosphategroup, along with its new desulfo analog. The evaluation by observing the interaction of CB1 and CB2 receptors of these analogs compounds was done using binding test. The binding test showed that desulfohaplosamate was low selectivity to the CB2 receptors,and the semi-synthetic derivative with cleaved ring B showed a completely loss of affinity for both receptors [9].
The isocopalane diterpenes coelodiol and coeloic acid were the oxidative degradation activitywhich in the *Coelocarteria cfr. singaporensis*. Both compounds were reported active against MKN-45 cell line (with IC_{50} 20 and 40 µg/mL respectively [10]. The 20,24-bishomo-25-norscalaranes and 20,24-bishomoscalaranes have been yielded by the Indonesian marine sponge *Carteriospongia foliascens*. Those compounds were active inhibiting RCE-protease activity [11]. Merodìterpene halioxepine was yielded from *Haliclona* sp collected from Indonesia. This compound showed moderate activity against NBT-T2 cells line with IC50 4.8 mg/ml [12]. The strongylophorine, a meroditerpene compound was found in the *Petrosia corticata* and active as proteasome inhibitors. The hemiacetal strongylophorines-13/-14 was demonstrated as the strongest proteasome inhibitor with an IC_{50} 2.1 µM [13].
The study of bioactive compounds derived from *Axynissia* sp yielded two sesquiterpenoid bisabolen type, (+)-curcuphenol and (+)-curcudiol, The (+)-curcuphenol was confirmed active inhibiting protein kinase SRC with IC_{50} 7.8 µg/mL, while (+)-curcudiol was active inhibiting FAK with IC_{50} 9.2 µg/mL [14]. The other reported sesquiterpene, 5-epi-smenospongorine, and nine quinone phenols were reported derived from the Indonesian sponge *Dactylospongia elegans*. Among ten evaluated compounds, six of them were showed activity for inducing K562 cells line. The structure-activity relationship study demonstrated that the amino group play a role in-inducing K562 cells [15]. Furospinosulin-1, isolated from Indonesia sponge showed selectifity antiproliferative against DU145 human prostate cancer cells line at concentration 1 to 100 µM. The in *vivo* test of furospinosulin-1 confirmed that this compound active against S180 cells line at 10–50 mg/kg BW [16]. Two unique sesterterpenes, hyattellactones A and B, also reported produced by sponge *Hyattella* sp. Hyattellactone A was reported inhibiting PTP1B activity with IC_{50} as 7.45 µM, while hyattellactone B showed much-less activity [17].

Chemical separation of active compounds derived from*Diacarnus megaspinorhabdosa* yielded norterpene derivatives such as, diacarperoxides D to G, and diacardiol A. Those compounds were active against L.5178Y, HeLa and PC12 cell lines with EC_{50} values 0.06-4.8; 0.6-10; and 0.8-10 μg/mL respectively [18].

3.2. Alkaloid

The first alkaloid group with reported having anticancer activity produced from Indonesian sponges is aaptamine or benzonaphthyridine. This compound was decided as chemotaxon of Aaptos genus especially found in *Aaptos suberitoides.* The investigation of this species collected from Carita bay West Java found aaptamine that active against human osteosarcoma (MG63luc+) [19]. Still in the same species but different sampling location from Ambon coastal area generated aaptamine, demethyl aaptamine, and 5-benzoyldemethylaaptamine. These compounds inhibited the growth of L5178Y cells line with IC50 8.3;0.9; and 5.5 μM, respectively [20] also HeLa cells [21]. Isoaaptamine isolated from Indonesian *Aaptos* sp. strongly inhibited murine leukimia cell line P338 with IC_{50} 0.28 μg/mL. Modification in C9 from this isoaaptamine generated compounds that showed strong activity against leukemia cell line K-5629. Those compounds were 9-O-4-trifluoromethoxybenzoylisoaaptamine and 9-O-4-tert butylbenzoylisoaaptamine, , 9-O-4-pentylbenzoylisoaaptamine with GI_{50} values are 1.66 ,0.05 and 1.9 μM, respectively.[22].
The second sponge derived alkaloid group is manzamine. Manzamine is β-carboline pentacyclic alkaloid that reported potential pharmacological activities, including anticancer activity [23]. Recently there are about 100 of manzamines analogs were reported produce by marine sponge with the sampling location along the Red Sea until the Indonesian [24]. The first manzamine from Indonesian sponges was reported by Ohtani research group [25]. The another manzamine group, kauluamine isolated from *Prianos* sp., reported having immunosuppressive activity. Other related compounds were pre-neo-kauluamine, neo-kauluamine, and manzamine A isolated from *Acanthostrongylophora ingens* that was active asas proteasome inhibitor activity [26]. Manzamine A derived from Manado *Acanthostrongylophora* demonstrate inhibiting cervical cancer cell C33A and HeLa with IC_{50} 2.1 and 4 μM, respectively [27]. Manzamine A also reported remarkable activity against cervical cancer (C33A, HeLa, SiHa, and CaSki) with very low and non-cytotoxic (IC50 up to 4 μM) [46]
Investigation of potential active compounds from *Agelas linnaei* and *Agelas nakamurai* generated 24 alkaloid group that illustrating bromopyrrole and another diterpene alkaloids. *A. linnaei* was reported generating 16 bromopyrrole groups., Agelanesins A–D with the uniqe iodinated tyramine revealing cytotoxic against L5178Y cells line with IC_{50} from 9.25 to 16.76 μM. Another study to *A. nakamurai* exhibiting eight alkaloids including agelasine D and its oxime derivative, also longamide C. Agelasine D and its derivatives showing cytotoxic against L5178Y with IC_{50} 4.03 and 12.5 μM [28]. Another brominated alkaloid was bromotyrosine group, the resulted work of Aplysinellidae generated seven bromotyrosine derivatives, purpuramine M–N (1–2) and araplysillin VII–XI (3–7). The pharmacological activity evaluation of purpuramine M was reported that this compound inhibiting A2780S, A2780SCP5, and U251MG cancer cell lines with IC_{50} values 20 μM, 40 μM, and 50 μM, respectively. In additional the guanidino moiety course decreasing the activity against these cell line[29] .
Three new compounds 3-bromofascaplysin, 14-bromoreticulatine, and 14-bromoreticulatate were reported produce by *Fascaplysinopsis reticulata* that collected from Pahena in the Pulau-Pulau Region Indonesia. These compounds were screened for several cell lines, and the most potential anticancer was 3-bromofascaplysin against melanoma cell line (MALME-3M) with IC_{50} value 0.45 μM. The activity of this compound against several colon, melanoma, ovarian, renal, and breast cancer cell lines showed moderate until strong activity with an IC_{50} in the range of 0.45-4.4 μM [30].
The pentacyclic alkaloids, papuamine and haliclonadiamine, that isolated from Indonesian sponge *Haliclona* sp inhibited cell proliferation of six human cancer cell lines with IC_{50} values of 0.93–1.50 and 1.00–4.44 mM, respectively [31].

The other sponge's alkaloid hyrtioreticulins A and B were potential inhibited ubiquitin-activating enzyme (E1) with IC_{50} values of 0.75 and 11 μg/mL, respectively. Both of those compounds were isolated from the Indonesian *Hyrtios reticulatus* [32]. The Indonesian sponge *Leucetta chagosensis* contained naamine group. naamineA and naamine G showed a weak cytotoxic against mouse lymphoma (L5178Y) and human cervix carcinoma (HeLa) cell lines [33]. In contrast, naamidines H and I were presented the strong cytotoxic against HeLa cells (IC_{50} 5.6 and 15)[34].

3.3. Peptide

Chemical investigation of *Siliquariaspongia mirabilis* reported the cytotoxic theopapuamides A-C that active against human colon carcinoma (HCT-116) cells with IC_{50} values from 2.1to 4.0 μg/m [35]. The peptide spongiacidin C was generated from the *Stylissa massa* that showed inhibiting USP7. with an IC_{50} of 3.8 μM [36]. A very potential HeLa cell inhibitor stylissamide X was reported produce by *Stylissa* sp. sponge. The IC50 of sstylissamide X against HeLa cells migration was in the ranges of 0.1–10 μM [37].

The study of Indonesian *Theonella swinhoei* was reported two tridecadepsipeptides theonellapeptolide and sulfinyltheonellapeptolide. Both of these compoundsdemonstrated moderate antiproliferative activity against HepG2, a hepatic carcinoma cell line with an IC_{50} of 1.5μM [38].

The chlorinated peptides, sintokamides A - E were isolated from the Indonesian *Dysidea* sp. Sintokamide A active inhibited prostate cancer cells [39].

3.4. Others

The polybrominated diphenyl ether compound derived from sponge Indonesian *Lamellodysidea herbacea* showing active moderate cytotoxic against two human cancer cell lines, HCT-15 (colon) and Jurkat (T-cell lymphoma) cells with IC_{50} values 12 and 9.5 μM respectively [31].

The extraction of lipid contained in *Lendenfeldia* sp. Sponge yielded furanolipid, and three homoscalarane sesterterpenes . This naphthalene dimer and three homoscalarane not only inhibited hypoxia induced HIF-1 activation withIC_{50} of 0.64–6.9 μM, but also reduced the viability of T47D and MDA-MB- 231,the breast cancer cells line. One of the homoscalarane confirmed the most potent active as unique tumor cell line selectivity to the NCI 60-cell line [40]. The manadic acids A and B as the fatty acids derivative have been isolated from the Indonesian sponge *Plakortis* sp. The evaluation of acticancer activity showed that both compounds moderately active against various tumor cell lines such as P388, A-549, HT 29, and MEL 28 with IC50 in the range of 0.5-5 μg/mL[41].

Indonesian *Plakortis nigra* have been yielded plakorstatins 1 and 2, exhibited the growth of the P388 lymphocytic leukemia cell line with ED_{50} of 1.1 and 0.91 μg/mL, respectively [42].The most potential sponge from Indonesia *Theonella* cf. *swinhoei,* derived *polyketide b*itungolides A-F,that inhibit phosphatase VHR [43].

4.The potential anticancer compounds

Based on the publication between 1995-2020 on Indonesian marine sponge, there were several potential anticancer compounds in many groups such as alkaloid, terpenoid, steroid, peptide, polyketide, lipid, etc. Most of the anticancer compounds derived from the Indonesian marine sponges were alkaloid (23 compounds), while the others were steroid & terpenoid (16 compounds), peptide (8 compounds), and others (14 compounds). Anticancer compounds derived from Indonesian sponges with remarkable activity (IC_{50} <1 μM or μg/mL) is described in table 1.

Table 1. Anticancer compounds derived from Indonesian Sponges with remarkable activity.

Groups	Compounds	Cell Lines/enzyme inhibitor	IC50	Source	References
Alkaloid	demethylaaptamine	L5178Y murine leukemia	0.9 μM	*Aaptos suberitoides*	[20]
	Isoaaptamine	P338	0.28 μg/mL	*Aaptos* sp	[22]
	3-bromofascaplysin	MALME-3M	0.45 μM	*Fascaplysinopsis reticulata*	[30]
	Hyrtioreticulins A	the ubiquitin-activating enzyme (E1)	0.75 μg/mL	*Hyrtios reticulatus*	[32]
Peptide	Stylissamide X	HeLa cells	0.1 μM	*Stylissa* sp	[37]
Terpenoid	Sigmosceptrellin B	L.5178Y	0.55 μg/mL	*Diacarnus megaspinorhabdosa*	[18]
	Diacarperoxide S	L.5178Y	0.88 μg/mL	*Diacarnus megaspinorhabdosa*	

The structure of the potential compounds could be seen in Figure 1

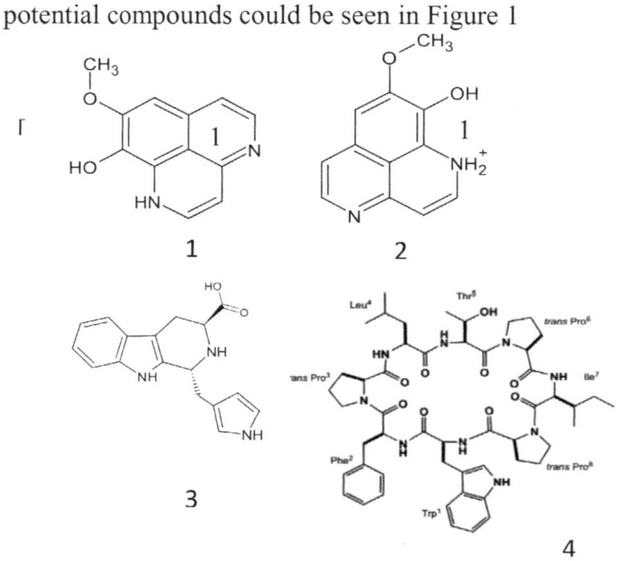

Figure 2. The structure of potential compounds: demethylaaptamine (1), isoaaptamine (2), Hyrtioreticulins A (3), stylissamide X (4)

The hydroxyl group of demethylaaptamine (1) plays an important role in the anticancer activity, methylation of this group decreased the activity [20]. Isoaaptamine showed strong activity against murine leukemia cell line P338. The para substitution of a phenyl ring attached via an ester linkage to the hydroxyl position of isoaaptamine (2) determined the selectivity toward the target [22]. Hyrtioreticulins A (3), also showed strong inhibition toward the E1-ubiquitin intermediate formation in a dose-dependent manner with IC_{50} values of 0.75 μg/mL. The imidazole moiety attached in C1 and trans position plays an important role in inhibiting E1 enzyme [32].
Cyclic peptides such as several compounds isolated from marine origin including stylissamide X (4) have been reported active as anti-infective and anticancer [37].

5. Conclusion

In conclusion, The Indonesian marine sponges derived the potential anticancer compounds such as demethylaaptamine, Isoaaptamine, 3-bromofascaplysin, 3-bromofascaplysin, Hyrtioreticulins A,

Stylissamide X, Sigmosceptrellin B Sigmosceptrellin B, and Diacarperoxide. Most of them were identified from the alkaloid group. Further investigation such as *in silico* studies is needed to find more structure-function information and specific target receptor.

Acknowledgment
The authors appreciate the support from the Research Center for Oceanography and Research Center For Biotechnology Indonesian Institute of Science for providing the access to the online library.

References

[1] TIMES Indonesia, Kanker Penyebab Kematian Tertinggi, Ini Faktanya |, 11 Februari 2020. https://www.timesindonesia.co.id/read/news/250486/kanker-penyebab-kematian-tertinggi-ini-faktanya
[2] Hu Y, Chen J, Hu G, Yu J, Zhu X, Lin Y, Chen S, Yuan J 2015 *Marine Drugs* 13p 202-221. https://doi.org/10.3390/md13010202
[3] Blunt JW, Carroll AR, Copp BR, Davis RA, Keyzers RA, and Prinsep MR 2018 *Natural product reports 35* **1** 8–53. https://doi.org/10.1039/c7np00052a
[4] Koopmans M; Martens D; and Wijffels RH 2009 *Mar. Drugs* 2009 **7p** 787–802
[5] Fusetani N, Sugawara T, and Matsunaga S 1992 *Journal of Organic Chemistry* 57(14):3828-32.
[6] Soest V 1989 *Netherlands Journal of Sea Research* **23** 223-230
[7] Hanif N, Murni A, Tanaka C, and Tanaka J 2019 *Mar. Drugs* **364** 2-69 doi:10.3390/md1706036
[8] Gallimore W, Cabral C, Kelly M, & Scheuer P J 2008 *Natural Product Research* **22** 1339–43.
[9] Chianese G, Fattorusso E, Taglialatela-Scafati O, Bavestrello G, Calcinai B, Dien HA, and Di Marzo V 2011 *Steroids* **76** 998–1002.
[10] Fattorusso E, Romano A, Taglialatela-Scafati O, Bavestrello G, Bonelli P, and Calcinai B 2006 *Tetrahedron Letters* **47** 2197–2200.
[11] Williams D E, Hollander I, Feldberg L, Frommer E, Mallon R, Tahir A, and Andersen R J 2009 *J. Nat. Prod.* **72** 1106–1109
[12] Trianto A,de Voodg NJ, and Tanaka J 2014. *Journal of Asian Natural Products Research* **16** 163–8.
[13] Noda A, Sakai E, Kato H, Losung F, Mangindaan REP, De Voogd N. J, and Tsukamoto S. 2015 Bioorganic and Medicinal Chemistry Letters, **25** 2650–2653.
[14] Hertiani T, Edrada-Ebel R, Kubbutat M, van Soest RWM, and Proksch P 2008 *Majalah Farmasi Indonesia* **19**78–85.
[15] Aoki S, Kong D, Matsui K, Rachmat R, & Kobayashi M. 2004 *Chemical & Pharmaceutical Bulletin* **52** 935–7.
[16] Arai M, Kawachi T, Setiawan A, & Kobayashi M 2010 *ChemMedChem* **5**1919–26
[17] Abdjul D B, Yamazaki H, Takahashi O, Kirikoshi R, Mangindaan REP, and Namikoshi M 2014 *Bioorganic and Medicinal Chemistry Letters* **25**904–907
[18] Ibrahim SRM, Ebel R, Wray V, Müller WEG, Edrada-Ebel R, and Proksch P 2008 *J. Nat. Prod.* **71** 1358–1364.
[19] Aoki S, Kong D, Suna H, Sowa Y, Sakai T, Setiawan A, and Kobayashi M 2006 *Biochemical and Biophysical Research Communications* 342101–106.
[20] Pham CD,† Hartmann R,‡ Werner EG, Müller, Voogd N, Lai D, and Proksch P 2013 *J. Nat. Prod.* **76** 103−106 .dx.doi.org/10.1021/np300794b
[21] Tsukamoto S, Kawabata T, Kato H, Ohta T, Rotinsulu H, Mangindaan R. E. P,and Namikoshi, M 2010 *J. Nat. Prod.* **70** 1658–1660.
[22] . Gul W, Hammond NL, Yousaf M, Bowling JJ, Schinazi RF, Wirtz SS, and Hamann MT 2006 *Bioorganic and Medicinal Chemistry* **14**8495–8505.
[23] Rao KV, Donia MS, Peng J, Garcia-Palomero E, Alonso D, Martinez A, …and Hamann MT 2006 *J. Nat. Prod.* **69** 1034–1040
[24] Ashok P, Ganguly S,and Murugesan S. 2014 *Drug Discov. Today* http://dx.doi.org/10.1016

[25] Ohtani II, Ichiba T, Isobe M, Borges MK, and Scheuer 1995 J. Am. Chem. Soc. **117** 10743–10744
[26] El-Desoky AH, Kato H, Eguchi K, Kawabata T, Fujiwara Y, Losung F, Mangindaan REP, de Voogd NJ, Takeya M, Yokosawa H, and Tsukamoto S 2014 *Journal of Natural Products* *77*1536-1540. DOI: 10.1021/np500290a
[27] Karan D, Dubey S, Pirisi L, Nagel A, Pina I, Choo YM, and Hamann MT 2020 *Journal of Natural Products* **83** 286-295 DOI: 10.1021/acs.jnatprod.9b00577
[28] Hertiani T, Edrada-Ebel R, Ortlepp S, van Soest RWM, de Voogd NJ, Wray V, and Proksch P, 2010 *Bioorganic & Medicinal Chemistry*, **18**1297–311.
[29] Dai J, Parrish SM, Yoshida WY, Yip MLR, Turkson J, Kelly M, & Williams P 2016. *Bioorganic & Medicinal Chemistry Letters*, 26(**2**): 499–504.
[30] Segraves NL, Robinson SJ, Garcia D, Said SA, Fu X, Schmitz FJ, Pietraszkiewicz H, Valeriote FA, and Crews P 2004 *J. Nat. Prod.* **67**: 783-792
[31] Yamazaki H, Sumilat D, Kanno S, Ukai K, Rotinsulu H, Wewengkang D S, … and Namikoshi M 2013 *Journal of Natural Medicines* **67** 730–5.
[32] Yamanokuchi R, Imada K, Miyazaki M, Kato H, Watanabe T, Fujimuro M, and Tsukamoto S 2012 *Bioorganic and Medicinal Chemistry*, 20(**14**): 4437–4442.
[33] Hassan W, Edrada R, Ebel R, Wray V, Berg A, Soest R Van, … and Proksch P 2004 *J. Nat. Prod.* **67**817–822
[34] Tsukamoto S, Kawabata T, Kato H, Ohta T, Rotinsulu H, Mangindaan R E P, … and Namikoshi M 2007 *J. Nat. Prod* **70** 1658–1660.
[35] Plaza A, Bifulco G, Keffer JL, Lloyd JR, Baker HL, and Bewley CA 2009 *J. Org. Chem.* **74**504–512.
[36] Yamaguchi M, Miyazaki M, Kodrasov MP, Rotinsulu H, Losung F, Mangindaan REP, de Voogd NJ, Yokosawa H, Nicholson B, and Tsukamoto S 2013. *Bioorganic & Medicinal Chemistry Letters*, **23**3884-3886 https://doi.org/10.1016/j.bmcl.2013.04.066.
[37] Arai M, Yamano Y, Fujita M, Setiawan A, and Kobayashi M 2012 *Bioorganic & Medicinal Chemistry Letters* **22**1818–21. http://doi.org/10.1016/j.bmcl.2011.10.023.
[38] Sinisi A, Calcinai B, Cerrano C, Dien H, Zampella A, D'Amore C, …and Taglialatela-Scafati, O 2013 *Journal of Organic Chemistry* **9** 1643–51.
[39] Sadar MD, Williams DE, Mawji N R, Patrick B O, Wikanta T, Chasanah E,…and Andersen RJ 2008 *Organic Letters* **10**4947–4950.
[40] Dai J, Liu Y, Zhou Y, & Nagle D G 2007 *J. Nat. Prod.* **70** 1824–1826.
[41] Ichiba T, Scheuer PJ, and Kelly-borges M 1995. *Tetrahedron*, **51**12195–12202.
[42] Pettit GR, Nogawa T, Knight JC, Doubek DL, and Hoopert JNA 2004 *Journal of Natural Products* **67**1611–1613.
[43] Sirirath S, Tanaka J, Ohtani II, Ichiba T, Rachmat R, Ueda K, …and HigaT 2002 *J. Nat. Prod.* **65** 1820–1823
[44] Indonesian Ministry of Health Jan. 13th, 2020 www.kemenkes.go.id/article/view/20011400002/
[45] Yea J, Zhoua F, Ammar M.Q. Al-Kareefa AMQ and Wanga H 2015 *Journal of Asian Natural Products Research* **17** 64–88, http://dx.doi.org/10.1080/10286020.2014.970535
[46] Karan D, Dubey S, Pirisi L, Nagel A, Pina I, Choo YM, Hamann MT.2020 *J Nat Prod.* **83**286-295. doi: 10.1021/acs.jnatprod.9b00577.

IOP Conf. Series: Materials Science and Engineering 1192 (2021) 012013 doi:10.1088/1757-899X/1192/1/012013

Engineering of Marine-derived Antimicrobial Peptides (mAMPs) into Improved Anti-infective Drug Leads: A Mini-review

L Sukmarini

Research Center for Biotechnology, Indonesian Institute of Sciences (LIPI), Jl. Raya Bogor Km. 46, Cibinong 16911, Bogor, West Java, Indonesia

E-mail: linda.sukmarini@lipi.go.id

Abstract. Marine-derived antimicrobial compounds possess chemical diversity varying from peptides, fatty acids to terpenes, alkaloids, and polyketides. These compounds, especially of peptide origin called antimicrobial peptides (AMPs), are present in the majority of marine organisms, including microbes (bacteria and fungi), invertebrates (molluscs, echinoderms, and sponges), vertebrates (fish and mammals), and plants (marine algae). They are defined by small molecular weight (less than 10 kDa), a net positive charge, and amphipathic structures. Moreover, due to their profound in vitro antimicrobial and cytotoxic activities and a low risk for resistance development, naturally occurring marine-derived AMPs (mAMPs) have been used as drug design templates for a large variety of semi-synthetic or synthetic AMPs, some of which have reached clinical trials. This mini-review aims to discuss AMPs from marine sources, mainly emphasizing the engineering of these peptides with improved pharmacological properties to develop drug candidates. Some selected recent examples of these engineered mAMPs as anti-infective drug leads are herein highlighted.

1. Introduction

Antimicrobial peptides (AMPs) belong to a group of peptide compounds exerting antimicrobial activity in vitro. These peptides are relatively small (comprising 5-60 amino acids or less than 10 kDa), have a net positive charge, and form amphipathic structures in a hydrophobic environment. In nature, AMPs are used as a vital tool for survival in many organisms as a part of the innate immune defense. Nowadays, AMPs are gaining increasing attention for promising drug leads because of their wide spectrum of activities against bacteria, fungi, parasites, viruses, and even cancer cells [1-7], and more importantly, they are less likely to develop resistance. These peptides mainly act on the lipid membrane structure of the incursive pathogens, creating a barrier in attaining resistance. Hence, AMPs may overcome antimicrobial resistance over conventional antibiotics [8-10].

The marine habitat has been regarded as one of the wealthiest reservoirs for AMPs. Its extreme and harsh conditions, such as a broad range of microbes (including the pathogenic microorganisms) as well as high salt concentration, drive marine organisms to produce various unique and potent bioactive molecules as their first line of defenses, in particular AMPs [11, 12]. Marine-derived AMPs (mAMPs) are produced from various marine species, including marine microorganisms (e.g., bacteria, fungi), invertebrates (e.g., molluscs, echinoderms, and sponges), plants (marine algae), and vertebrates (e.g., fish and mammals) [13]. Their chemical structure were found to vary from the terrestrial counterparts derivatives and often present novel and more diverse structures. Marine AMPs can be differed by their biochemical characteristics and structural features into linear α-helical (e.g., mixinidin), extended linear with the high content of one amino acid (glycine, histidine, proline, or tryptophan) (e.g., arasin), hairpin-like β-sheet or α-helical/β-sheet combined structure (e.g., hepcidins), and cyclic (primarily derived from sponges and tunicates; e.g., discodermin A) peptides. These mostly gene-encoded ribosomal peptides are extensively modified to adapt to a harsh marine environment, especially the high salinity. The salt-resistant feature of mAMPs allows peptides to retain their biological functions (particularly their anti-infective properties) in relatively high-salt environments, such as gastrointestinal fluid, saliva, serum, or other body fluids. Apparently, natural AMPs from marine organisms show tremendous prospects to be evolved into therapeutics, including anti-infective drug leads [14-16].

In spite of the propitious biological activity of the AMPs, they have some drawbacks towards the challenges for clinical applications, such as limited stability at certain pH; hemolytic side effect; elevated production cost; limited information on toxicity, pharmacodynamics, and pharmacokinetic properties; rapid degradation by protease; and cations-related reduced activity. However, a peptide can be altered in different approaches so that the expected properties, including effectiveness and stability, can be improved while the undesirable characteristics are attenuated. In addition, they can be optimized through the modification of their primary sequences. The rational design has received tremendous interest as well as depicts a significant revolution in the development of AMPs-based drug leads or drugs that have higher activity, less cytotoxicity, as well as feasible manufacturability [1, 17-19].

Several studies on AMPs from marine species have been extensively reviewed so far [13, 14, 18, 20-24]. However, this mini-review focuses on the engineering of these marine peptides into improved desired properties, particularly anti-infective. Their chemical modifications are also highlighted. Some selected recent engineered mAMPs as anti-infective drug leads emphasizing on antibacterial and antifungal agents during 2015-2020 are documented and listed herein.

2. Modifications in naturally occurring marine-derived AMPs (mAMPs)

Post-translational modifications (PTMs) are prevalently found in naturally occurring AMPs as these alterations are critical to the diverse chemical structure and biological function of natural AMPs. As mentioned earlier, mAMPs undergo various PTMs (Table 1). Some of the PTMs are either specific to mAMPs or can be shared with terrestrial AMPs.

Most of these chemical modifications direct the proper peptides folding into various chemical scaffolds that are necessary to recognize the bacterial surfaces and membranes target. Moreover, to aid the design of mAMPs with improved stability as well as efficacy to be applied in therapeutics, PTMs can maintain protease activity and physiological salt concentration [33]. These features, therefore, become an advantage for peptide drug candidates.

Deciphering the relationship between PTMs and peptide properties has inspired peptide engineering. Some PTMs strategies have been found to apply in peptide engineering. A number of studies have corroborated substantial evidence that antibacterial potencies can be generated by modifying synthetic AMPs. Peptide engineering plays an essential role in developing naturally occuring mAMPs into an improved peptide as well as a new variation of therapeutic compounds. Herein, some of the recent engineered AMPs from marine sources against bacteria and fungi will be discussed below.

Table 1. Selected PTMs present in naturally occurring mAMPs

Modifications	AMPs	Marine species origin	References
Bromination	cathelicidins	hagfish *Myxine glutinosa*	Shinnar *et al.*, 2003 [25]
	strongylocins	hagfish *Strongylocentrotus droebachiensis*	Li *et al.*, 2008 [26]
	jaspamide	sponge *Jaspis* sp.	Zabriskie et al., 1986 [27]
Unnatural amino acids (a) 6-bromotryptophan (bromination), dihydroxylysine, dihydroxyarginine, 3,4-dihydroxyphenylalanine	styelin	tunicate *Stylia cava*	Taylor *et al.*, 2000 [28]
(b) 3-methylisoleucine	polydiscamide A	sponge *Discodermia* spp.	Gulavita *et al.*, 1992 [29]
N- or C-terminal modification (a) N-terminal blocking by formyl group and C-terminal lactonization with threonine	halicylindronide D	sponge *Halichondria cylindrata*	Li *et al.*, 1995 [30]
(b) Oxidatively decarboxylated aromatic C-terminus	halocyamines	ascidian *Halocynthia roretzi*	Azumi *et al.*, 1990 [31]
(c) C-terminal amidation	clavanins	Hemocytes of sea squirt *Styela clava*	van Kan *et al.*, 2002 [32]

3. Engineered antibacterial mAMPs

It has been reported that high antimicrobial (antibacterial) activity is related to a free N-terminal amino group and amidated C-terminal on a peptide structure. The proper equilibrium of amphipathicity (cationic, e.g., arginine and aromatic, e.g., tryptophan residues), hydrophobicity (the hydrophobic interaction with bacterial membrane), cationicity (the number of positively-charged residues), and a free N-terminal amino group and an amidated C-terminus constitute a favourable motif in AMPs [34-36].

A novel cationic peptide with high polyalanine content, namely *Pa*-MAP 1.9, has been rationally designed based on a synthetic multifunctional peptide analogue *Pa*-MAP. The *Pa*-MAP is derived from HPLC-8, a peptide obtained originally from the polar fish *Pleuronecetes americanus* [37]. This newly analogue synthetic mAMPs seems to be a promising anti-infective drug leads against the pathogenic Gram-negative infection. It showed greater antibacterial properties primarily against Gram-negative strains, including *E. coli, Kleibsiella pneumoniae,* and *Enterococcus faecalis*. It has been suggested that adopting the conformation of α-helical, which may favour the insertion into the lipid bilayer as well as the orientation mimicking the bacterial membrane, can facilitate a bacterial membrane interaction [34].

IOP Conf. Series: Materials Science and Engineering 1192 (2021) 012013 doi:10.1088/1757-899X/1192/1/012013

A bioengineering approach has been applied to develop β-hairpin mAMPs arenicin-1 derived from marine lugworm *Arenicola marina* as a template for antimicrobials with potent activity and less toxicity [38]. The shortened 17-residue analogue of arenicin-1, namely ALP1 [39] (Figure 1), was obtained by heterologous expression in the *Escherichia coli* strain. It has been reported that the ALP1 analogue demonstrates potent antibacterial activity towards both Gram-positive and Gram-negative pathogens, including multidrug resistance strains, no cytotoxicity at high peptide concentration (up to 50 μM), as well as good tolerance to physiological or high salt concentrations. These features perform the peptide a favourable chemical temp for further designing antibacterial agents with more potent and nontoxicity properties [38, 39]. In addition, recently, Orlov *et al.* (2019) [40] redesigned arenicin-1 through branching or strand rearrangement, specific amino acids substitution, as well as structure cyclization or linearization. Those generated analogues make both the outer and inner *E. coli* membrane cells permeable, indicating the membranolytic mechanism of action. Their findings also suggest that the arenicin scaffold could be acceptable as a model for refining new antibacterial compounds.

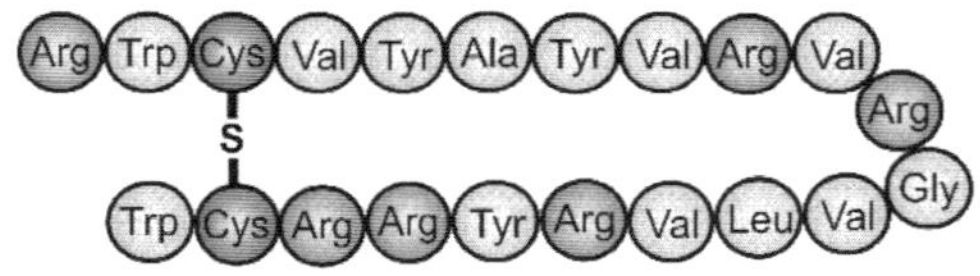

arenicin-1

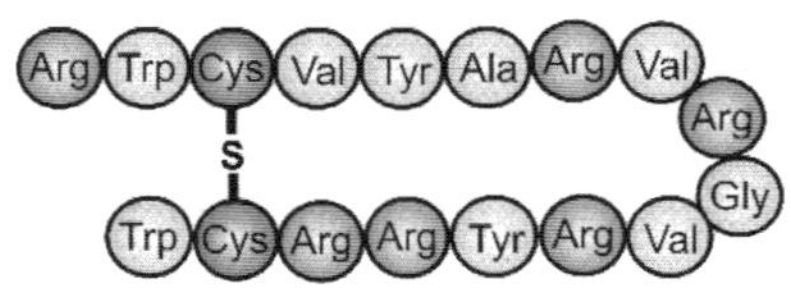

ALP-1

Figure 1. Chemical structures of antibacterial peptide arenicin-1 and its ALP1 analogue.

Moreover, the other member of the arenicin family, NZ17074 (N1), which is a variant of arenicin-3 (GFCW**N**VCVYRNGVRVC**H**RRCN; changes vs. arenicin-3 in bold; Tyr5Asn, Tyr17His), also exhibits potent antimicrobial activity towards Gram-negative bacteria and fungi, and is currently entering preclinical evaluations [41]. However, due to its significant degree of cytotoxicity in normal eukaryotic cells, N1 has been modified by replacing multiple residues with hydrophobic/cationic residues (alanine/lysine) and by removing disulfide bond, generating several analogues to reduce cell toxicity and generate more therapeutically molecules. Two chemically synthetic analogues, N2 (Gly1Ala, Gly12Ala; a 'rocket' analogue bearing two disulfide bonds) and N6 (Cys3Ala, Cys20Ala; a 'kite' analogue bearing a Cys7-Cys16 disulfide bond), demonstrated potent activity against *E. coli* and *Salmonella enteritidis,* with N2 showing higher activity than N1. The increased flexibility in the N- and C- terminus of both newly synthesized analogues led to lower haemolytic and cytotoxic effects than N1 [42, 43].

Furthermore, the improved antibacterial activity of the synthetic N6 analogues against intracellular *S. typhimurium* has demonstrated a novel platform of cell-penetrating peptides (CPPs)-cleavable linker-AMP as an efficient AMPs delivery and a C-terminal modification as an effective strategy to largely improve peptides stability. Those analogues, the CPP-linker-N6, the C-amidated N6, and CPP-linker-C-amidated N6, had almost no hemolysis and cytotoxicity but with increased antibacterial activity. However, further investigations are required to be applied clinically [44].

A very recently promising lead arenicin-3 derivative for further clinical evaluation, named AA139 analogue, has also been successfully synthesized by Cooper and his coworkers [45]. They developed the synthetic peptide AA139 (Val8Ala, Tyr9Arg, Val13Ala) by structure-guided optimization by combining the structure-activity/toxicity relationship evaluation with the membrane models and NMR solution structures. The peptide analogue showed reduced mammalian toxicity and propitious efficacy in various pathogenic Gram-negative in vivo models.

4. Engineered antifungal mAMPs

Pathogen resistance and drug toxicity have restrained the treatments for fungal infections. From previous studies, the antifungal mAMP Cm-p1 (SRSELIVHQR) has been isolated from a marine mollusc *Chenchritis muricatus*. Moreover, to improve its antifungal activity, Cm-p5 (SRSE-LIVHQRLF) has been engineered by adding two more residues, leucine and phenylalanine to the C-terminus of Cm-p1. Besides antifungal property towards human fungal pathogens such as *Trichophyton mentagrophytes, T. rubrum, Candida parapsilosis*, and *Cryptococcus neoformans*, this modification has considerably increased fungistatic effect towards pathogenic *Candida albicans* (EC = 146 μg mL^{-}1; MIC = 10 μg mL^{-1}). Moreover, this newly engineered analogue exhibited low toxicity towards mammalian cell cultures and low haemolytic. In silico study and isothermal titration calorimetry showed Cm-p5 binds to a bilayer membrane of phosphatidylserine and possessed affinity from the least to the highest level to ergosterol, mammalian phospholipid, and fungal phospholipid, respectively, but no affinity to chitin. However, this analogue was unfortunately incapable of regulating the kidney fungal burden in a systemic candidiasis mice model [46, 47].

Cm-p5 analogue, however, is also not suitable for therapeutic application because the analogue is readily degraded by proteases. Two arginine residues in the peptide design seems to lead to trypsin cleavage, and the free N-terminal can be prone to be attacked by aminopeptidase. As the structure-activity relationship is determined, the covalent modification is expected to enhance stability. Therefore, based on the following rational design study of this peptide, glutamic acid and histidine residues are essential in the helix stabilization, allowing Vicente *et al.* (2019) to synthesize a cyclic helical-stabilized analogue of Cm-p5, named CysCysCm-p5, by changing those residues for cysteine [48] (Figure 2). The new cyclic CysCysCm-p5 has shown greater fungal control capacity (than conventional antibiotic fluconazole), low cytotoxicity, and a stabilized α-helix and disulfide bridges that may promote metabolic stability and activity in vivo. Thus, this monomer shows a favourable systemic antifungal drug lead.

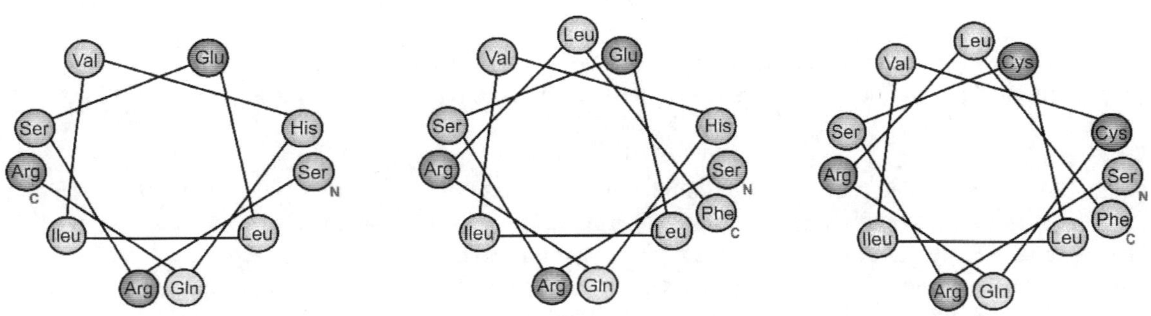

(a) Cm-p1 **(b) Cm-p5** **(c) cyclic CysCys-Cm-p5**

Figure 2. Helical projections of antifungal peptide Cm-p1 (a) and its analogues Cm-p5 (b) and CysCys-Cm-p5 (c). Adapted from Vicente *et al.*, [48].

In addition, further modified monomers of CysCys-Cm-p5, parallel and anti-parallel dimers were generated by intramolecular disulfide bond formation. Recently, Rosenau and his coworkers [49] have demonstrated that these new derivates initially exhibited semi-inhibitory concentration (10-21 μg mL^{-1} toward *C. aurius* biofilm in vitro that did not observe in the previous cyclic Cm-p5.

Another recent engineered antifungal mAMPs is the *N*-methylated analogue of a marine bacteria-derived high proline content cyclotetrapeptide [50]. The non-methylated naturally occurring cyclotetrapeptide was previously obtained from the seaweed *Diginea* sp.-associated *Pseudomonas* sp. and *Pseudoalteromonas* sp. [51]. The newly synthesized methylated analogue exhibited increased antifungal activity towards *Microsporum audouinii* and *T. mentagrophytes* (dermatophyte pathogens) compared to its non-methylated natural peptide and considerable activity against *C. albicans*. Besides modulating the biological properties or activities, the *N*-methylation of the cyclopeptides has shown several advantages. These advantages include increasing the hydrophobic interaction (via minimizing the hydrogen-bond donors), improving the efficacity (via improving receptor selectivity), and improving the oral bioavailability (via avoiding inter-and intra-molecular hydrogen bonds formation) [52].

5. Concluding remarks

AMPs from marine organisms are a promising treatment for antibiotic resistance. These peptides show advantageous post-translational modifications that enable coping with harsh marine environmental conditions such as high salinity. The modifications aim to increase the half-life of mAMPs in life systems, endowing improved stability and bioavailability to protease resistance in vivo. These characteristics lead to mAMPs as fascinating chemical models for the design of drugs or pharmaceuticals. Elucidation of the modification-peptide property correlation prompts peptide manipulation. Further continued structural alterations of the naturally occurring mAMPs are needed to obtain (semi)synthetic peptides with increased pharmacokinetics as well as bioactivity, which will be a tremendous battle against microbial infections.

References

[1] Mahlapuu M, Björn C and Ekblom J *Crit. Rev. Biotechnol.* 2020 **20** 978-992
[2] Galdiero S, Falanga A, Berisio R, Grieco P, Morelli G and Galdiero M 2015 *Curr. Med. Chem.* **22** 1665-1677
[3] De Ullivarri MF, Arbulu S, Garcia-Gutierrez E and Cotter PD 2020 *Front. Cell. Infect. Microbiol.* **10** 105
[4] Giovati L, Ciociola T, Magliani W and Conti S 2018 *Future Med. Chem.* **10** 2569-2572
[5] Hsieh I-N and Hartshorn L 2016 *Pharmaceuticals* **9** 53
[6] Felício MR, Silva ON, Gonçalves S, Santos NC and Franco OL 2017 *Front. Chem.* **5** 5
[7] Yaghoubi A, Khazaei M, Ghazvini K, Movaqar A, Avan A, Hasanian AM and Soleimanpour S 2021 *Int. J. Pept. Res. Ther.* **27** 1057-1067
[8] Baltzer SA and Brown MH 2011 *J. Mol. Microbiol. Biotechnol.* **20** 228-235
[9] Browne K, Charaborty S, Chen R, Willcox MDP, Black DS, Walsh WR and Kumar N 2020 *Int. J. Mol. Sci.* **21** 7047
[10] Huan Y, Kong Q, Mou H and Yi H 2020 *Front. Microbiol.* **11** 582779
[11] Choudary A, Naughton LM, Montánchez I, Dobson ADW and Rai DK 2017 *Mar. Drugs* **15** 272
[12] Cheung RCF, Ng TB and Wong JH 2015 *Mar. Drugs* **13** 4006-4043
[13] Kang HK, Seo CH and Park Y 2015 *Mar. Drugs* **13** 618-654
[14] Otero-González AJ, Magalhães BS, Garcia-Villarino M, López-Abarrategui C, Sousa DA, Dias SC and Franco OL 2010 *FASEB J.* **24** 1320-1334
[15] Falanga A, Lombardi L, Franci G, Vitiello M, Iovene MR, Morelli G, Galdiero M and Galdiero S 2016 *Int. Mol. Sci.* **17** 785
[16] Koehbach J and Craik DJ 2019 *Trends Pharmacol. Sci.* **40** 517-528
[17] Mahlupu M, Håkansson J, Ringstad L and Björn C 2016 *Front. Cell. Infect. Microbiol.* **6** 194
[18] Semreen MH, El-Gamal MI, Abdin S, Alkhazraji H, Kamal L, Hammad S, El-Awady F, Waleed D and Kourbaj L 2018 *Saudi Pharm. J.* **26** 396-409
[19] Avan I, Hall CD and Katritzky AR 2014 *Chem. Soc. Rev.* **43** 3575-3594

[20] Ponnappan N, Budagavi DP, Yadav BK and Chugh A 2015 *Probiotics & Antimicro. Prot.* **7** 75-89
[21] Bertrand B and Munoz-Garay 2019 *Int. J. Pept. Res. Ther.* **25** 1441-1450
[22] Falanga A, Lombardi L, Franci G, Vitiello M, Iovene MR, Morelli G, Galdiero M and Galdiero S 2016 *Int. J. Mol. Sci.* **17** 785
[23] Sperstad SV, Haug T, Blencke H-M, Styrvold OB, Li C and Stensvåg 2011 *Biotechnol. Adv.* **29** 519-530
[24] Destoumieux-Garzón D, Rosa RD, Schmitt P, Barreto C, Vidal-Dupiol J, Mitta G, Gueguen Y and Bachére E 2016 *Phil. Trans. R. Soc. B.* **371** 20150300
[25] Shinnar AE, Butler KL and Park HJ 2003 *Bioorg. Chem.* **31** 425-436
[26] Li C, Haug T, Styrvold OB, Jørgensen TØ and Stensvåg K 2008 *Dev. Comp. Immunol.* **32** 1430-1440
[27] Zabriskie TM, Klocke JA, Ireland CM, Marcus AH, Molinski TF, Faulkner DJ, Xu C and Clardy 1986 *J. Am. Chem. Soc.* **108** 3123-3124
[28] Taylor SW, Craig AG, Fischer WH, Park M, Lehrer RI and Steylin D 2000 *J. Biol. Chem.* **275** 38417-3846
[29] Gulavita NK, Gunasekera SP, Pomponi SA and Robinson EV 1992 *J. Org. Chem.* **57** 1767-1772
[30] Li HY, Matsunaga S and Fusetani N 1995 *J. Med. Chem.* **38** 338-343
[31] Azumi K, Yokosawa H and Ishii S 1990 *Biochemistry* **29** 159-165
[32] van Kan EJM, Demel RA, Breukink E, van der Bent A and Kruijff 2002 *Biochemistry* **41** 7529-7539
[33] Wang G 2012 *Curr. Biotechnol.* **1** 72-79
[34] Falanga A, Nigro E, De Biasi MG, Daniele A, Morelli G, Galdiero S and Scudiero O 2017 *Molecules* **22** 1217
[35] Ong ZY, Wiradharma N and Yang YY 2014 *Adv. Drug Deliv. Rev.* **78** 28-45
[36] Cardoso P, Glossop H, Meikle TG, Aburto-Medina A, Conn CE, Sarojini V and Valery C 2021 *Biophys. Rev.* **13** 35-69
[37] Cardoso MH, Ribeiro S, Nolasco DO, de la Fuente-Nùñez C, Felicio MR, Gonçalves S, Matos CO, Liao LM, Santos NC, Hancock REW, et al. 2016 *Sci. Rep.* **6** 21385
[38] Panteleev PV, Bolosov IA, Balandin SV and Ovchinnika TV 2015 *J. Pept. Sci.* **21** 105-113
[39] Panteleev PV, Bolosov IA and Ovchinnikova TV 2016 *J. Pept. Sci.* **22** 82-91
[40] Orlov DS, Shamova OV, Elisveev IE, Zharkova MS, Chakchir OB, Antcheva N, Zachariev S, Panteleev PV, Kokryakov VN, Ovchinnikova TV *et al.* 2019 *Mar. Drugs* **17** 376
[41] Wang X, Wang X, Teng D, Zhang Y, Mao R, Xi D and Wang J 2014 *Appl. Microbiol. Biotechnol.* **98** 7387-7389
[42] Yang N, Wang X, Teng D, Mao R, Hao Y, Zong L, Feng X and Wang J 2016 *Process Biochem.* **51** 734-739
[43] Yang N, Liu X, Teng D, Li Z, Wang X, Mao R, Wang X and Wang J 2017 *Sci. Rep.* **7** 3392
[44] Li Z, Teng D, Mao R, Wang X, Hao Y, Wang X and Wang J 2018 *J. Med. Chem.* **61** 7991-8000
[45] Elliot AG, Huang JX, Neve S, Zuegg J, Edwards IA, Cain AK, Boinett CJ, Barquist L, Lundberg CV, Steen J, *et al.* 2020 *Nat. Commun.* **11** 3184
[46] López-Abarrategui C, McBeth C, Mandal SM, Sun ZJ, Heffron G, Alba-Menéndez A, Migliolo L, Reyes-Acosta O, García-Villarino M, Nolasco DO, *et al.* 2015 *FASEB J.* **29** 3315-3325
[47] González-Garcia M, Valdés ME, Freitas CG, Menéndez AA, López-Abarrategui C, San Juan-Galán J, Campos Diaz S, Luiz Franco O and Otero-González AJ 2017 *Rev. Cubana Med. Tropical* **69** 1-15
[48] Vicente FEM, González-Garcia M, Pico ED, Moreno-Castillo E, Garay HE, Rosi PE, Jimenez AM, Campos-Delgado JA, Rivera DG, Chinea G, *et al.* 2019 *ACS Omega* **4** 19081-19095
[49] Kubiczek D, Raber H, Gonzalez-García M, Morales-Vicente F, Staendker L, Otero-Gonzalez AJ and Rosenau F 2020 *Antibiotics* **9** 363
[50] Dahiya R, Kumar S, Khokra L, Gupta SV, Sutariya VB, Bhatia D, Sharma A, Singh S and Maharaj

S 2018 *Mar. Drugs* **16** 305
[51] Rungprom W, Siwu ERO, Lambert LK, Dechsakulwatana C, Barden MC, Kokpol U, Blanchfield JT, Kita M and Garson MJ 2008 *Tetrahedron* **64** 3147-3152
[52] Asfaw H, Laqua K, Walkowska AM, Cunningham F, Martinez-Martinez MS, Cuevas-Zurita JC, Ballell-Pages L and Imming P 2017 *PloS ONE* **12** e0176088

Reinforcement of Polylactic acid (PLA) bio-composite with lignin from oil palm empty fruit bunches (OPEFB) for 3D printing application

M S Pairon[1], F Ali[1,5], H Anuar[2], F Ahmad[1], J Suhr[3] and M E S Mirghani[4]

[1]Department of Biotechnology Engineering, Faculty of Engineering, International Islamic University Malaysia (IIUM), 53100 Gombak, Malaysia
[2]Department of Manufacturing and Material Engineering, Faculty of Engineering, International Islamic University Malaysia (IIUM), 53100 Gombak, Malaysia
[3]Department of Polymer Science and Engineering, School of Mechanical Engineering, Sungkyunkwan University (SKKU), Seoul, South Korea
[4]International Institute for Halal Research and Training (INHART), International Islamic University Malaysia (IIUM), 53100 Gombak, Malaysia

[5]Corresponding author email: fathilah@iium.edu.my

Abstract. Polylactic acid (PLA) has been used as an additive material in 3D printing due to its toxic-free and environmentally friendly property. Lignin with complex and branched chemical structures had been used as a filler to improve the mechanical strength of PLA. The availability of oil palm empty fruit bunches (OPEFB) in Malaysia had made this material a good source for lignin extraction. Thus, in this research, we aim to study the mechanical strength of PLA bio-composite material with reinforcement of lignin from OPEFB. The lignin was extracted by 1,4-dioxane with hydrochloric acid as a catalyst. The recovery of lignin from extraction solvent was done by precipitation. The lignin was successfully extracted from OPEFB with 9.04% of lignin extraction yield. The PLA/lignin bio-composite filament with 0.1% (w/w) lignin was performed by filament extruder and then used for 3D printing. The prepared sample in the form of filament and 3D printed material was characterized for mechanical strength and surface morphology. The Young's modulus, ultimate strength and elongation at break of the PLA/lignin bio-composite sample had increased 11%, 7% and 10% respectively. After the filament was used for 3D printing, the reduction of mechanical strength had been observed. The morphology by scanning electron microscopy (SEM) confirmed the presence of lignin on the fracture surface of PLA bio-composite material. Apart from that, the interlayer adhesion had been observed in the 3D printed PLA/lignin bio-composite that caused the drop in mechanical strength of the material.

Keywords: Polylactic acid, lignin, oil palm empty fruit bunch (OPEFB), 3D printing, bio-composite material.

1. Introduction

3D printing is the construction of a physical object from additive materials with a graphical construction by using computer-aided design [1]. 3D printing, which was first invented by Charles W. Hull in 1986 under the name stereolithography, is increasingly being used nowadays in medical applications [2]. Bone restoration, healing, and recovery are some of the medical uses of 3D printing [2]. The prototype model was constructed by the 3D printer in the manufacturing industry before going into full production [3]. Low-melting-point polymers were used as additive materials in 3D printing, including polylactic acid (PLA), polyamide (PA), acrylonitrile butadiene styrene (ABS), and polycarbonate (PC). PLA had been commonly used as an additive material because it does not caused irritation and can be used in a variety of applications [4]. Apart from that, biodegradability of PLA was also taken into consideration in material selection because of environmental concerns [4].

Liu et al. [4] had stated that PLA had a simple linear molecular chain structure microscopically. In conjunction to that, PLA had a limited mechanical strength compared to ABS, PA and PC that had a graph molecular chain [4]. This limitation could reduce the capability of using PLA as additive material in 3D printing. To improve the mechanical strength of PLA, the possible use of bio-composites materials with lignocellulose reinforcements had been studied. Furthermore, the reinforcement of complex structure lignin and cellulose from lignocellulose could modify the simple linear molecular chain of the PLA. The availability of oil palm empty fruit bunches (OPEFB) in Malaysia had triggered the selection of this raw material as a source of lignocellulose extraction [5]. In comparison to the cellulose, lignin is the best lignocellulose portion to be reinforced into PLA. The crystallinity of cellulose can help in increasing the storage modulus of the polymer and improve its mechanical strength [6]. However, owing to its hydrophobicity, PLA can mix homogeneously with lignin but not with cellulose. During the mixing and blending phase with PLA, hydrophilic cellulose appears to have poor dispersion and clump together [7].

1,4-dioxane had been selected as extraction solvent to extract the lignin from OPEFB. Different from sodium hydroxide (alkaline solvent) and ethanol, 1,4-dioxane could extract lignin with minor modification and was nearly similar towards the native structure of lignin [8][9]. This study will cover the reinforcement of dioxane lignin from OPEFB into PLA for 3D printing. Also, the comparative study of mechanical strength will be covered in the filament and 3D printed PLA/lignin bio-composite material.

2. Experimental

2.1. Material

OPEFB used in this study obtained from Tennamaram Palm Oil Mill, Batang Berjuntai. Absolute undenatured ethanol >99.8% (AR grade) purchased from Sciefex and benzene (AR grade) purchased from R&M chemical had been used for the pre-treatment of OPEFB. 1,4-dioxane (AR grade) as extraction solvent was purchased from Acros Organics with the addition of hydrochloric acid as acid catalyst purchased from Bendosen. Sodium sulfate anhydrous (AR grade) from Systerm had been used during lignin recovery. PLA granules used for PLA/lignin bio-composite preparation were supplied from NatureWorks (USA) Ingeo with grade 2003D.

2.2. Lignin extraction

The soil in OPEFB fiber was removed by washing with flowing tap water and drying in an oven at 80°C. Crushers and grinders were used to reduce the size of the dried OPEFB fiber. The water-soluble extractives were removed using cold distilled water for 8 hours, filtered, and dried overnight in an oven at 80°C. The steps were repeated by using cold ethanol-benzene mixture (1:2 (v/v)) to remove water-insoluble extractive in the OPEFB. Lignin was extracted for 90 minutes in an N_2 atmosphere using a 1,4-dioxane solution as extraction solvent and 0.3M HCl concentration as an acid catalyst. The liquid fraction (filtrate) was concentrated at 40°C with reducing pressure using a rotatory vacuum evaporator. After that, 1% (w/v) Na_2SO_4 was added into the condensed liquid fraction in a ratio of 7:1 (v/v), followed by filtration using a Buchner funnel with water as a washing medium. The extracted lignin was dried in an oven at 40°C until it reached a constant weight. Finally, the extracted lignin was purified by washing and rinsing with diethyl ether and dried. The dried lignin was transformed into lignin powder by using the grinder. Equation (1) was used to measure the lignin extraction yield from OPEFB.

$$\text{Lignin yield (\%)} = \frac{\text{Extracted lignin (g)}}{\text{OPEFB}_{\text{dry weight}}\text{ (g)}} \times 100\% \quad (1)$$

2.3. Preparation of PLA/lignin bio-composite sample

The PLA and PLA/lignin bio-composite samples were prepared in filament and 3D printed form. 100g of PLA was mixed with 0.1g of extracted lignin powder before inserted into the feeder in the Filabot FB00655 EX6 filament extruder. The temperature used was 40°C at feed, 180°C at back, 190°C at the middle, and 170°C at front of the filament extruder. The extruded filament was cooled by a cooling fan with a 75% speed and the filament diameter was adjusted to be in average of 1.75mm. The filament was 3D printed by using Creality Ender-3 Pro 3D printer at nozzle temperature of 225°C and bed temperature of 70°C. The sample was sliced by Cura software with 100% infill and line infill pattern.

2.4. Characterization

2.4.1. Universal Test Machine (UTM). Young's modulus (MPa), ultimate tensile strength (MPa), and elongation at break (%) of the samples obtained from Universal Testing Machine (UTM) (Shimadzu, AGS-X, Japan). Filament samples were tested with crosshead speed 2.5 mm/min, load cell of 5 kN, and 25mm gauge length. 3D printed samples were prepared according to ASTM D-638 type-I with crosshead speed 50 mm/min, load cell of 5 kN, and 57mm gauge length. The samples were repeated 3 times for an average value.

2.4.2. Scanning Electron Microscope (SEM). The fracture surface of the PLA and PLA/lignin bio-composite from mechanical test was investigated using Scanning Electron Microscopy (JEOL, JSM-5600, UK). The samples were coated with palladium sputter and observed at a magnification of x50, x500, and x1500 with 7 kV accelerating voltage.

3. Result and discussion

3.1. Lignin extraction yield

1.13g of lignin was extracted from 12.5g of OPEFB by using 1,4-dioxane as extraction solvent. Based on equation (1), the lignin extraction yield of 9.04% was extracted from the OPEFB. With comparison from the study by Saha et al. [10], the extraction of lignin by using 1,4-dioxane from debarked bamboo waste achieved 15% yield. The difference in the extraction yield was due to the different type of lignocellulose biomass used for lignin extraction. Different type of biomass could extract different quality of the extracted lignin [11]. Apart from that, lignin extraction method was another factor that affect the lignin extraction yield. The lignin extraction yield from OPEFB from previous study by using sodium hydroxide (alkaline treatment) was 15% and ethanol solution (organosolv treatment) was 13% [12][13]. Each extraction method had provided different structural of lignin that could be used for their application [8]. Despite that, the mechanisms of extraction for each treatment was quite similar. β-O-4 linkage was one of the types of lignin linkage that became the critical steps in depolymerize lignin in lignocellulose biomass [14].

3.2. Universal Test Machine (UTM)

Figure 1, figure 2 and figure 3 showed the reinforcement of lignin changed the mechanical strength of the filament and 3D printed samples in terms of the Young's modulus (MPa), ultimate tensile strength (MPa) and elongation at break (%). Overall, the filament and the 3D printed sample had shown a different value because the samples were tested in different sizes and dimensions. After the reinforcement of lignin, Young's modulus increased from 238.58 MPa to 263.02 MPa for the filament and decreased from 1848 MPa to 1797 MPa for the 3D printed sample (figure 1). In figure 2, the ultimate

IOP Conf. Series: Materials Science and Engineering 1192 (2021) 012014 doi:10.1088/1757-899X/1192/1/012014

tensile strength of the PLA filament also increased with lignin reinforcement from 15.83 MPA to 16.95 MPa (figure 2(a)). However, when the filament was used for 3D printing, the ultimate tensile strength of PLA/lignin bio-composite had decreased from 68.27 MPa to 61.24 MPa (figure 2(b)). A similar finding also had been obtained from a previous study, where the reinforcement of lignin had reduced the Young's modulus and ultimate tensile strength of the 3D printed PLA/lignin bio-composite [15]. From figure 3, a slight change with no major difference on elongation at break when the lignin was introduced into the PLA. The value increased from 7.27% to 8.00% for filament and decreased from 6.12% to 5.68% for the 3D printed sample.

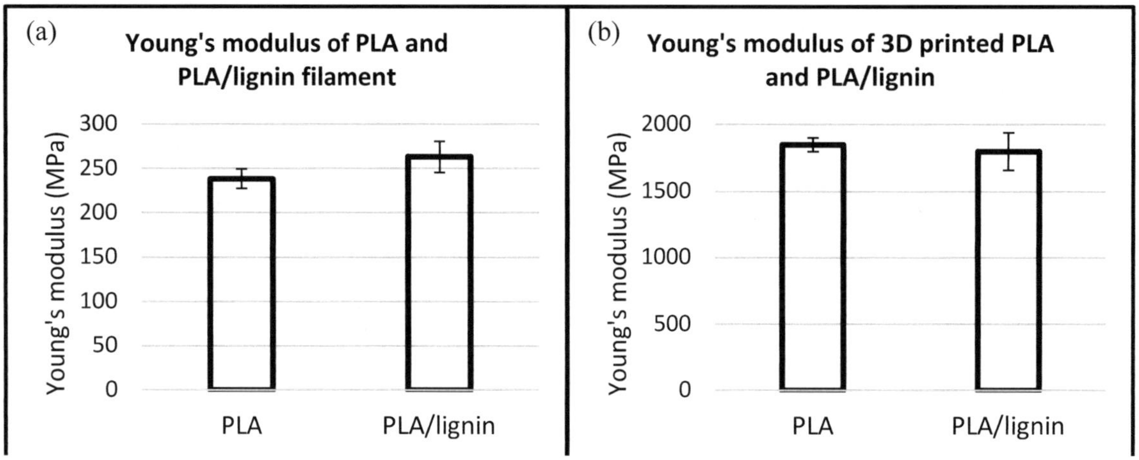

Figure 1. Young's modulus of PLA and PLA/lignin bio-composite in form of (a) filament and (b) 3D printed.

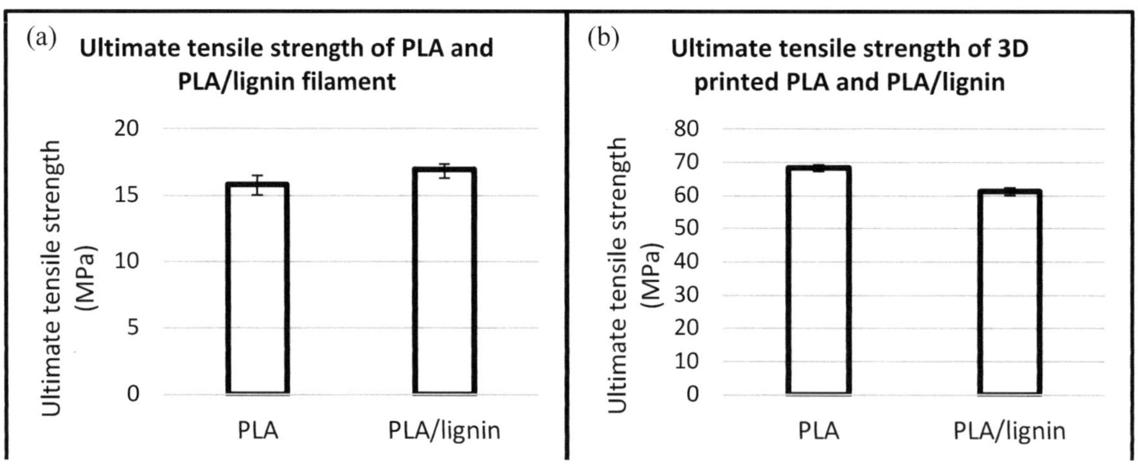

Figure 2. Ultimate tensile strength of PLA and PLA/lignin bio-composite in form of (a) filament and (b) 3D printed.

The reinforcement of lignin was proved to increase the mechanical strength of the PLA. Even with 0.1% (w/w) lignin reinforcement, Young's modulus, ultimate strength, and elongation at break of the PLA/lignin bio-composite filament had all increased by 11%, 7%, and 10%. Theoretically, the lignin filler had changed the molecular structure of PLA into branched polymer and increased its mechanical strength. During the extrusion, the terminal COOH group of PLA reacts with the phenolic and aliphatic hydroxyl (OH) group of lignin by chemical bonding [16]. The intermolecular of lignin-lignin, PLA-PLA, and PLA-lignin had influenced the tensile strength in PLA/lignin bio-composite [17]. The influence of lignin filler within PLA matrix will be explained further with SEM morphology. As the filament was used for 3D printing, the drops in mechanical strength was observed. This was due to the poor interlayer adhesion between the layer during 3D printing [15]. The relationship between viscosity, crystallize temperature and crystallization phase had played a major role for a good interlayer adhesion in 3D printing process [18].

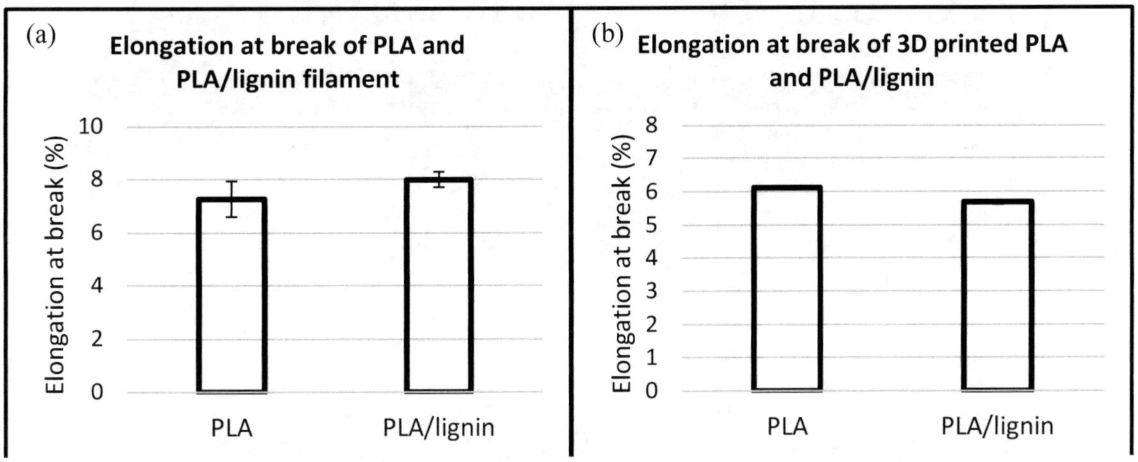

Figure 3. Elongation at break of PLA and PLA/lignin bio-composite in form of (a) filament and (b) 3D printed.

3.3. Scanning Electron Microscope (SEM)
Figure 4(a) and figure 4(b) showed the fracture surface of the PLA filament and PLA/lignin bio-composite filament from the SEM observation respectively. Results indicated that both samples had shown brittle fracture with a little plastic deformation. Figure 4(c) showed the presence of lignin that acted as rigid filler in increasing the mechanical strength of lignin [17]. Similar lignin morphology observed by [16], where the presence of cavities was also observed between the lignin and PLA in the composites. In resisting the stress during tensile test, some of the lignin had been pulled out from the sample and left a lignin pore as observed in figure 4 (d).

With little plastic deformation, the 3D printed PLA (figure 4 (e)) and 3D printed PLA/lignin bio-composite (figure 4 (f)) also had shown the brittle fracture during tensile test. However, a poor interlayer adhesion in 3D printed PLA/lignin bio-composite was observed within its structure compare to 3D printed PLA. The poor interlayer adhesion was related to the decreasing in Young's modulus, ultimate tensile strength and elongation at break of the 3D printed PLA/lignin bio-composite. Similar observation also had been found in the previous study, where the reinforcement of lignin could cause poor interlayer adhesion for 3D printed samples [15]. Triangular void observed in the figure 4(g) in 3D printed PLA/lignin bio-composite also caused from the poor interlayer adhesion [18]. To have a better interlayer adhesion, the nozzle and bed temperature during the printing should be adjusted in the future study. Since the reinforced lignin was only 0.1% (w/w) concentration, it was hardly observed under SEM. However, similar lignin pore was observed in figure 4(h) that resulted from the pulled-out lignin.

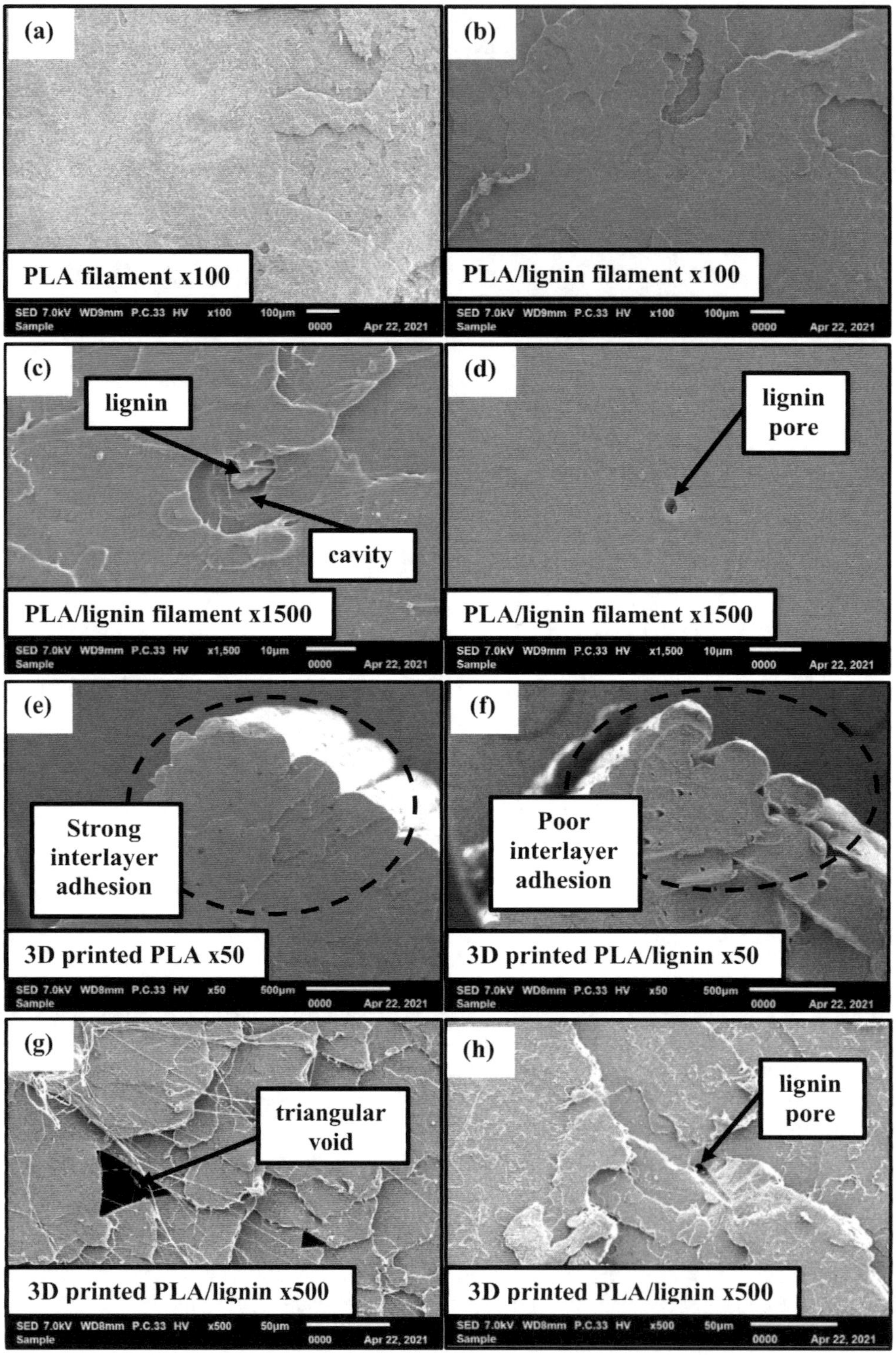

Figure 4. SEM micrograph at fracture surface of (a) PLA filament at x100, (b) PLA/lignin bio-composite filament at x100, (c)(d) PLA/lignin bio-composite filament at x1500, (e) 3D printed PLA bio-composite at x50, (f) 3D printed PLA/lignin bio-composite x50 and (g)(h) 3D printed PLA/lignin bio-composite at x500

4. Conclusion

Lignin had been successfully extracted from OPEFB by using 1,4-dioxane with an extraction yield of 9.04%. In this study, the reinforcement of lignin was proved to increase the Young's modulus, ultimate tensile strength and elongation at break by 11%, 7% and 10%. Although with only 0.1% of lignin reinforcement, the SEM micrograph observed the reinforced lignin supporting PLA/bio-composite filament on the fracture region. However, further studies are needed for this PLA/lignin bio-composite filament to be used in the 3D printing application. This was due to the decreasing in mechanical strength caused by the poor interlayer adhesion of the 3D printed PLA/lignin bio-composite. Also, optimized lignin composition with a perfect printing temperature was another factor that should be considered in future studies.

Acknowledgement

The authors would like to thank Ministry of Education Malaysia for the Fundamental Research Grant Scheme, Grant Number FRGS/1/2019/TK05/UIAM/02/1 and Research & Business Foundation, Sungkyunkwan University (SPI20-009-0009) for the financial support of this research. Also, the authors would like to thank Kulliyyah of Engineering for providing the laboratory facilities.

References

[1] Lee T C, Ramlan R, Shahrubudin N, Lee T C and Ramlan R 2019 *Proc. Manuf.* **35** 1286–1296
[2] Maroulakos M, Kamperos G, Tayebi L, Halazonetis D and Ren Y 2019 *J. Dent.* **80** 1–14
[3] Attaran M 2017 *Bus. Horiz.* **60** 677–688
[4] Liu Z, Wang Y, Wu B, Cui C, Guo Y and Yan C 2019 *Int. J. Adv. Manuf. Tech.* **102** 2877-2889
[5] Ling J H, Lim Y T, Leong W K, Euniza J and Sia H T 2019 *J. Civ. Eng. Forum* **5** 289-300
[6] Wang Q, Sun J, Yao Q, Ji C, Liu J and Zhu Q 2018 *Cellulose* **25** 4275-4301
[7] Mokhena T C, Sefadi J S, Sadiku E R, John M J, Mochane M J and Mtibe A 2018 *Polym.* **10** 1363
[8] Lu Y, Lu Y C, Hu H Q, Xie F J, Wei X Y and Fan X 2017 *J. Spectrosc.* **2017** 8951658
[9] Xu C C, Mou H, Lei M, Huang J, Li W and Huang H 2018 *BioResources* **13** 8578-8592
[10] Saha M, Saynik P B, Borah A, Malani R S, Arya P, Shivangi and Moholkar V S 2019 *Bioresour. Technol. Rep.* **5** 206-211
[11] Gunasekaran V, Ramesh S, Sathiasivan K, Shankar M, Rajesh M and Tamilarasan K 2020 *Chem* **74** 273-283
[12] Hassan N S and Badri K H 2014 *AIP Conf. Proc.* **1614** 433-438
[13] Pasma S A, Daik R, Ramli S, Maskat M Y and Zulfakar M H 2019 *BioResources* **14** 8879-8891
[14] Chio C, Sain M, and Qin W 2019 *Renewable Sustainable Energy Rev.* **107** 232–249
[15] Tanase-Opedal M, Espinosa E, Rodriguez A and Chinga-Carrasco G 2019 *Materials* **12** 3006
[16] Mimini V, Sykacek E, Hashim S N A, Holzweber J, Hettegger H, Fackler K, Potthast A, Mundigler N and Rosenau T 2019 *J. Wood Chem. Technol.* **39** 14-30
[17] Spiridon I, Leluk K, Resmerita A M and Darie R N 2015 Compos. B. **69** 342-349
[18] Prasong W, Ishigami A, Thumsorn S, Kurose T and Ito H 2021 *Polymers* **13** 740

IOP Conf. Series: Materials Science and Engineering 1192 (2021) 012015 doi:10.1088/1757-899X/1192/1/012015

Microwave-Assisted Synthesis of Polylactic Acid-diol for Polyurethane as Biodegradable Packaging Material

N I M Razali[1], F Ali[1,4], A S Azmi[1], T N M T Ismail[2], M E S Mirghani[3] and M F Omar[1]

[1]Department of Biotechnology Engineering, Kulliyyah of Engineering, International Islamic University Malaysia, Jalan Gombak, 53100 Kuala Lumpur, Selangor, Malaysia

[2]Advance Oleochemical Technology Department, Malaysia Palm Oil Berhad, 8, Persiaran Institusi, Kawasan Institusi Bangi, 43000 Kajang, Selangor, Malaysia.

[3]International Institute for Halal Research and Training (INHART), International Islamic University Malaysia, Jalan Gombak, 53100 Kuala Lumpur, Selangor, Malaysia

[4]Coresponding Author's Email: fathilah@iium.edu.my

Abstract. Since its discovery, plastic has been a part of human life and is widely employed in our daily lives. Excessive use of plastic has raised pollution rates around the world, with plastic ending up in landfills or the sea, posing a threat to both terrestrial and aquatic life. Considering this problem, the widespread use of polyurethanes (PUs) in many industries has resulted in unavoidable PUs pollution in everyday life. A reaction involving prepolymer, isocyanate, and polyol can be used to make PUs. Petroleum-based polyol and vegetable oil-based polyol are the two types of polyols available. Isocyanate will become the hard domain of the polymer in the PUs polymer chain, while polyol will become the soft domain. Polylactic acid-diol is the prepolymer used to make PU (PLA-diol). PLA-diol was previously made using a traditional heating approach, which takes a long time. To overcome this traditional method, microwave-assisted synthesis is proposed to synthesize the PLA-diol. The synthesis process involved synthesizing PLA-diol at different microwave power (450W – 900W) and at different reaction time (1 hour – 2 hours). The peak of hydroxyl group in synthesized PLA-diol was characterized via the Fourier Transform Infrared Spectroscopy (FTIR) characterization to determine the functional groups of PLA-diol and gel permeation chromatography (GPC) characterization was done to determine the molecular weight of PLA-diol. The resulting PLA-diol will then be used to synthesis biodegradable PUs in the subsequence study.

1. Introduction

Polyurethanes (PUs) is one of the most consumed and used plastic in the world. According to statistic, 18 million tons of PUs are produced in 2016 [1]. In fact, in 2020, PU production is expected to reach 22 million tons [2]. PU is a polymer consists of organic units, combined by carbamates (urethanes) links. PU polymers are the resulting product of a step growth polymerization reaction between an isocyanates and alcohol [3]. Isocyanates can be in the form of di-isocyanates or poly-isocyanates whereas alcohol can be in the form of a diol or macro-diol. The resulting reaction between these two together with a chain extender will produce PU with a soft segment and a hard segment, where polyol as the soft segment and isocyanate as the hard segment.

IOP Conf. Series: Materials Science and Engineering 1192 (2021) 012015 doi:10.1088/1757-899X/1192/1/012015

Biopolymers such as polylactic acid (PLA), cellulose and polyhydroxybutyrate (PHB) have been used in producing polymers. PLA specifically has become center of interest in producing PU. PLA are used as the diol in PU synthesis. PLA can be polymerized from its monomer which is lactic acid (LA) through polycondensation and ring opening polymerization. PLA polymerized from polycondensation method typically will have low molecular weight whereas PLA polymerized from ring opening polymerization (ROP) method will have high molecular weight [4]. ROP allows higher control of polymerization hence remains as the most widely used polymerization method for synthesis of well-defined materials [5].

To obtain PLA diol, polycondensation of LA can be done in two different ways, which are through conventional heating method and microwave-assisted heating method. In conventional heating method, the heating time required to achieve reaction temperature is lengthy. As reported by Bakibaev et al., through conventional method, polycondensation require up to 16 hours of time for the process to be completed with 160°C reaction temperature [6]. Another study done by Ismail et al., 6 hours reflux time was needed to obtain PLA diol [7].

Microwave-assisted polymerization is one of the suggested alternatives to replace conventional heating method. Despite of being newly studied, microwave-assisted polymerization shows promising results of synthesizing PLA diol. By using microwave-assisted method, Bakibaev and his team manage to synthesis PLA diol at 350W microwave power with only 10 minutes reaction time [6]. In another microwave-assisted single-step polycondensation of LA, PLA diol with weight-average molecular weight of 1800 Da were produced after being irradiated with 40W microwave power, at reaction temperature of 200°C and with 30 minutes of reaction time [8]. These two studies proved that microwave-assisted polymerization of LA is possible to be done and the reaction time is significantly reduced.

Microwave irradiation is defined as radiation having a frequency ranging from 0.3 to 300 GHz. Most synthetic and home microwaves are built to function in frequencies between 2 GHz and 8 GHz to prevent interference with contemporary telecommunication and radar systems, with residential microwaves commonly operating at 2.45 GHz. [9]. To fulfil the research needs, microwave instrumentation suppliers have moved towards combinatorial or high throughput medium governs by two philosophies. These two philosophies are the actual type of the microwave itself, known as single mode (monomode) and multimode microwave reactors. Microwave heating mechanism is based on dielectric heating [10], [11]. Dielectric heating can be defined as the ability of some polar substances (liquids and solids) to absorb and convert microwave irradiation into heat [10]. The ability of these materials is closely associated with its dielectric properties and can be determined using loss tangent, tan δ which can be expressed in quotient tan $\delta = \varepsilon'/\varepsilon''$. The dielectric loss, denote as ε'' indicates the efficiency in which electromagnetic is converted into heat whereas dielectric constant (ε') indicates the ability of molecules to be polarized by the electric field, allowing reasonable amount of energy to be absorbed. Therefore, to obtain high absorption efficiency for fast heating, a reaction medium with high tan δ value is needed. However, the value of ε' should be taken into consideration too since high tan δ value does not necessarily means the substances would allow reasonable energy absorption since it is depending on ε' value.

Conventional method of producing PLA-diol requires production time up to 6 hours. Therefore, a new method which is microwave assisted synthesis of PLA-diol will be introduced in this study. The hypothesis is that by using microwave irradiation, less time will be needed to activate the reaction in PLA-diol synthesis. If this method shows a promising future, the conventional method of synthesizing PLA-diol can be replaced with a more versatile and shorter production time method.

2. Experimental Procedures

2.1. Material & Instrumentation

2.1.1. Material. L-lactide (98%), 1,4-butanediol (99%), stannous octoate (95%), and dichloromethane (99%) were purchased from Sigma Aldrich. Toluene (99.8%) were used as the solvent in the reaction

IOP Conf. Series: Materials Science and Engineering 1192 (2021) 012015 doi:10.1088/1757-899X/1192/1/012015

whereas methanol was used as it is in the purification process. Both toluene and methanol are from analytical grade and were used as it is.

2.1.2. Instrumentation. The microwave used in the polymerization experiment was a modified domestic microwave oven, SAMSUNG MC455THRCSR (2450 MHz, 900W), with different power level (100W – 900W). The microwave was modified so that a 250 mL three neck flask, an overhead stirrer, a reflux condenser, and a nitrogen gas flow system could be introduced into the cavity of the microwave oven where appropriate additional safety steps were taken to avoid radiations leakages.

2.2. Synthesis of PLA-diol via Microwave assisted synthesis

Prior to experiment, an adequate amount of L-lactide was dried in vacuum oven at 45°C for a few hours to remove existing moisture. 1g of L-lactide (6.94 x 10^{-3} mol) was mixed with 42 mL anhydrous toluene in a 250mL three-neck round bottom flask. The mixture was purged with nitrogen gas for a few minutes. Next, 1,4-butanediol (4.16 x 10^{-4} mol) was added to the mixture while still being purged by nitrogen gas. Finally, stannous octoate (3.47 x 10^{-5} mol) was added into the mixture. After that, the flask was placed into microwave oven cavity. The flask was securely connected to an overhead stirrer, a reflux condenser, and a nitrogen gas flow system. All the connections were checked to be secured and safe before operating the microwave synthesizer. The microwave synthesizer was operated at 450W for 2 hours. After reaction completed, the solvent was removed via separation process. The product was dissolved in dichloromethane and then precipitated in methanol. The remaining moisture in the precipitate was removed by vacuum oven, yielding white powder. The experiment was repeated with different microwave operating conditions so that an optimum condition could be identified. Figure 1 illustrated the schematic of the experimental procedure.

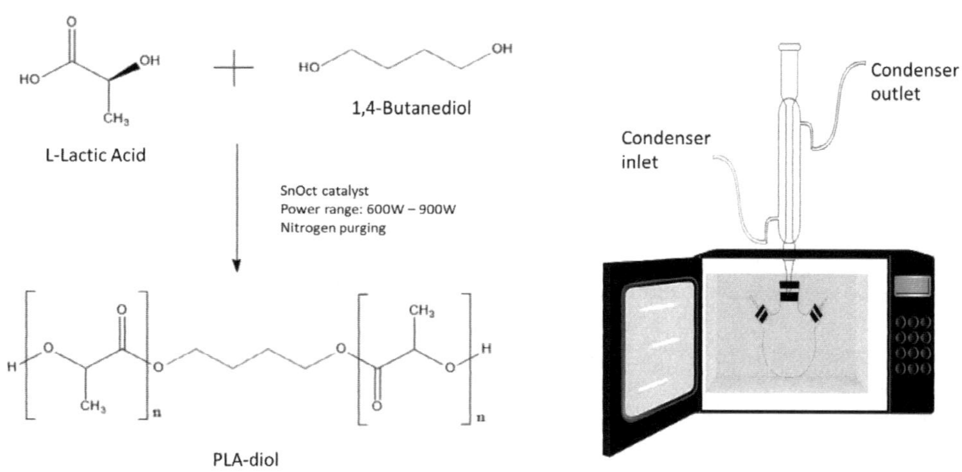

Figure 1. Synthesis pathway and the experimental set up of the experiment.

2.3. Characterization

The resulting PLA-diol samples were characterized for gel permeation chromatography (GPC) to determine its molecular weight. During GPC analysis, the elution phase was tetrahydrofuran (THF) and was calibrated with polystyrene standard. The peak of hydroxyl group in PLA-diol was characterized by Fourier Transform Infrared Spectroscopy (FTIR) spectra, taken at 650 cm^{-1} – 4000 cm^{-1} spectral resolution.

3. Results and discussion

3.1. Microwave-assisted synthesis of PLA-diol optimization

PLA diol was synthesized via microwave-assisted synthesis at different microwave power and reaction time to find optimum synthesis reaction condition. Table 1 illustrates the optimization runs done during the experiment and the resulting PLA diol obtained in weight.

Table 1. Optimization run for PLA-diol microwave-assisted synthesis.

Microwave Power (W)	Reaction time (mins)	Weight (g)
600	45	N/A
	60	N/A
	90	0.54
	120	0.65
900	45	0.52
	60	0.93
	90	1.21
	120	0.19

The weight from reactions from 600W at 45 minutes and 60 minutes could not be computed because the samples were dissolved in methanol during precipitation process which means PLA-diol did not form. From the resulting PLA-diol weight, it can be said that PLA-diol was successfully synthesized via microwave-assisted synthesis starting from 600W at 90 minutes. On the other hands, it clearly showed that when reacted at 900W for 90 minutes, the weight of PLA-diol produced was the highest. However, when reacted at 900W for 120 minutes, the weight of PLA-diol produced was the least among all runs. The weight of PLA-diol produced at 120 minutes, 900W power was lost due to extensive evaporation of water molecule and intensified polycondensation process [12]. Therefore, it can be concluded that the reaction between l-lactide, 1,4-butanediol and stannous octoate was at optimum condition when reacted at 900W for 90 minutes. Figure 2 shows the relation between microwave power, reaction time and weight of PLA-diol produced.

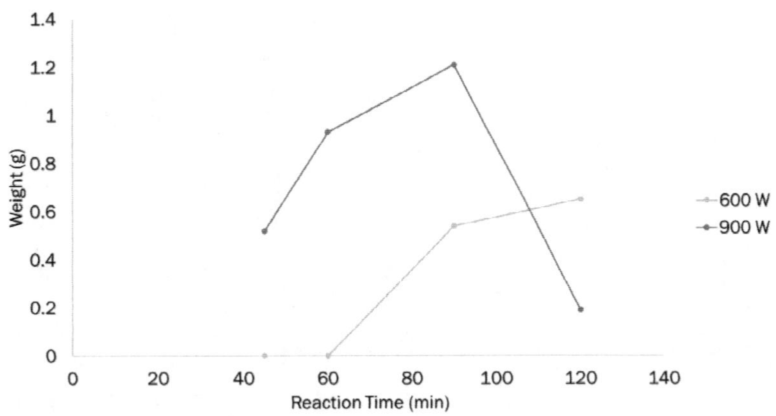

Figure 2. The relation between the weight of PLA-diol produced with microwave power and reaction time.

3.2. Gel Permeation Chromatography (GPC)

Figure 3 shows GPC chromatogram. GPC was done on samples for molecular weight determination. The samples were eluted in THF, and the system was calibrated with polystyrene standard. From the chromatogram, the samples showed peaks at the retention time of 28 minutes.

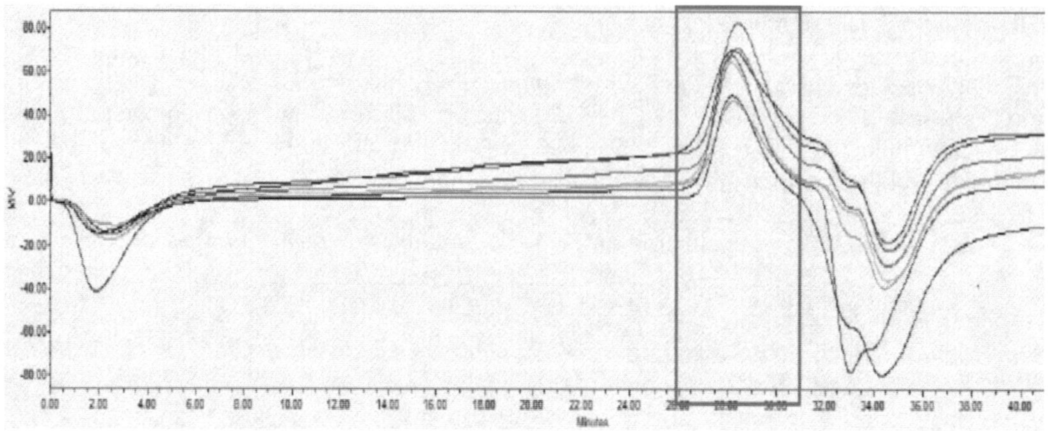

Figure 3. GPC chromatogram for synthesized PLA-diol.

3.3. Fourier Transfer Infrared Spectroscopy (FTIR)

Figure 4 shows FTIR spectra for PLA-diol synthesized via microwave-assisted synthesis method. The broad peak of hydroxyl group of PLA-diol produced can be observed at 3500 cm^{-1}. A similar peak for PLA-diol was reported by previous studies [7], [13]–[15]. This high hydroxyl absorption peak is expected to disappear when reacted with isocyanate in polyurethane synthesis. [7], [13]. The peak at 2995 cm^{-1} indicated -CH_3 end groups and the peak at 1759 cm^{-1} indicated C=O end groups. A similar finding was reported by Luo et al., [16]. From the FTIR spectra, it can be concluded that PLA-diol was successfully synthesized via microwave-assisted synthesis.

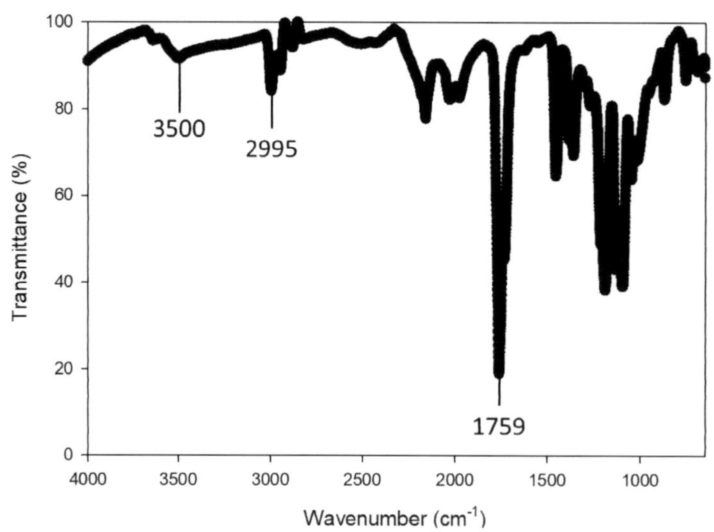

Figure 4. FTIR spectra for PLA-diol synthesized via microwave-assisted synthesis method.

4. Conclusion

PLA-diol from reaction between l-lactide, 1,4-butanediol and stannous octoate was successfully synthesized via microwave-assisted synthesis method. From the optimization process, it was found that to achieve maximum amount of PLA-diol synthesized, the reaction should take place at 900W for 90 minutes reaction time, compared to 6 hours of reaction time of synthesis via conventional method. This showed a great reaction time reduction. GPC chromatogram showed the produced PLA-diol was able to separate and showing peaks at 28 minutes retention time. On the other hands, FTIR spectra confirmed the formation of PLA-diol by showing broad hydroxyl group at 3500 cm^{-1}. In conclusion, PLA-diol was able to be synthesized via newly studied method, known as microwave-assisted synthesis. This also concluded that, even with modified domestic microwave oven, synthesis process could be carried out when appropriate safety measures were taken into consideration.

Acknowledgement
Our gratitude goes to the Malaysian Ministry of Education (MOE) for funding this research through the Fundamental Research Grant Scheme (FRGS) project, FRGS/1/2019/TK05/UIAM/02/1. Also, we would like to extend our thanks to Kulliyyah of Engineering IIUM for providing laboratory facilities for this research.

References
[1] Furtwengler P, Boumbimba R M and Avérous L 2018 Elaboration and Characterization of Advanced Biobased Polyurethane Foams Presenting Anisotropic Behavior *Macromol. Mater. Eng.* **303**, pp 1–11
[2] Akindoyo J O, Beg M D H, Ghazali S, Islam M R, Jeyaratnam N and. Yuvaraj A R 2016 Polyurethane types, synthesis and applications-a review *RSC Adv.* **6** pp 114453–114482
[3] Sonnenschein M F 2014 Polyurethanes: Science, Technology, Markets, and Trends," *Polyurethanes Sci. Technol. Mark. Trends* **9781118737** pp 1–417
[4] Ali F B and Ismail N 2018 Polymerization Methods and Characterizations for Poly(Lactic Acid) (PLA) Based Polymers *Multifaceted Protoc. Biotechnol.* pp 139–150
[5] Dechy-Cabaret O, Martin-Vaca B. and Bourissou D 2004 Controlled ring-opening polymerization of lactide and glycolide *Chem. Rev.* **104**, pp 6147–6176
[6] Bakibaev A A *et al.* 2015 Polymerization of Lactic Acid Using Microwave and Conventional Heating *Procedia Chem.* **15** pp 97–102
[7] Ismail M N, Rafidi N, Awale R J and Ali F 2017 Preparation and characterization of polylactic acid based polyurethane for environmental friendly packaging materials *J. Adv. Res. Fluid Mech. Therm. Sci.* **38**, pp 16–21
[8] Nagahata R, Sano D, Suzuki H and Takeuchi K 2007 Microwave-assisted single-step synthesis of poly(lactic acid) by direct polycondensation of lactic acid *Macromol. Rapid Commun.* **28** pp 437–442
[9] Kappe C O 2004 Controlled microwave heating in modern organic synthesis," *Angew. Chemie - Int. Ed.* **43** pp 6250–6284
[10] Mallakpour S and Rafiee Z 2011 New developments in polymer science and technology using combination of ionic liquids and microwave irradiation *Prog. Polym. Sci.* **36** pp 1754–1765
[11] Sosnik A, Gotelli G and Abraham G A 2011 Microwave-assisted polymer synthesis (MAPS) as a tool in biomaterials science: How new and how powerful *Prog. Polym. Sci.* **36** pp 1050–1078
[12] Zahari S M S N S, Mansor M H, Azman H H and Rosli D 2020 Uncatalysed Polycondensation of Lactic Acid to Polylactic Acid under Microwave Irradiation: Effect of Microwave Power *J. Phys. Conf. Ser.* **1551**
[13] Ali F B, Kang D J, Kim M P, Cho C H and Kim B J 2014 Synthesis of biodegradable and flexible, polylactic acid based, thermoplastic polyurethane with high gas barrier properties *Polym. Int.* **63** pp 1620-1626
[14] Hiltunen K, Härkönen M, Seppälä J V and Väänänen T 1996 Synthesis and characterization of lactic acid based telechelic prepolymers *Macromolecules* **29** pp 8677–8682
[15] Zhu K J, Xiangzhou L andShilin Y 1990 Preparation, characterization, and properties of polylactide (PLA)–poly(ethylene glycol) (PEG) copolymers: A potential drug carrier *J. Appl. Polym. Sci.* **39** pp 1–9
[16] Luo W, Li S, Bei J and Wang S 2002 Synthesis and characterization of poly(L-lactide)-poly(ethylene glycol) multiblock copolymers *J. Appl. Polym. Sci.* **84** pp 1729–1736

Chitin Fiber from Mushroom as Reinforcement for Biobased Polymer

M I M Zin[1], N S Shamsudin[1], F Ali[1], W M F W Nawawi[1,2]

[1]Department of Biotechnology Engineering, Kulliyyah of Engineering, International Islamic University Malaysia (IIUM), P.O. Box 10, 50728, Gombak, Kuala Lumpur, Malaysia

[2]Corresponding Author's Email: wanmohdfazli@iium.edu.my

Abstract. This project aimed to study the reinforcement effect of fungal chitin at different loading on chitin/PLA composite. The chitin nanofibers were extracted from three mushroom species (oyster mushroom *Pleurotus ostreatus,* shiitake mushroom *Lentinula edodes,* enoki mushroom *Flammulina velutipes*) and used as a reinforcement element in PLA. The chitin/PLA composite was fabricated using a solvent-casting method followed by the hot-compress molding method. In the solvent-casting method, the chitin nanofibers were dispersed in PLA/chloroform mixture and the mixture was left for solvent evaporation. The solvent-free chitin/PLA thin film was then filled in dog bone mold before proceeded with hot-compress molding at 190°C and 70 bar. The samples with different chitin loading were tested with tensile test to study the mechanical performance of nanocomposite. The chitin/PLA composite from oyster mushroom shows the optimum result (σ= 43 MPa, E= 12 MPa) at 5% chitin loading. The increment of the chitin loading leads to a decrease in both strength and strain. However, for the samples from enoki and shitake mushrooms, the optimum chitin loading is 10% with 55 MPa and 56 MPa tensile strength, respectively. This study suggests the potential of fungal chitin as reinforcement in PLA.

1. Introduction

Chitin is a natural fiber that has been recently studied for its beneficial properties as reinforcing materials in biopolymers [1-3]. Its is abundantly found in crustacean shell, insect cuticle and cell wall of fungi. Fungal chitin has gained attention as the alternative to chitin source since it can provide continuous and consistent supply [4, 5]. Since fungal chitin has excellent properties such as biodegradability and biocompatibility, it can be used as a reinforcing element for biobased materials such as polylactic acid (PLA) [6].

PLA is the most promising biodegradable bioplastic which is considered as eco-friendly material to replace conventional plastics, especially in the packaging industry. However, its brittleness and poor stability limit its industrial applications. Therefore, modification of PLA with the addition of plasticizers, coupling and antistatic agents was done as a strategy to improve its performance so that it can compete with the conventional polymers used in the packaging industry [3]. Coltelli, et al. [7] discovered that incorporating PLA with chitin nanofibrils increased the elongation of the polymer by 180%. The chitin loading in PLA affects the mechanical properties of PLA composite [1, 8, 9] Another study by Maskur, et al. [10] reported that the incorporation of 5 wt% crustacean chitin resulted in higher tensile strength and modulus. However, no study has been done yet on the effect of fungal chitin loading in PLA.

Therefore, this study demonstrates the effect of fungal chitin loading on the mechanical properties of chitin/PLA composite. The fungal chitin in this study was extracted from oyster (*Pleurotus ostreatus),* Shiitake (*Lentinula edodes)* and Enoki mushroom (*Flammulina velutipes)* before the chitin powder was mixed with PLA using chloroform. The mechanical properties of chitin/PLA composite were studied using a universal tensile machine.

2. Materials and Methods

2.1. Materials

Three different mushrooms which are oyster (*Pleurotus ostreatus),* Shiitake (*Lentinula edodes)* and Enoki mushroom (*Flammulina velutipes)* were used as fungal chitin sources in this study. The mushrooms were frozen at least overnight at -20°C. Sodium hydroxide (Merck, pellets) was used for chitin extraction process. The extracted chitin and PLA (Nature Works, Ingeo 3251D and Ingeo 3052D) were mixed using chloroform (Merck).

2.2. Chitin Nanofiber Extraction

The extraction was carried out according to a procedure performed by Nawawi, et al. [11]. The frozen mushrooms were thawed and washed with distilled water to remove any impurities before blending in a kitchen blender (Vita Mixer Innofood, SX766) for 5 minutes to break the cell wall. Distilled water was added to the blended mushrooms and the suspension was heated at 85°C for 1 hour to remove any water-soluble components. The water-soluble components were removed by filtration process using a cotton cloth. A vacuum pump (VALUE, VE 115N) was used to facilitate the filtration process. Then, the filter cake was soaked in 1 M NaOH (65°C, 3 hours) to remove protein, lipid and alkaline-soluble polysaccharides before neutralizing with distilled water until pH 7 was indicated by universal test paper. The neutralized chitin was freeze-dried to remove the excess moisture and ground into powder (150 μm) for further use.

2.3. Chitin/PLA Composite Fabrication

Prior to chitin/PLA composite fabrication, the PLA was dried overnight at 70°C to remove moisture content. Then, the fabrication process started by first dispersing the chitin powder (5%, 10%, 15% and 20%) in chloroform before dispersing in PLA/chloroform mixture. The mixture was then stirred at high speed using the magnetic hot plate (IKA®, G-MAG HS 7) until the chitin was homogenously dispersed within the mixture. The mixture was poured into glass petri dish covered with aluminium foil and left in the fume hood until the chloroform was fully evaporated, leaving behind chitin/PLA film composite. The film composite was ground and filled into the dog bone mold cavities (Type IV, ASTM D638) that were coated with a releasing agent (Loctite 770-NC Frekote) to ease the removal process. The step was proceeded with compression molding by hot press (SUN RUN, SPH-900) at 190°C and 70 bar for 15 minutes. As a control, neat PLA sample was prepared by adding the PLA granule until it fully occupied the dog bone mold cavities.

3. Results and Discussion

3.1. Chitin/PLA Composite Fabrication

The agglomeration of chitin during composite fabrication was minimized by grinding freeze-dried chitin into powder form. The smaller size of chitin powder increases the contact area with PLA/chloroform mixture which contributes to a homogenous dispersion within PLA polymeric matric. Still, stirring was required to promote a homogenous dispersion of chitin since the chitin is not easily solubilized in a common solvent, such as chloroform [12]. During the solvent casting method, the chloroform was slowly evaporated by covering the glass petri dish with aluminium foil. This step was important to prevent bubble formation in the film which would affect the preparation of tensile test specimens via compression molding. Figure 1 shows the chitin/PLA film with different chitin loading. The accumulation spots of chitin become more visible as higher chitin loading is introduced. This formation describes the properties of chitin that attracted and crystallized when less solvent was present. The higher chitin loading in PLA induces more interaction between chitin nanocrystals and the solvent.

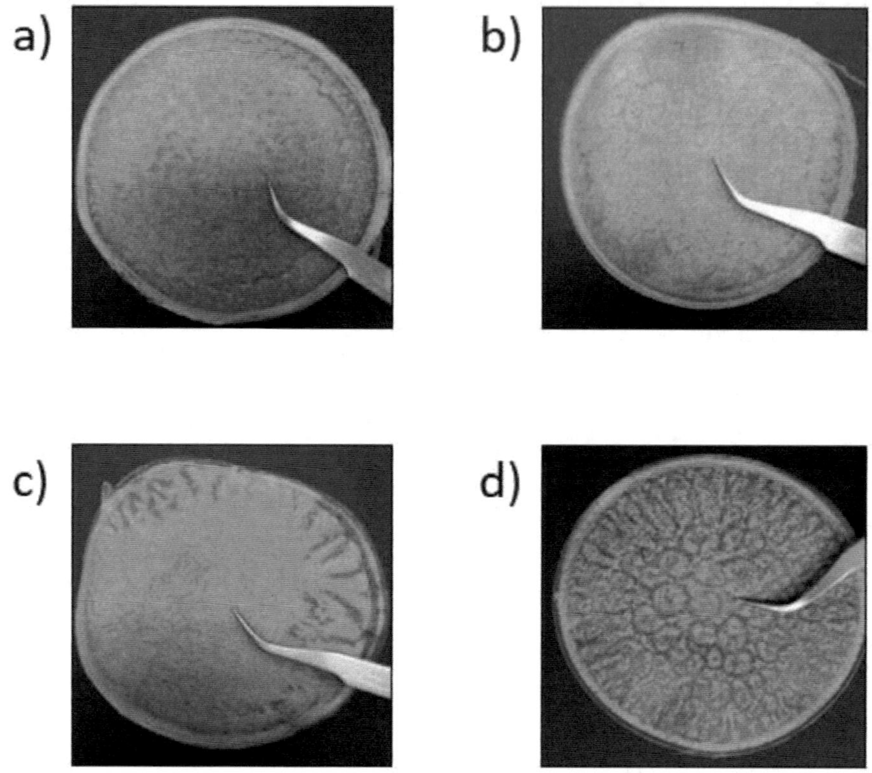

Figure 1. Chitin/PLA composite with (a) 5 wt%, (b) 10 wt%, (c) 15 wt% and (d) 20 wt%.

Figure 2 shows the tensile specimen of neat PLA. No bubble formation was observed within the tensile specimen. This was achieved by pre-dried PLA and grinding the film before tensile specimen preparation. The formation of bubbles in the samples acts as a weak point for crack propagation to initiate. Thus, will affect the mechanical properties.

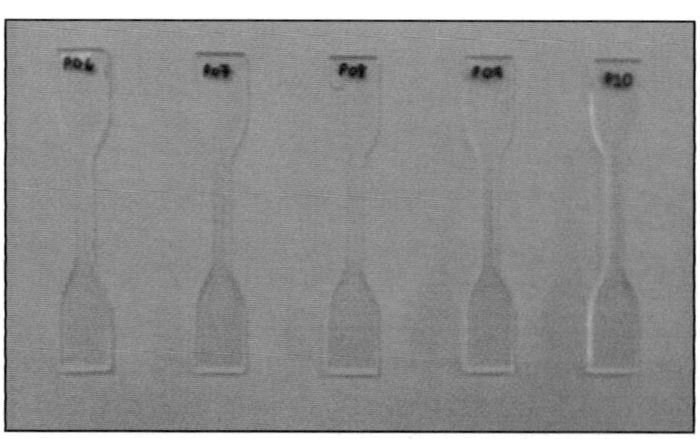

Figure 2. Tensile specimen of neat PLA.

3.2. Mechanical Properties of Chitin/PLA composite

The mechanical properties of oyster chitin/PLA composite were compared to neat PLA (Ingeo 3052D). While for shitake and enoki chitin/PLA composite, PLA (Ingeo 3251D) was used since we are unable to purchase the same product due to technical limitations. However, the effect of fungal chitin loading in PLA still could be discussed except that we could not compare the result for oyster with shitake and enoki chitin/PLA composite. Table 1 shows the mechanical properties of oyster chitin/PLA at different chitin loading.

Table 1. Mechanical properties of oyster chitin/PLA at different chitin loading.

Sample	Tensile strength (MPa)	Young modulus (MPa)	Tensile strain (%)	Tensile toughness (MJ/m^2)
NP1	17.76 ± 7.83	12.86 ± 0.5	1.56 ± 0.73	0.42 ± 0.36
CP1O5	43.04 ± 2.45	12.15 ± 0.14	4.35 ± 0.45	0.91 ± 0.08
CP1O10	41.52 ± 0.54	19.15 ± 9.24	3.16 ± 1.13	0.73 ± 0.26
CP1O15	30.37 ± 0.86	12.07 ± 0.51	2.77 ± 0.05	0.44 ± 0.01
CP1O20	34.75 ± 4.10	12.54 ± 0.3	3.10 ± 0.40	0.59 ± 0.15

*The numbers represent the chitin loading percentage in PLA.

The strength and toughness of PLA were improved when oyster chitin was added. The composite containing 5 wt % of oyster chitin recorded the highest tensile strength and toughness with the value of 43 MPa and 0.91 MJ/m^2, respectively. The result was in agreement with a study by Maskur, et al. [10] when they investigate the effect of crustacean chitin loading in PLA. A lower chitin loading promotes a better homogenous dispersion within PLA matrix. Since chitin is less hydrophilic, lower chitin loading makes it more compatible with the hydrophobic PLA matrix [8]. As the result, the affinity of chitin and PLA matrix contributes to the properties of composite [2]. Figure 3 shows the trends of tensile strength when more chitin loading was introduced. When the chitin loading was increased to 10, 15 and 20 wt%, the tensile strength and strain were decreased. At higher chitin loading, the compatibility of chitin with PLA is weaker and the chitin tends to agglomerate. Therefore, the bigger chitin particle gives a negative impact on the strength of the composite. However, there is no reduction of strain to failure observed, proposing that no damage on fibers occurred during the extraction process.

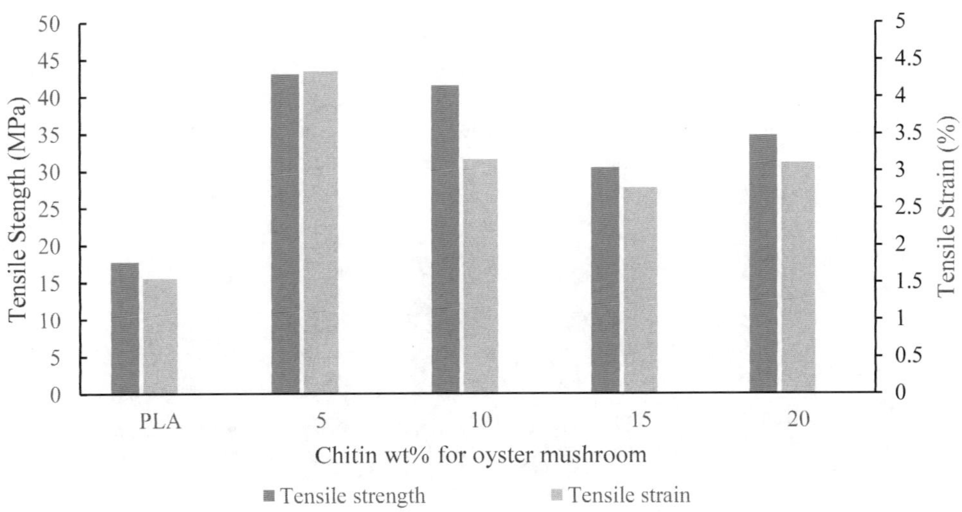

Figure 3. Tensile strength and strain of chitin/PLA nanocomposite for oyster mushroom.

Table 2 presents the mechanical properties of enoki and shitake chitin/PLA composite. For both enoki and shitake chitin/PLA composite samples, the highest tensile strength and toughness were exhibited when 10 wt% chitin was added into PLA. The result showed that PLA (3251D) promotes better chitin compatibility up to 10 wt% chitin loading compared to PLA (3052D) which only up to 5 wt% chitin loading. An increase in fiber content in a composite system theoretically would increase the tensile strength [9, 13, 14]. However, the compatibility between the fiber with PLA matrix also contributes to the tensile strength. Therefore, if better compatibility at higher fiber content could be maintained, the tensile strength probably would be improved further.

Table 2. The mechanical properties on chitin/PLA nanocomposite of Enokii and Shiitake mushroom.

Sample	Tensile strength (MPa)	Young modulus (MPa)	Tensile strain (%)	Tensile toughness (MJ/m2)
NP2	48.63 ± 11.83	12.22 ± 0.42	5.97 ± 2.22	1.90 ± 1.08
CP2E5	53.18 ± 4.98	12.37 ± 0.20	6.06 ± 0.87	1.91 ± 0.45
CP2E10	55.87 ± 0.56	10.27 ± 4.16	7.21 ± 0.34	2.33 ± 0.10
CP2E15	53.97 ± 0.71	9.78 ± 0.48	7.27 ± 0.59	2.22 ±0.24
CP2E20	48.77 ± 0.22	13.68 ± 0.04	4.68 ± 0.07	0.68 ± 0.67
CP2S5	52.05 ± 1.19	12.13 ± 0.21	5.35 ± 0.19	1.58 ± 0.10
CP2S10	56.85 ± 2.01	11.85 ± 0.45	6.46 ± 0.50	2.17 ± 0.25
CP2S15	51.33 ± 4.47	13.02 ± 0.20	5.15 ±0.52	1.53 ± 0.29
CP2S20	47.27 ± 4.78	13.86 ± 0.36	4.37 ± 0.66	1.20 ± 0.33

*CP2E and CP2S indicate the samples from chitin enoki and shiitake respectively. The numbers represent the chitin loading percentage in PLA.

The highest modulus was recorded at 20 wt% chitin loading for both enoki and shitake chitin/PLA samples. According to Rizvi, et al. [9], the modulus is expected to increase when the nanofiller content increases up to some limit. Differ from tensile strength, the modulus is probably not affected by the compatibility of chitin and PLA (3251D) since the result showed a different pattern. Nevertheless, the tensile strain of enoki chitin/PLA composite showed a slight increase up to 15 wt% of chitin before it plummeted at 20 wt% (Figure 4). The tensile strain for enoki chitin/PLA at 15 wt% showed no significant difference with the one at 10 wt% chitin loading. Therefore, we can claim that the tensile strain pattern for both enoki and shitake chitin/PLA composite was quite similar.

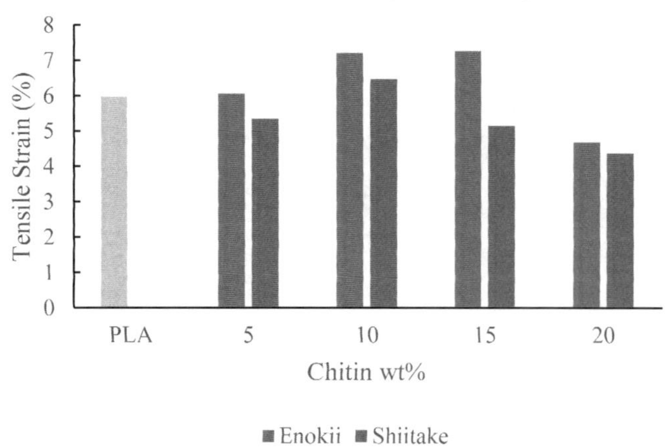

Figure 4. Tensile strain of enoki and shitake chitin/PLA nanocomposite at different chitin loading.

4. Conclusions

The use of fungal chitin nanofiber as a reinforcement element in PLA has a positive effect to enhance its mechanical properties. The chitin/PLA (3052D) composite has optimum tensile strength and toughness when 5 wt% oyster chitin was incorporated into the PLA. However, for enoki and shitake chitin/PLA (3251D) composite, the optimum tensile strength and toughness were recorded at 10 wt% chitin loading. The overall result suggested the potential of fungal chitin nanofiber as a reinforcement element in the development of sustainable material.

Acknowledgements

The authors acknowledge the Ministry of Education Malaysia and International Islamic University Malaysia (FRGS/1/2018/TK05/UIAM/03/3) for providing financial support to this project.

References

[1] Mohd Asri S E A, Zakaria Z, Hassan A and Mohamad Kassim M H 2020 Effect of Chitin Source and Content on Properties of Chitin Nanowhiskers Filled Polylactic Acid Composites *IIUM Engineering Journal* **21** pp 239-255

[2] Nakagaito A, Kanzawa S and Takagi H 2018 Polylactic Acid Reinforced with Mixed Cellulose and Chitin Nanofibers—Effect of Mixture Ratio on the Mechanical Properties of Composites *Journal of Composites Science* **2**

[3] Herrera N, Salaberria A M, Mathew A P and Oksman K 2016 Plasticized polylactic acid nanocomposite films with cellulose and chitin nanocrystals prepared using extrusion and compression molding with two cooling rates: Effects on mechanical, thermal and optical properties *Composites Part A: Applied Science and Manufacturing* **83** pp 89-97

[4] Dhillon G S, Kaur S, Brar S K and Verma M 2013 Green synthesis approach: extraction of chitosan from fungus mycelia *Critical Reviews in Biotechnology* **33** pp 379-403

[5] Yousefi N, Jones M, Bismarck A and Mautner A 2020 Fungal chitin-glucan nanopapers with heavy metal adsorption properties for ultrafiltration of organic solvents and water,*Carbohydrate Polymers* **253**

[6] Li J, Li J, Feng D, Zhao J, Sun J and Li D 2017 Comparative Study on Properties of Polylactic Acid Nanocomposites with Cellulose and Chitin Nanofibers Extracted from Different Raw Materials *Journal of Nanomaterials* **2017** pp 1-11

[7] Coltelli M B *et al.* Chitin Nanofibrils in Poly(Lactic Acid) (PLA) Nanocomposites: Dispersion and Thermo-Mechanical Properties *Int J Mol Sci* **20**

[8] Nakagaito A N, Yamada K, Ifuku S, Morimoto M and Saimoto H 2013 Fabrication of Chitin Nanofiber-Reinforced Polylactic Acid Nanocomposites by an Environmentally Friendly Process *Journal of Biobased Materials and Bioenergy* **7** pp 152-156

[9] Rizvi R, Cochrane B, Naguib H and Lee P C 2011 Fabrication and characterization of melt-blended polylactide-chitin composites and their foams *Journal of Cellular Plastics* **47** pp 283-300

[10] Maskur M H, Wan Nawawi W M F, Ali F and Alkhatib M F R 2020 Crustacean chitin nanomaterial as reinforcement for bio-based polymer *Advances in Materials and Processing Technologies* pp 1-12

[11] Nawawi W, Lee K Y, Kontturi E, Murphy R J and Bismarck A 2019 Chitin Nanopaper from Mushroom Extract: Natural Composite of Nanofibers and Glucan from a Single Biobased Source *ACS Sustainable Chemistry & Engineering* **7** pp 6492-6496

[12] Nawawi W, Jones M, Murphy R J, Lee K Y, Kontturi E and Bismarck A 2020 Nanomaterials Derived from Fungal Sources-Is It the New Hype? *Biomacromolecules* **21** pp 30-55

[13] Ku H, Wang H, Pattarachaiyakoop N and Trada M 2011 A review on the tensile properties of natural fiber reinforced polymer composites *Composites Part B: Engineering* **42** pp 856-873

[14] Rojo E, Alonso M V, Oliet M, Del Saz-Orozco B and Rodriguez F 2015 Effect of fiber loading on the properties of treated cellulose fiber-reinforced phenolic composites *Composites Part B: Engineering* **68** pp 185-192

A Mini Review on the Antibacterial Activity of Roselle (*Hibiscus sabdariffa* L.) Phytochemicals

A Prasetyoputri*, S I Rahmawati, A Atikana, F N Izzati, Y Hapsari, E Septiana, Bustanussalam and M Y Putra

Research Centre for Biotechnology, Indonesian Institute of Sciences (LIPI). Jl. Raya Bogor KM 46, Cibinong, Indonesia 16911

E-mail: anggia.prasetyoputri@uq.net.au

Abstract. Roselle (*Hibiscus sabdariffa* L.) has been shown to have various bioactivities with therapeutic benefits. These bioactivities are owing to the different kinds of phytochemicals, which include anthocyanins, phenolic acids, and flavonoids. This mini review aims to summarize the reported antibacterial activity of roselle phytochemicals in the literature in the past decade (2011-2021). The results revealed that roselle extracts from various extraction methods were shown to have antibacterial activity against clinical isolates and food pathogens, including multidrug resistant bacteria. Furthermore, there is evidence that roselle extract showed potential synergy with antibiotics. Overall, phytochemicals in roselle have the potential as an antibacterial for different beneficial applications.

Keywords: *Hibiscus sabdariffa* L.; phytochemicals; antibacterial activity; bioactive compounds; synergy

1. Introduction

In light of antibiotic resistance and limited new antibiotics entering the clinic, various efforts are undertaken to find sources of new antimicrobials. As many plants are used as traditional medicine, plant-derived compounds are therefore a promising source of new antimicrobial molecules [1]. Several main phytochemicals that contribute to the antimicrobial activities of plant extracts include flavonoids, alkaloids, terpenes, polyphenols, and coumarins [1,2]. These phytochemicals exert varying mechanisms of action against bacteria, which include membrane disruption, complex formation with bacterial cell wall, or substrate deprivation [2].

One of medicinal plants that have been investigated for their antibacterial activity is *Hibiscus sabdariffa* L. (Hs). Distributed worldwide, Hs is also called roselle or karkade. Hs belongs to the Malvaceae family native to tropical Africa, though Hs also cultivated in several places such as in America and South-east Asia, including Indonesia. It is an annual herbaceous plant living in subtropical and tropical, dry, mountain climates. The calyces are harvested after around 3-4 months when the plants are mature [3]. The bright red colour of the fleshy calyces (Figure 1) is due to an abundance of anthocyanins [4]. The high concentrations of dietary fibre and polyphenol [5] and other nutrients such as high levels of vitamin C [3] led to Hs calyxes widespread use as herbal medicine.

Furthermore, the popular consumption of Hs is owing to its nice taste, colourful appearance, as well as its multiple medicinal benefits and use for culinary purposes.

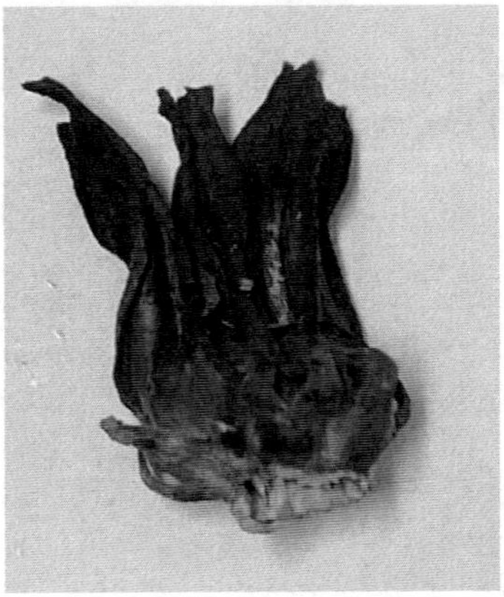

Figure 1. Hs dried calyx

Studies have shown that a wide range of therapeutic effects are exhibited by Hs, including antioxidant, antipyretic, anti-inflammatory, antibacterial, anticancer, immunomodulatory, antihypertensive, and hepatoprotective effects [4,6-9]. Furthermore, in insulin-resistant rats, both antihyperglycemia and antihyperlipidemia activities have been demonstrated, as well as evidence of reduced cardiac hypertrophy in hypertensive rats and patients [6]. Several the bioactivities as mentioned above are due to its rich source of high antioxidant polyphenols in its calyces [10]. Moreover, the dietary fibre from polysaccharide content in Hs was found to be useful in obesity management as the fibre can reduce body weight, adiposity, plasma total cholesterol and glucose [8].

2. Phytochemicals found in Hs

Hs is an attractive commodity for food and traditional medicine, therefore, phytochemicals that are present in Hs have been extensively studied [5,7,11]. The major constituents of Hs phytochemicals include anthocyanins, flavonoids, phenolics and organic acids [9,12]. These compounds are responsible for the many biological activities of Hs extracts.

Among many biological activities, Hs is known for its rich antioxidant properties, potentially owing to the diverse polyphenols that it produces, such as anthocyanins. High concentrations of anthocyanins can be found in Hs calyces [7]. These compounds are responsible for the calyces' red colour, which varies with the changing of its pH [7,13]. Among different anthocyanins, delphinidin-3-sambubioside and cyanidin-3-sambubioside are the major anthocyanins present in extracts of Hs calyxes, with cyanidin-3-glucoside and delphinidin-3-glucoside present in lesser amounts [7,13,14] (Figure 2). These two major anthocyanins are suggested to be the compounds contributing to the antihypertensive and anticholesterol activities of Hs extract [15]. Additionally, delphinidin-3-sambubioside has also demonstrated an apoptosis-inducing potential in cancer cells [12].

Polyphenols of the flavonol and flavanol type in simple or polymerised form have also been reported in Hs extracts, as well as a number of flavonoids [7]. Antioxidant, antihypertensive, antimicrobial, anti-inflammatory, anticancer and antidiabetic potentials have been demonstrated by the polyphenols and flavonoids from Hs [16]. Sabdaritrin, hibiscitrin (hibiscetin-3-glucoside), gossytrin, gossypitrin, quercetin, luteolin, chlorogenic acid, protocatechuic acid, pelargonidic acid, eugenol, and phytosterol (β-sitosterol and ergosterol) are some of the flavonoids that can be found in Hs extracts [7,11]. Moreover, chlorogenic acids, caffeic acids, quercetin compounds have also been found in Hs extract, with activities ranging from antioxidant, anti-inflammatory, antiparasitic and many more [9]. A number of bioactivities have been demonstrated by polyphenol compounds, for example, protocatechuic acid was shown to have anticancer activity due to its apoptosis-inducing potential [12].

Additionally, chemopreventive activity was associated with the flavonoids from aqueous Hs extract administration in Wistar rats [17].

In addition to phenolic acids, flavonoids, and anthocyanins, Hs extracts also contain a high percentage of organic acids, including hydroxycitric acid, hibiscus acid, malic, tartaric, and citric acids [9]. The composition of acid from different cultivars of Hs may vary as they are influenced by genetics, environment, ecology and harvest conditions [8]. A number of biological activities are attributed to these organic acids. For example, hibiscus acid was shown to have antihypertensive activity as it demonstrated to be a potent vasorelaxant in a rat aorta [18]. Moreover, hibiscus acid was also found to have antibacterial activity against multidrug resistant bacteria [19].

It is evident from the literature that Hs extracts and phytochemicals have a lot of potential with regards to their antioxidant, anticancer, and antihypertensive properties, as well as many other important biological activities [5,7,9,11,12]. However, their antibacterial activity has been investigated to a lesser extent. As plant-derived compounds are potential as sources for new antibacterial lead compounds [20], a summary of the antibacterial activities of Hs extracts and phytochemicals would be beneficial. Therefore, this mini review will attempt to summarise various studies on antibacterial activities of extracts derived from Hs in between 2011 and 2021. In addition to elaborating on the different pathogens that can be inhibited by bioactive compounds from Hs, this paper will also review the potential synergism of Hs extracts with antibiotics.

3. Antibacterial activity of Hs extract and phytochemicals

Antibacterial activity of Hs extracts has been reported against many bacterial strains, including multi-drug resistant strains (Table 1). A number of studies have also investigated the specific active compounds that may be responsible for the antibacterial activity of Hs extracts. Hs is known to primarily contain organic acids, flavonoids and phenolic acids, anthocyanins, and polysaccharides, as well as various volatile compounds [7].

Flavonoids, anthocyanins, terpenoids, and coumarins have been known to be active against various Gram-positive and Gram-negative bacteria [21], suggesting the presence of these compounds is the reason for antibacterial activity of Hs extracts. Indeed anthocyanins from various plants have been shown to exert antibacterial activity through disruption of bacterial cell wall and have activities against *Escherichia coli* and *Salmonella* [22]. Similarly, flavonoid compounds have been shown to have activity against pathogenic bacteria through various mechanisms, which include nucleic acid synthesis inhibition, biofilm inhibition, DNA gyrase inhibition, and inhibition of membrane function [23]. Interestingly, a study characterising 25 varieties of Mexican Hs found that a high total content of phenolic compounds that include flavonoids and anthocyanins does not always translate to good antibacterial activity [24]. This study suggested that certain phenolic composition within the plant extract and potential interaction with other minor compounds could contribute to the antibacterial activity of Hs extracts [24].

Phenolic rich extracts from Hs was found to contain various compounds, including anthocyanin, flavonoids, alcoholic compounds, triazine derivatives, and esters, which were suggested to contribute to their potent antibacterial activity [25]. Hs extracts also contain various organic acids, such as hibiscus acid [9] (Figure 3). Hibiscus acid was found to inhibit bacterial growth of *E. coli, Salmonella typhimurium, Pseudomonas aeruginosa, Staphylococcus aureus*, and *Vibrio cholerae* [19]. This mechanism was suggested to be a disruption in bacterial cell membrane permeability [19]. Additionally, the contribution of fatty acids from Hs calyces to its antibacterial activity has also been reported [26].

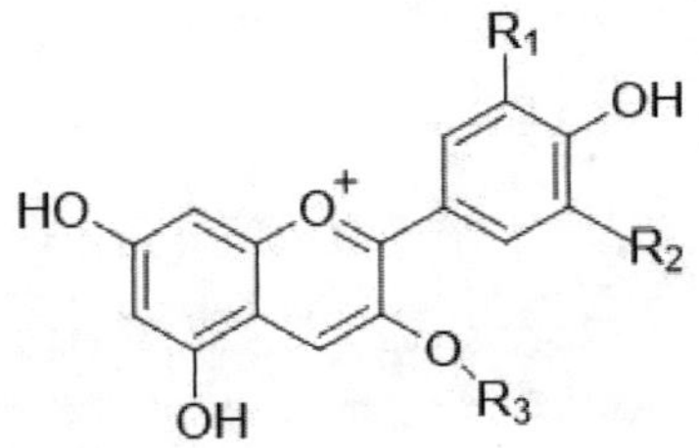

Figure 2. Chemical structure of anthocyanins found in Hs. Cyanidin-3-sambubioside (R1=OH; R2=H, R3=Sambubioside), delphnidin-3-sambubioside (R1=OH; R2=OH, R3=Sambubioside), cyanidin-3-glucoside (R1=OH; R2=H, R3=Glucose), and delphnidin-3-glucoside (R1=OH; R2=OH, R3=Glucose). Figure redrawn from [7].

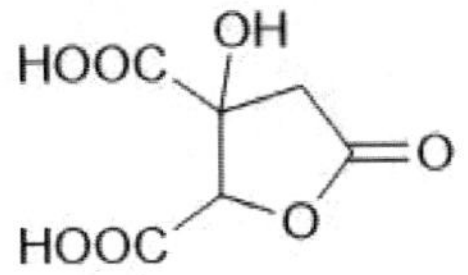

Figure 3. Chemical structure of hibiscus acid.

Table 1. Several reports on antibacterial activities of *H. sabdariffa* extracts *in vitro*

No.	Material preparations	Pathogen	Method of testing	Antibacterial activity	References
1.	Aqueous calyx extract	*Salmonella typhi*	Well diffusion method	Concentrations of 150, 250, 500 mg/mL produced inhibition zones of 18, 25, and 27 mm, respectively	[27]
2.	Methanol calyx extract	*E. coli* O157:H7 isolates from food, veterinary, and clinical samples	Disc diffusion method	Extract concentrations of 2.5%, 5%, and 10% were found to inhibit *Escherichia coli* O157:H7, with the 10% concentration being the most effective. Overall mean zones of inhibition were 8.9, 10.75, and 12.66 mm for 2.5%, 5%, and 10% concentrations, respectively	[28]
3.	Ethanol extract of 25 Mexican Hs varieties	*E. coli* (ATCC 25922) *Salmonella enteritidis* (ATCC 14028), *S. aureus* (ATCC 25923), and *Micrococcus luteus* (ATCC 9341)	Agar well diffusion method	All extracts exhibited antibacterial activity against the Gram-positive and Gram-negative strains tested. Greater antibacterial activity was observed against *S. aureus* and *M. luteus* compared to *E. coli* and *S. enteritidis*. Zone of inhibition range for *S. aureus*: 16-22 mm; *M. luteus*: 10-18 mm; *E. coli* and *S. enteritidis*: 10-16 mm.	[24]
4.	Methanol extract of Hs twigs	A total of 27 strains *of E. coli, Enterobacter. aerogenes, E. cloacae, Klebsiella pneumoniae, Providencia stuartii, P. aeruginosa*	A rapid p-Iodonitrotetrazolium chloride (INT) colorimetric assay	Hs extract was found to inhibit 18/27 (66.6%) bacteria tested (six *E. coli* strains, six *E. aerogenes* strains, one *E. cloacae* strain, four *K. pneumoniae* strains, and one *P. stuartii* strain, but no *P. aeruginosa* strain)	[29]
5.	Crude, phenolic rich ethanol extract	*S. aureus* DSM 1104, *Streptococcus pyogenes* ATCC 19615, *L. monocytogenes* LMG10470, *E. coli* LMG 8223, *K. pneumoniae* ATCC 43816 and *P. aeruginosa* LMG 8029	Agar well diffusion assay, MIC assay, as well as SEM and TEM microscopy	Using 2000 µg/mL extract, the inhibition zones against *S. aureus, Streptococcus pyogenes, Listeria monocytogenes, E. coli, K. pneumoniae, P. aeruginosa* ranged between 32 – 48 mm. MICs of phenolic rich extract against the above bacteria were the highest at 250 µg/mL concentration with inhibition zones ranging between 22- 30 mm. SEM and TEM images displayed malformations in bacterial cells in the presence of crude phenolic rich extract.	[25]
6.	Acetonic extract, hibiscus	*E. coli, Salmonella* Typhimurium	Gel diffusion with paper disc, MIC and	Testing with multidrug resistant *E. coli and Salmonella* Typhimurium, zones of inhibition of acetonic extract ranged between	[19]

	acid, and chromatographic fractions		MBC	9.8 – 12.6 mm, while hibiscus acid was 10.5- 16.0 mm. With the same pathogens as above, MICs of acetonic extract and hibiscus acid was 7 mg/mL and 4-7 mg/mL, respectively. Additionally, the MBCs of acetonic extract and hibiscus acid was 7-10 mg/mL and 5-7 mg/mL, respectively.	
7.	Aqueous methanol extract of leaves and stem	*S. aureus, P. aeruginosa*	Disc diffusion assay	Concentrations of 20, 40, 80, 160 mg/mL extract were tested. Inhibition zones of aqueous methanolic leaves extracts against S. *aureus* ranged between 5-11 mm and 6-12 mm for *P. aeruginosa.* Aqueous methanolic stem extracts of Hs were only active against *S. aureus* at >80 mg/mL and against P. *aeruginosa* at 160 mg/mL.	[30]
8.	Ethanolic extract	*E. coli* ATCC 25922, *L. monocytogenes* ATCC 51774, *S. aureus* ATCC 6538, *E. aerogenes* ATCC 13048 and *S.* Typhimurium ATCC 14028	Disc diffusion assay	Inhibition zone diameters for *E. coli*: 23.43 mm, *S. aureus*: 24.15 mm, *S.* Typhimurium: 23.85 mm, *E. aerogenes*: 8.49 mm, *L. monocytogenes*: 26.37 mm.	[31]
9.	Aqueous extract	*S. aureus, P. aeruginosa, E. coli, K. pneumoniae*	Disc diffusion assay	Dose-dependent inhibition was displayed. Extract concentrations of 25, 50, 100, and 200 mg/mL resulted in the following zones of inhibition, respectively: *S. aureus* (8.0, 12.0, 13.0, 13.5 mm), *K. pneumoniae* (10.5, 12.5, 14.0, 14.0 mm), *P. aeruginosa* (9.0, 11.0, 12.5, 17.0 mm), and *E. coli* (8.5, 9.0, 12.0, 13.5 mm).	[32]
10.	Infusions of dry calyces in hot water (infused for 7 mins at 75°C) and cold water (25°C, let to stand for 2h)	*S.* Typhimurium, *P. aeruginosa, E. coli* and *S. aureus*	Broth microdilution on 96-well plates to determine MIC and MBC	Cold tea was more effective than hot tea at inhibiting *P. aeruginosa* (MIC 62.5 mg/ml), *S.* Typhimurium, *E. coli* and *S. aureus* (MIC 125 mg/ml). With regards to MBC, cold hibiscus tea was also more effective than hot tea against *S.* Typhimurium and *P. aeruginosa* (MBC 125.0 mg/ml for both strains).	[33]

IOP Conf. Series: Materials Science and Engineering 1192 (2021) 012017 doi:10.1088/1757-899X/1192/1/012017

Bacterial pathogens, especially with multidrug resistance, are of interest in developing new antibiotics or in the discovery of antimicrobials. Plant extracts, including those derived from Hs, have the potential as antibacterial as shown by the different bioactive compounds that can exert antibacterial activity. The different pathogens that have been tested for the inhibition potential of Hs extracts are discussed below.

3.1. Human pathogens

Many studies have investigated the antibacterial potential of Hs extracts *in vitro* against bacterial pathogens of clinical importance. Aqueous methanolic extract has been tested against various pathogens with varying results. For example, aqueous methanolic extracts of Hs leaves were found to be more active against *S. aureus* and *P. aeruginosa* compared to extracts obtained from the stem [30]. Antibacterial activity of the leaves extract was also found to be concentration dependent, where the highest activity was found at the highest concentration tested of 160 mg/mL with an inhibition zone of 12.33 mm [30]. Methanolic extract of Hs was also tested against *E. coli* O157:H7 from different sources, including clinical isolates, with extract concentrations of 2.5%, 5%, and 10%. Here, the result also showed a concentration-dependent inhibition with 10% extract displaying the largest overall zone of inhibition of 12.66 mm [28]. Another study found that methanolic extract of Hs twigs were active against a number of *E. coli* strains, as well as against *E. cloacae*, *E. aerogenes*, and K. *pneumoniae* but not against the *P. aeruginosa* [29].

Ethanolic extracts of Hs are similarly showing *in vitro* antibacterial activity against a range of pathogens. Ethanolic extracts rich in phenolics were determined to not only have antibacterial activity *in vitro*, but also antifungal activity [25]. At 2000 μg/mL in an agar diffusion assay, crude phenolic rich extract was able to inhibit the growth of Gram-positive (S. *aureus, S. pyogenes, L. monocytogenes*) and Gram-negative bacteria (*E. coli, K. pneumonia, P. aeruginosa*) with inhibition zones ranging between 32 – 48 mm [25]. Interestingly, microscopic analysis revealed that in the presence of 50 μg/mL Hs extracts, cell deformation and other malformations were observed [25], suggesting that antibacterial activity of Hs extract affected the bacterial cell wall and/or membrane.

Ethanolic extracts of Hs was also found to inhibit *Bacillus subtilis in vitro* to a similar extent as the water extracts [34]. Other bacterial pathogens that could be inhibited by Hs extracts *in vitro* included oral bacteria *Streptococcus mutans, Streptococcus sanguinis,* and *Capnocytophaga gingivalis* [35], as well as *Fusobacterium nucleatum, Prevotella intermedia,* and *Porphyromonas gingivalis* [36], clinical isolates of *Acinetobacter baumannii* [37] and *Mycobacterium tuberculosis* [38]. Moreover, studies have shown that Hs extracts can inhibit biofilm formation [36,39]. A Drosophila infection model was also recently used to assess Hs extract activity against *S. aureus*, providing evidence that Hs extract antibacterial activity can be assessed *in vivo* [40]. It is therefore evident that Hs has a broad spectrum of antibacterial activity against pathogenic bacteria.

3.2. Foodborne pathogens

In addition to elucidating the antibacterial properties of Hs extracts against human pathogens, studies have also been conducted on the potential of Hs extracts against foodborne pathogens. This is very useful as Hs calyces are also consumed and addition of Hs to food may provide additional benefit as food additive or preservative and prevent contamination of foodborne bacteria.

Against bacteria obtained from food or other foodborne pathogens, Hs extract has shown antibacterial activity to a varying degree. In addition to *E. coli* O157:H7 from food samples [28], Hs extracts also inhibited a number of pathogens isolated from dairy products in a lawn diffusion assay [41]. Inhibition of food pathogens *E. coli, S. aureus, L. monocytogenes, E. aerogenes* and *S. Typhimurium* were also observed *in vitro* by ethanolic extracts of Hs [31]. With the exception of *E. aerogenes*, in that study, Hs extracts displayed the best antibacterial activity against all tested bacteria compared to other spice plants tested, providing further evidence of its potential as a food preservative.

Hs extracts have also been shown to inhibit a number of enteric pathogens, including clinical isolates [42]. Using broth microdilution, crude Hs extract was able to inhibit growth of six ATCC strains (MC 50-200 mg/ml) and 16 clinical isolates (MIC 25-200 mg/ml), with the lowest MIC was exhibited against clinical isolates of *Edwardsiella tarda* and *Yersinia pseudotuberculosis* [42].

In another study, ethanolic extract of Hs calyces was tested as preservative and was found to increase the duration of shelf-life of beef [43]. Here, the Hs phenolic extract was also determined to exert antibacterial activity against *E. coli, S. enterica* var Typhimurium, *S. aureus, L. monocytogenes,* and *B. cereus*. Moreover, the phenolic extract was separated into aqueous and organic phases, and a fraction from the organic phase that was rich in phenolic acids was found to have the lowest MIC and MBC against the tested bacteria compared to aqueous phase and other fractions containing anthocyanin [43].

The spectrum of Hs extract antibacterial activity was further demonstrated against seven food pathogens, namely B. *cereus* 10451, *S. aureus* 10786, *E. coli* GIM1.708, *S. enteritidis* 10982, *Vibrio parahaemolyticus* 17802, and *P. aeruginosa* (B) 10104 with both aqueous and ethanolic extracts displaying bacterial growth inhibition at 20% w/v [44]. MIC determination resulted in similar MICs between the aqueous and ethanolic extracts, with MICs of 0.625%-10% w/v for aqueous extracts and 2.5%-5% w/v for ethanolic extracts. Interestingly, measurement of cytoplasmic internal pH suggested that the plant extract affected the bacterial cell membrane [44], similar to what has been reported in other studies [25,35]. The broad-spectrum nature of Hs extracts against foodborne pathogens warrants further investigation of its use as food additive and preservatives.

4. Synergistic activity of Hs phytochemicals with antibiotics *in vitro*

In addition to finding sources of new antibacterial agents, efforts to alleviate antibiotic resistance issue include discovering molecules or agents that have synergistic activities to improve the activity of currently-available antibiotics, as an alternative strategy to tackle antibiotic resistance. Plant extracts are a potential source of compounds that may have synergistic activity with antibiotics against various pathogenic bacteria [1,45]. A number of studies have therefore attempted to elaborate on the potential synergistic activity of plant extracts against bacterial pathogens, including multi-drug resistant bacteria [30,46-50].

Synergy of Hs extract in inhibiting bacterial growth *in vitro* was demonstrated against *Helicobacter pylori* clinical isolates by agar dilution method using aqueous Hs extracts with clarithromycin and metronidazole [49]. Another study determined that Hs extract at concentrations of 3.13, 1.56, and 0.78 mg/mL displayed synergistic activity with ciprofloxacin at concentrations 1.25 and 0.625 μg/mL against *S. aureus*, but no synergistic activity was observed with higher concentrations of both ciprofloxacin and Hs extracts [51]. Against *M. tuberculosis*, Hs extracts were found to be a good combination to rifampicin in inhibiting growth of isoniazid/ethambutol resistant *M. tuberculosis*, but displayed antagonistic effects when combined with streptomycin, ethambutol and isoniazid [38].

The potential synergy of Hs extract with gentamicin has also been tested for wound healing [52]. In this study, methanol extract of Hs calyx was formulated as a cream in combination with gentamicin and tested on wounds on rats. Application of the cream containing a combination of 1% w/w extract and 1% w/w gentamicin showed reduced time required for wounds closure by five days compared to 1% w/w gentamicin alone [52]. Additionally, the cream formulation showed no significant changes in pH [52], demonstrating the potential of Hs extract as a formulation with antibiotics for topical applications.

Considering the different methods were used in determining synergy, caution must be taken to interpret synergy of Hs extracts with antibiotics. However, current results show potential of Hs extracts to improve the activity of antibiotics.

5. Conclusions

Hs extracts contain various phytochemicals that can inhibit growth of various Gram-positive and Gram-negative bacteria *in vitro*, including multidrug resistant strains. The suggested mode of action of antibacterial activity that has been reported seems to be the disruption of bacterial cell wall and/or membrane. The major components of bioactive compounds in Hs are primarily known and some of those have been shown to contribute to its antibacterial activity. Due to its large numbers of bioactive compounds, further investigation into the Hs extracts as a source of antibacterial lead compounds need to be performed, especially as the majority of studies were performed *in vitro*. Identification of specific compounds that are responsible for the antibacterial activity *in vitro* can assist in potentially selecting candidates for further studies *in vivo*. Additionally, the potential of such candidate compounds can be investigated further by synthesizing analogues to elucidate structure-activity relationship correlating to the antibacterial activity. This in turn may provide potential lead compounds that can be developed as new antibiotics. Alternatively, since synergy with antibiotics has also been demonstrated in a number of studies, Hs extracts or its bioactive compounds may also be explored further to improve antibiotic efficacy and alleviate antibiotic resistance. Information of synergy opens up the opportunity of Hs-derived compounds as antibiotic potentiators, which could help revive antibiotics that are losing their efficacy and therefore helping to invigorate the antibiotic pipeline.

6. References

[1] Savoia D 2012 *Future Microbiol* **7** 979-90
[2] Cowan MM 1999 *Clin Microbiol Rev* **12** 564
[3] Ismail A, Ikram EHK and Nazri HSM 2008 *Food* **2** 1-16
[4] Sindi HA, Marshall LJ and Morgan MR 2014 *Food Chem* **164** 23-9
[5] Guardiola S and Mach N 2014 *Endocrinología y Nutrición (English Edition)* **61** 274-95
[6] Si L-Y-N, Ramalingam A, Ali SS, Aminuddin A, Ng P-Y, Latip J, Kamisah Y, Budin SB and Zainalabidin S 2019 *EXCLI Journal* **18** 876-92
[7] Da-Costa-Rocha I, Bonnlaender B, Sievers H, Pischel I and Heinrich M 2014 *Food Chem* **165** 424-43
[8] Moyano G, Sáyago-Ayerdi SG, Largo C, Caz V, Santamaria M and Tabernero M 2016 *J Funct Foods* **21** 1-9
[9] Izquierdo-Vega JA, Arteaga-Badillo DA, Sánchez-Gutiérrez M, Morales-González JA, Vargas-Mendoza N, Gómez-Aldapa CA, Castro-Rosas J, Delgado-Olivares L, Madrigal-Bujaidar E and Madrigal-Santillán E 2020 *Biomedicines* **8** 100
[10] Si LY-N, Ali SAM, Latip J, Fauzi NM, Budin SB and Zainalabidin S 2017 *Life Sci* **191** 157-65
[11] Bedi PS, Bekele M and Gure G 2020 *Int Res J Pure Appl Chem* **21** 41-54
[12] Laskar YB and Mazumder PB 2020 *Biomed Pharmacother* **127** 110153
[13] Grajeda-Iglesias C, Figueroa-Espinoza MC, Barouh N, Baréa B, Fernandes A, de Freitas V and Salas E 2016 *J Nat Prod* **79** 1709-18
[14] Herranz-López M, Fernández-Arroyo S, Pérez-Sanchez A, Barrajón-Catalán E, Beltrán-Debón R, Menéndez JA, Alonso-Villaverde C, Segura-Carretero A, Joven J and Micol V 2012 *Phytomedicine* **19** 253-61
[15] Hopkins AL, Lamm MG, Funk JL and Ritenbaugh C 2013 *Fitoterapia* **85** 84-94
[16] Riaz G and Chopra R 2018 *Biomed Pharmacother* **102** 575-86
[17] Gheller AC, Kerkhoff J, Vieira Júnior GM, de Campos KE and Sugui MM 2017 *Sci World J* **2017** 9392532
[18] Zheoat AM, Gray AI, Igoli JO, Ferro VA and Drummond RM 2019 *Fitoterapia* **134** 5-13
[19] Portillo-Torres LA, Bernardino-Nicanor A, Gómez-Aldapa CA, González-Montiel S, Rangel-Vargas E, Villagómez-Ibarra JR, González-Cruz L, Cortés-López H and Castro-Rosas J 2019 *Antibiotics* **8**
[20] Shin J, Prabhakaran V-S and Kim K-s 2018 *Microb Pathog* **116** 209-14
[21] Barbieri R, Coppo E, Marchese A, Daglia M, Sobarzo-Sánchez E, Nabavi SF and Nabavi SM 2017 *Microbiol Res* **196** 44-68
[22] Ma Y, Ding S, Fei Y, Liu G, Jang H and Fang J 2019 *Food Control* **106** 106712
[23] Adamczak A, Ożarowski M and Karpiński TM 2020 *J Clin Med* **9**

[24] Borrás-Linares I, Fernández-Arroyo S, Arráez-Roman D, Palmeros-Suárez PA, Del Val-Díaz R, Andrade-Gonzáles I, Fernández-Gutiérrez A, Gómez-Leyva JF and Segura-Carretero A 2015 *Ind Crops Prod* **69** 385-94
[25] Abdel-Shafi S, Al-Mohammadi A-R, Sitohy M, Mosa B, Ismaiel A, Enan G and Osman A 2019 *Molecules* **24** 4280
[26] Sultan FI, Khorsheed AC and Mahmood A-RK 2014 *IJRRAS* **19** 140-49
[27] Mohammad GJ 2020 *Int J Drug Deliv Technol* **10** 217-21
[28] Fullerton M, Khatiwada J, Johnson JU, Davis S and Williams LL 2011 *J Med Food* **14** 950-56
[29] Djeussi DE, Noumedem JAK, Seukep JA, Fankam AG, Voukeng IK, Tankeo SB, Nkuete AHL and Kuete V 2013 *BMC Complem Altern M* **13** 164
[30] Kumar S and Anna Sheba L 2019 *Asian J Pharm Clin Res* **12** 198-201
[31] Akarca G, Tomar O, Güney İ, Erdur S and Gök V 2019 *J Food Sci Technol* **56** 5253-61
[32] Sulaiman FA, Kazeem MO, Waheed AM, Temowo SO, Azeez IO, Zubair FI, Adeyemi TA, Nyang A and Adeyemi OS 2014 *J Taibah Univ Sci* **8** 315-22
[33] Paraíso CM, Januário JGB, Mizuta AG, dos Santos SS, dos Santos Magon TF, Ogawa CYL, de Oliveira Silva JV, Sato F, Visentainer JV, Mikcha JMG *et al.* 2021 *J Food Meas Charact* 10.1007/s11694-021-00936-4
[34] Jung E, Kim Y and Joo N 2013 *J Sci Food Agric* **93** 3769-76
[35] Baena-Santillán ES, Piloni-Martini J, Santos-López EM, Gómez-Aldapa CA, Rangel-Vargas E and Castro-Rosas J 2020 *J Med Food* **24** 67-76
[36] Sulistyani H, Fujita M, Miyakawa H and Nakazawa F 2016 *Asian Pac J Trop Med* **9** 119-24
[37] Abdallah EM 2016 *J Acute Dis* **5** 512-16
[38] Fauziyah PN, Sukandar EY and Ayuningtyas DK 2017 *Sci Pharm* **85**
[39] Sartini S, Djide MN and Nainu F In *Correlation Phenolic Concentration to Antioxidant and Antibacterial Activities of Several Ethanolic extracts from Indonesia*, Journal of Physics: Conference Series, 2019/10; IOP Publishing: 2019; p 072009.
[40] Nainu F, Djide MN, Subehan, Sartini, Roska TP, Salim E and Kuraishi T 2020 *HAYATI J Biosciences* **27** 306-13
[41] Lim H-W, Seo K-H, Chon J-W and Song K-Y 2020 *J Dairy Sci Biotechnol* **38** 37-44
[42] Jantrapanukorn B, Pongpraritt S, Powthong P and Pheungphu T 2017 *J Appl Pharm Sci* **7** 119-22
[43] Márquez-Rodríguez AS, Nevárez-Baca S, Lerma-Hernández JC, Hernández-Ochoa LR, Nevárez-Moorillon GV, Gutiérrez-Méndez N, Muñoz-Castellanos LN and Salas E 2020 *Foods* **9**
[44] Gonelimali FD, Lin J, Miao W, Xuan J, Charles F, Chen M and Hatab SR 2018 *Front Microbiol* **9**
[45] Stefanović OD, Synergistic Activity of Antibiotics and Bioactive Plant Extracts: A Study Against Gram-Positive and Gram-Negative Bacteria. In *Bacterial Pathogenesis and Antibacterial Control*, Kırmusaoğlu, S., Ed. IntechOpen: United Kingdom, 2018, DOI: 10.5772/intechopen.72026.
[46] Farooqui A, Khan A, Borghetto I, Kazmi SU, Rubino S and Paglietti B 2015 *PloS one* **10** e0118431-e31
[47] Maia NL, de Barros M, de Oliveira LL, Cardoso SA, dos Santos MH, Pieri FA, Ramalho TC, da Cunha EFF and Moreira MAS 2018 *Front Microbiol* **9**
[48] EL-Zawahry YA, Reda FM and Azazy WM 2013 *Life Sci* **10** 3477-89
[49] Hassan STS, Berchová K, Majerová M, Pokorná M and Švajdlenka E 2016 *Pharm Biol* **54** 1736-40
[50] Haroun MF and Al-Kayali RS 2016 *Iran J Basic Med Sci* **19** 1193-200
[51] Ezenwanyi NE 2016 *Br JAppl Sci Technol* **15** 1-8
[52] Builders PF, Kabele-Toge B, Builders M, Chindo BA, Anwunobi PA and Isimi YC 2013 *Indian J Pharm Sci* **75** 45-52

IOP Conf. Series: Materials Science and Engineering 1192 (2021) 012018 doi:10.1088/1757-899X/1192/1/012018

Enzymatic Glycerolysis of Palm Kernel Olein and Palm Kernel Stearin in different Ratios for Monolaurin Synthesis

N Ngatirah[1,3], C Hidayat[2], ES Rahayu[2] and T Utami[2,4]

[1]Study Program of Food Science, Faculty of Agricultural Technology, Universitas Gadjah Mada, Yogyakarta, Indonesia
[2]Department of Food and Agricultural Product Technology, Faculty of Agricultural Technology, Universitas Gadjah Mada, Yogyakarta, Indonesia
[3]Department of Agricultural Product Technology, Faculty of Agricultural Technology, Institut Pertanian Stiper, Yogyakarta, Indonesia

[4]Corresponding Author's Email: tyas_utami@ugm.ac.id

Abstract: Monolaurin is known as a monoglyceride of lauric acid, which act as an emulsifier and antimicrobial. One potential of monolaurin raw materials is palm kernel oil (PKO), which can be fractionated into palm kernel olein (PKOo) and palm kernel stearin (PKS). This study aims to determine the enzymatic glycerolysis influence of PKOo and PKS mixture in different ratios and time, on monoacylglycerol (MAG), diacylglycerol (DAG), and triacylglycerol (TAG) profiles. The Glycerolysis process of PKOo-PKS fat blends were carried out in different ratios of (60:40, 40:60, and 20:80 w/w) using commercial lipases (Lypozime RM IM) at 50°C, with an oil:glycerol molar ratio of 1:4, in a tert-butanol solvent system, between 3 to 24 hours. The results showed that the glycerolysis of PKOo and PKS in different ratios yielded varying product reactions with similar compositions (1.8-3.9% MAG, 73.2-76.4% DAG, and 20.7-24.2% TAG). A higher and not significant MAG was observed during the period of 24 hours. Also, the monolaurin was obtained at the ratio of 40:60 PKOo:PKS, within the glycerolysis time of 3 to 24 hours, at approximately 2.18±0.59% and 3.47±0.62%, respectively. The FTIR analysis also showed that the monolaurin sample was identical to the standard type, with the formation of OH group at wave number and cluster C=O ester of 3368.64 cm^{-1} and 1734.03 cm^{-1}, respectively.

1. Introduction

Monolaurin is a monoglyceride of lauric acid, which is naturally found in breast milk, at 5.8% of fat. It serves as an essential fat and immune system boosters for infants [1] has two hydroxyl and one lauryl groups. Based on these classification, monolaurin is categorized as non-ionic emulsifier containing hydrophilic and hydrophobic groups, in its molecular structure [2]. While as an emulsifier, it is used as food industry-based emulsion and in the pharmaceutical industry. This compound is also capable of acting as an antimicrobial agent [3], and food preservative.

The production of commercial monolaurin uses lauric acid esterification with chemical catalysts. Meanwhile, the use of these catalysts also possess several drawbacks, such as occurring at high temperatures (220-260°C). The energy consumption is also observed to be higher and not environmentally friendly, as the product is further found to be darker in colour [4]. Therefore, the development of monolaurin synthesis started through the use of enzymatic catalysts. Based on the enzymatic use of lipase in this process, several advantages are being detected. These includes the occurrence of reaction at a lower temperature as well as the production of higher quality and purity.

The sources of monolaurin are known as lauric acid, methyl lauric, as well as coconut and palm kernel oils (PKO). Meanwhile, the use of lauric acid and methyl lauric are known to have some disadvantages, due to their unavailability. This causes the relatively high price of the sources, due to only being obtained from hydrolysis or transesterification of coconut or PKO. Therefore, the most potential monolaurin samples are derived from both coconut and PKO. Meanwhile, the palm kernel oil has higher productivity advantages than other sources in Indonesia.

Monolaurin synthesis with PKO and coconut oil also uses ethanolysis [5,6], hydrolysis- continued esterification [7], or Glycerolysis [8–10] reactions. Meanwhile, the use of ethanol is feared to still be contained in the product, during the ethanolysis method. The use of oil in the hydrolysis-continued esterification reaction is also observed to take longer, leading to the production of responsive inhibitors. Also, the triglyceride glycerolysis reaction is the most economical method of monolaurin synthesis. Based on the use of specific lipase at sn 1 and 3, this method is known to produce 2-monoglyceride products.

Triacylglycerol (TAG) is found to be the main composition in PKO, with its value approximately at 95% fat. The main TAG composition of PKO is also known to be 20.46%, 17.77%, and 16.88% of lauric-lauric-lauric (LaLaLa/trilaurin), lauric-lauric-myristic (LaLaM), and capric-lauric-lauric (CaLaLa), respectively [11]. This oil also contains 50% lauric acid [12]. Based on being a monolaurin raw material, the use of PKO has several disadvantages, i.e., the inconsistency of quality, which is affected by the processing and storage condition of the kernel. The PKO is further fractionated to 60-70% Palm Kernel Olein (PKOo) and 30-40% Palm Kernel Stearin (PKS), with the amount of lauric acid at 39.7-48% and 56-59.7%, respectively [12]. Based on the amount of the acid composition, PKOo and PKS potentially serve as the raw material of monolaurin synthesis. Although the availability of PKOo is higher, its lauric acid content is found to be lower. Therefore it is necessary to mix both PKOo and PKS, in order to increase the value of lauric acid and consistent quality of fat mixture. The mixture also produces a significant change in trilaurin and lauric acid, with the acidic compound at the sn-2 position. Also, the PKOo:PKS ratios of 60:40, 40:60, and 20:80 is found to produce the trilaurin and lauric acid values of $23.72 \pm 0.6\%$ and $55.33 \pm 0.41\%$, $24.58 \pm 0.56\%$ and $57.01 \pm 0.15\%$, as well as $26.93 \pm 0.19\%$ and $59.92 \pm 0.09\%$, respectively. Based on the position of the acid at sn-2, the values of PKOo and PKS at 60:40, 40:60, and 20:80, are 42.29%, 50.71%, and 51.84%, respectively [13]. Furthermore, differences in the values of trilaurin and lauric acid present in the mixture are found to affect the amount of monolaurin produced. Based on the use of specific lipase at sn 1 and 3 during the process of glycerolysis, the trilaurin present in the fat mixture is converted into monolaurin (a mixture of 2 and 1 monolaurin) and dilaurin, as intermediate products. Meanwhile, the influence of PKOo and PKS mixture on the composition of monoacylglycerol (MAG), diacylglycerol (DAG), and triacylglycerol (TAG), is still unknown.

The amount of monolaurin produced is also affected by the time of glycerolysis, which is an essential factor used in producing MAG and DAG [14]. The increased reaction time also improves the amount of monolaurin. Also, glycerolysis is found to reach dynamic equilibrium at 8 hours, as the product tends to be constant with higher reaction time [15]. Based on different ratios and time, this study aims to determine the influence of PKOo and PKS mixture on MAG, DAG, and TAG profiles, during enzymatic glycerolysis.

2. Materials and Methods

2.1. *Materials*

The PKOo, PKS, and refined glycerin were obtained from PT Wilmar, Indonesia. Also, Lipozyme RM IM and Molecular sieves were purchased from Sigma-Aldrich (St. Louis, MO, USA). The tert-butanol was also obtained from Merck KGaA (Darmstadt, Germany).

2.2. *Preparation of fat blends*

The materials were heated at 70°C for 30 mins, in order to homogenize all oil fractions before mixing. Based on the results, the liquefied products were mixed at PKOo:PKS ratios of 60:40, 40:60, and 20:80 (w/w), homogenized at 70°C for 15 mins, and further stored in a refrigerator.

2.3. *Monolaurin synthesis*
A 10 g mixture of PKOo and PKS in different ratios (60:40; 40:60; and 20:80) was mixed with glycerol, in the proportion of 1:4. The tert-butanol was further added to the mixture, with the ratio of 1.5 for oil and molecular sieve of 12% glycerol. Furthermore, the mixture was placed in the batch stirred tank reactor at 50°C, and mix at a speed of 200 rpm. Lipozyme RM IM with 5% w/w of oil was further added to the mixture, and reacted for 3 and 24 hours. The separation of the product from the enzyme and glycerol was carried out through centrifugation at 3000g for 5 mins, and then extracted from the solvent using a rotary evaporator. This product also analyzed the composition of MAG, DAG, and TAG, by using gas chromatography (GC). According to Nitbani et al. [5] the products of glycerolysis were obtained through the isolation or purification process of monolaurin. These products dissolved in hydroalcoholic (ethanol:water=8:2), with a ratio of 1:9 v/v. The mixture was further extracted, by using n-hexane with a balance of 1:3 v/v. This mixture dissolved in the hydroalcoholic phase, and then evaporated to obtain monolaurin, which was weighed and analyzed using FTIR. Additionally, the experiment was conducted with duplicates.

2.4. *Statistical analysis*
The completely randomized design was used as an experimental design. The experiment was conducted with duplicates. The resulting data is then performed to an analysis of variance and if there is a significant difference between the treatment followed by a Duncan Multiple Range Test (DMRT) level 5%.

3. Result and Discussion

3.1. *Compositions of MAG, DAG, and TAG*
Based on different ratios, the composition of MAG, DAG, and TAG from the enzymatic glycerolysis of PKO-PKS mixture are shown in Figure 1.

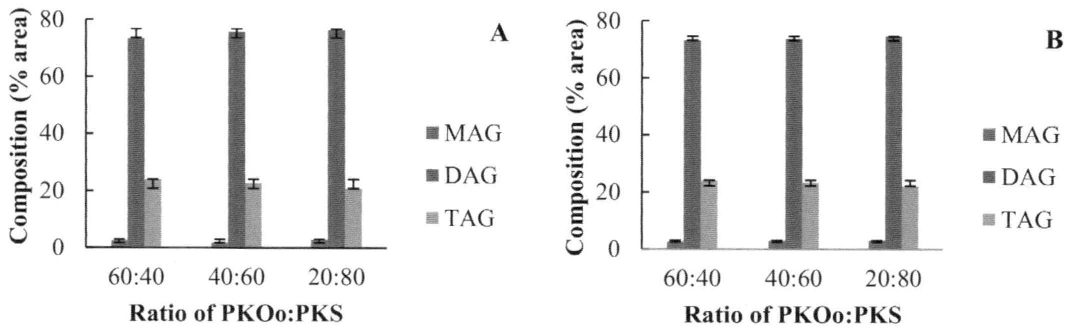

Figure 1. The composition of MAG, DAG, and TAG results of PKOo: PKS glycerolysis with different ratios (60:40; 40:60; and 20:80) at 50 °C, oil: glycerol molar ratio of 1:4, solvent amount 1.5x oil, enzyme concentration 5% of oil weight, stirring speed 200 rpm with a time of 3 hours (A) and 24 hours (B).

Based on Figure 1, the composition of MAG, DAG, and TAG at different ratios (60:40, 40:60, and 20:80), did not show a significant difference. The reaction time of 3 and 24 hours also produced MAG, DAG, and TAG compositions that did not have significant differences. Also, the TAG conversion results in several DAG (73-74%) and MAG approximately produced 3-3.19%. This was due to the use of the sn-1,3 enzyme, which specifically decomposed fatty acids, and produced 2,3-DAG and 1,2-DAG at sn 1 and 3 positions, respectively. Also, the values of DAG produced was found to be high, as it was assumed not to be totally hydrolyzed into MAG by the lipozyme RM IM enzyme. Furthermore, glycerolysis was a complicated process, which included hydrolysis, esterification, and isomerization of MAG and DAG. Based on the initial 5 h, hydrolysis dominated the process, accompanied by esterification and isomerization [16]. The higher molar fraction of TAG compared to glycerol was found to also enhance the synthesis of dilaurin. Based on the higher production of DAG, the free fatty acids did not totally react to the glycerol. This was because most glycerol were found to react with fatty acids obtained from TAG, in order to produce monoacylglycerol during glycerolysis reactions [17]. Theoretically, 1 mole of TAG and 2 moles of glycerol provided 3 moles of MAG during this process [18]. However, the yield of monoacylglycerol depends on equilibrium conditions, due to several occurrences [19]. Glycerolysis was also influenced by the concentration of enzymes, molar ratio of oil:glycerol, amount of solvent, and the temperature of the response [20]. Also, the molar ratio of fat:glycerol determined the glycerolysis products that were produced. Based on the excess and limited use of glycerol in this reaction, MAG and DAG were found to be high, respectively [21]. According to the equilibrium law, an increase in the glycerol content shifted the balance towards the production of MAG [18]. Based on the study of Solasea et al.[18] a glycerol:oil mole ratio of 3:1 was selected as optimum value for MAG production. The low yield of MAG was also caused by low concentrations of enzymes. This study was also found to use an enzyme concentration of 5%, compared to the 5-10% of total reactants (oil and glycerol) used in previous studies. The MAG yield also increased with higher lipase concentration, which was due to the greater number of active sites [22]. Solasea et al.[22], also found that the glycerolysis of sardine oil was carried out using Lipozyme 435, at a load of 5 and 10 wt%. This was further conducted with a glycerol:oil mole ratio of 3:1, in order to produce MAG at 90 and 80%, for the lipase load of 10 and 5%, respectively.

The conversion of TAG to MAG and DAG at 3 and 24 h showed no significant difference on all PKOo and PKS ratios (Figure 2). This was due to the decreasing activities of the lipase enzymes. Excess glycerol also inhibited the enzymatic activity due to its polar nature, which often led to inactivation [23]. Also, the glycerol content directly affected the lipase activity, due to its influence on the polarity of the reaction medium [18]. Based on the studies of Mustafa et al. [24], there was no further increase in the percentage of conversion after 3 hours of reaction. This was observed in the esterification of lauric acid and glycerine into glycerine laurate, using the lipozyme RM IM enzymes. Furthermore, the conversion percentage dramatically increased in the initial 30 mins of the reaction, as approximately 70% of the original fatty acids were converted into esters [24]. Based on the use of Novozymes 435 (10% oil weight) at 65°C for 7 hours, the enzymatic glycerolysis of PKO and glycerol at a molar ratio of 1:1 produced 4.00 ± 0.09% MAG, 88.41 ± 0.19% DAG, and 7.60 ± 0.10% TAG, after purification [8]. The reaction reached dynamic equilibrium at the eighth hour, as well as essentially maintained the type and content of products formed in the reversible response constant [15]. Based on the reaction times above 36 hours, the amount of monolaurin observed was not growing [10]. The glycerolytic reaction of VCO and glycerol reported no significant increase in the amount of MAG and DAG, due to using the Lipase Candida antartica (Novozymes 435) that was continuous for approximately 48 h [25].

3.2. *The amount of free fatty acids*

The amount of free fatty acids produced during the enzymatic glycerolysis of PKOo and PKS mixtures are shown in Table 1.

Table 1. The average of free fatty acids (%).

Time of glycerolysis (hour)	PKOo: PKS Ratios*		
	60:40	40:60	20:80
3	0.85±0.14[a]	0.76±0.01[a]	0.69±0.09[a]
24	0.97±0.13[a]	0.89±0.11[a]	0.87±0.10[a]

*value is average ± SD (n=2). Average on row or column, followed by the same letter, indicates a not significant difference in the Duncan Multiple Range Test (DMRT).

Based on Table 1, the free fatty acids produced during the glycerolysis process were insignificantly different. This showed that the hydrolysis of oil was very tiny, leading to the minimal production of these compounds. Free fatty acids were also the result of oil hydrolysis, which was carried out by the water derived from glycerol, in the reaction system. Although several studies added water to glycerol in order to increase the amount of MAG, it was also found to enhance free fatty acids [18]. Based on the use of Lypozime RM IM (10% oil) at 50°C for 20 h, the glycerolysis sardine oil and glycerol at a mole ratio of 1:3 increased FFA by 5.3 and 32.5%, for WGC (water glycerol content) at 7 and 24%, respectively. Furthermore, the highest MAG yield was obtained, due to the addition of 12% water in glycerol [18]. Based on the use of Lipozyme TL IM (7% w/w oil and 5 ml alcohol) at 60°C and 200 rpm for 24 h, the MDG (mono-diglyceride) synthesis using RBDPO (refined bleached deodorized palm oil) and glycerol with a molar ratio of 1:5, produced 40.16% MAG, 49.92% DAG, 10% TAG, and 0.16% FFA [14].

3.3. Yield of monolaurin

Glycerolysis products are separated by using hydroalcoholic solvents, in order to obtain monolaurin. The result of monolaurin yield produced is shown in Table 2.

Table 2. The yield of monolaurin (%).

Time of glycerolysis (hour)	PKOo: PKS ratios		
	60:40	40:60	20:80
3	1.96±0.81^{a}	2.18±0.59^{a}	1.24±0.02^{a}
24	2.91±1.49^{a}	3.47±0.62^{a}	2.05±1.26^{a}

*value is average ± SD (n=2). Average on row or column, followed by the same letter, indicates a not significant difference in the Duncan Multiple Range Test (DMRT) 5%.

Based on Table 2, the yield of monolaurin was insignificantly different. Meanwhile, a PKOo:PKS ratio of 40:60 obtained the highest yield at 3.47%, during a period of 24 h. The low output of monolaurin produced in this study was mainly caused by the glycerolytic reaction of the DAG product (Figure 1). Also, the yield obtained was lower, compared to previous studies. Based on the use of Lypozime TL IM (5% oil) at 35°C for 12 h, the synthesis of 2-monolaurin from coconut oil produced a yield of 35% through ethanolysis, at oil:ethanol ratio of 1:4 [5]. The synthesis of 1-monolaurin at lauric acid and glycerol at 1:1 also produced a yield of 27.89%, based on the use of a 5% pTSA catalyst at 130°C for 6 h [26]. Furthermore, the synthesis of lauric acid and glycerol with mole ratio of 1:1 produced a yield of 17.25%, based on using *the Rhizopus* spp. lipase at 50°C for 72 h [27]. Based on using a 20% Carica papaya lipase at 45°C for 36 h, the monolaurin synthesis from coconut oil with crude glycerol at 1:8 in ethanol also produced a yield of 58.35% [10].

3.4. Analysis of FTIR (Fourier Transform Infra Red) monolaurin

Analysis of FTIR monolaurin after purification is shown in Figure 2. According to Figure 2, the monolaurin sample was analyzed by FTIR, after purification. It also indicated that the OH group occurred at a wavenumber, C=O ester, C-O-C, CH Stretch vibration, as well as Methyl and methylene group of 3368.64 cm^{-1}, 1734.03 cm^{-1}, 1174.05-993.03 cm^{-1}, 2853.21-2922.01 cm^{-1}, and 1461.04-1380.46 cm^{-1}, respectively. The wavenumber of 719.77 cm^{-1} also indicated the rocking vibration from $(CH2)_n$. Furthermore, the FTIR analysis showed that the sample monolaurin was identical to the standard monolaurin, which was found to be in line with previous studies. Widiyarti et al., also found a new group at wavenumbers of 3224.98 cm^{-1} and 3290.56 cm^{-1}. This indicated that the asymmetric stretching vibration of the hydroxyl and carbonyl groups were similar at wavenumber of 1730.15 cm^{-1} [28]. Also, Sangadah et al. obtained the C=O ester and OH clusters at wavelengths of 1748 cm-1 and 3650-3200 cm^{-1}, respectively [29].

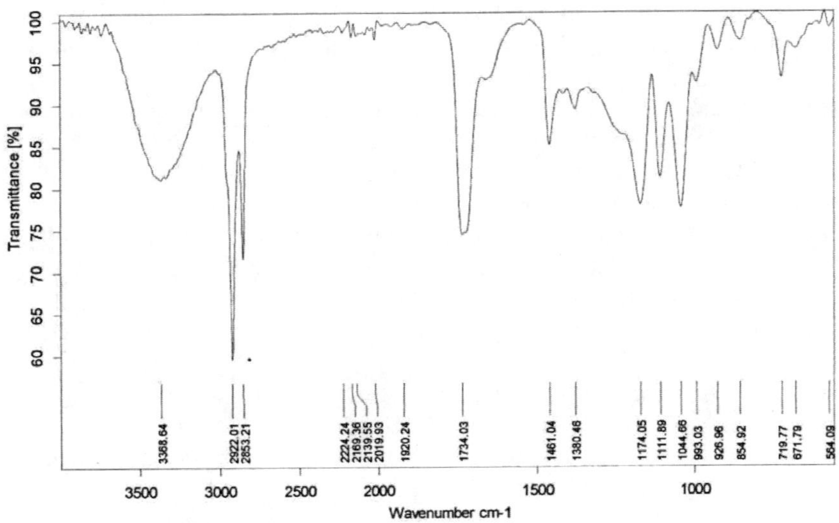

Figure 2. Analysis of FTIR monolaurin from glycerolysis with a PKOo-PKS fat blend with ratio of 40:60, glycerolysis time of 24 hours, at 50 °C, the molar ratio of oil: glycerol 1:4, amount of solvent 1.5x oil, enzyme concentration 5% w/w oil, stirring speed 200 rpm.

4. Conclusion

It is concluded that the glycerolysis of PKOo-PKS mixtures in different ratios produced similar compositions (1.8-3.9% MAG, 73.2-76.4% DAG, and 20.7-24.2% TAG). The reaction at a period of 24 h produced higher MAG, which were insignificant. Also, the PKOo:PKS ratio of 40:60 at a period of 3 and 24 h obtained monolaurin yields of 2.18±0.59% and 3.47±0.62%, respectively. The FTIR analysis also showed that the sample of monolaurin was identical to the standard type, with the formation of OH group and C=O ester having wavenumbers of 3368.64 cm^{-1} and 1734.03 cm^{-1}, respectively.

Acknowledgments

The authors are grateful to the Directorate of General Education, Ministry of Education and Culture, Indonesia, for funding this study through the doctor's dissertation research scheme (research and amendment of contract number 6/E1/KP.PTNBH/2020 and 6/AMD/E1/KP.PTNBH/2020).

References

[1] Lieberman S Enig M G and Preuss H G 2006 A review of monolaurin and lauric acid: Natural virucidal and bactericidal agents *Altern. Complement. Ther.* **12** p 310–314

[2] Nitbani F O Juminaa Siswanta D and Solikhah E N 2015 Reaction path synthesis of monoacylglycerol from fat and oils *Int. J. Pharm. Sci. Rev. Res.* **35** p 126–136

[3] Jumina J Lavendi W Singgih T Triono S Steven Kurniawan Y and Koketsu M 2019 Preparation of Monoacylglycerol Derivatives from Indonesian Edible Oil and Their Antimicrobial Assay against Staphylococcus aureus and Escherichia coli *Sci. Rep.* **9** p 1–8

[4] Rarokar N R Menghani S Kerzare D and Khedekar P B 2017 Progress in Synthesis of Monoglycerides for Use in Food and Pharmaceuticals *J. Exp. Food Chem.* **03** p 1–6

[5] Nitbani F O Jumina Siswanta D Sholikhah E N and Fitriastuti D 2016 Synthesis and antibacterial activity of 2-monolaurin *Orient. J. Chem.* **32** p 3113–20

[6] Khasbullah F Murhadi and Suharyono, 2013 The Study of functional characteristics of ethanolysis product of CPO (Crude Palm Oil) and PKO (Palm Kernel Oil) mixture at level two ethanolysis reaction *J. Teknol. Ind. dan Has. Pertan.* **18** 1 p 13–27

[7] Nandi S Gangopadhyay S and Ghosh S 2004 Production of Medium Chain Glycerides and Monolaurin from Coconut Acid Oil by Lipase-Catalyzed Reactions *J. Oleo Sci.* **53** 10 p 497–501

[8] Saberi A H Kee B B Oi-Ming L and Miskandar M S 2011 Physico-chemical properties of various palm-based diacylglycerol oils in comparison with their corresponding palm-based oils *Food Chem.* **127** p 1031–1038
[9] Zha B Chen Z Wang L Wang R Chen Z and Zheng L 2014 Production of glycerol monolaurate-enriched monoacylglycerols by lipase-catalyzed glycerolysis from coconut oil *Eur. J. Lipid Sci. Technol.* **116** p 328–335
[10] Pinyaphong P Sriburi P and Phutrakul S, 2012 Synthesis of Monoacylglycerol from Glycerolysis of Crude Glycerol with Coconut Oil Catalyzed by Carica Papaya Lipase **50200** p 506–511
[11] Chai X *et al.*, 2018 Non-triglyceride components modulate the fat crystal network of palm kernel oil and coconut oil *Food Res. Int.* **105** 423–431
[12] Nuzul A I 2013 Characteristics of malaysian palm kernel and its products *J. Oil Palm Res.* **25** p 245–252
[13] Ngatirah N Hidayat C Rahayu E S and Utami T, 2020 Doctor's dissertation research report *Fac. Agric. Technol. Gadjah Mada Univ. Yogyakarta.*
[14] Zakwan Julianti E and Lubis Z 2017 Production mono-diglyceride (MDG) from refined deodorized palm oil (RBDPO) by enzymatic process *Int. Food Res. J.* **24** p 56–59
[15] Diao X Guan H Kong B and Zhao X 2017 Preparation of diacylglycerol from lard by enzymatic glycerolysis and its compositional characteristics *Korean J. Food Sci. Anim. Resour.* **37** p 813–822
[16] Tan T and Yin C 2005 The mechanism and kinetic model for glycerolysis by 1,3 position specific lipase from Rhizopus arrhizus *Biochem. Eng. J.* **25** p 39–45
[17] Cheirsilp B and H-Kittikul A 2007 A mathematical model approach to a glycerolysis reaction for monoacylglycerol production *WIT Trans. Modelling Simul.* **46** p 225–232
[18] Solaesa G Sanz M T Melgosa R Bucio S L and Beltrán S 2015 Glycerolysis of sardine oil catalyzed by a water dependent lipase in different tert-alcohols as reaction medium *Grasas y Aceites* **66**
[19] Choong T S Y *et al.* 2018 Kinetic study of lipase-catalyzed glycerolysis of palm olein using Lipozyme TLIM in solvent-free system *PLoS One* **13** p 1–13
[20] Mamtani K Shahbaz K and Farid M M 2021 Glycerolysis of free fatty acids: A review *Renew. Sustain. Energy Rev.* **137** p. 110501
[21] Subroto E 2020 Monoacylglycerols and diacylglycerols for fat-based food products: A review *Food Res.* **4** p 932–943
[22] Solaesa Á G Sanz M T Falkeborg M Beltrán S and Guo Z 2016 Production and concentration of monoacylglycerols rich in omega-3 polyunsaturated fatty acids by enzymatic glycerolysis and molecular distillation *Food Chem.* **190** 17743 p 960–967
[23] Satyawali Y Cauwenberghs L Maesen M and Dejonghe W 2021 Lipase catalyzed solvent free synthesis of monoacylglycerols in various reaction systems and coupling reaction with pervaporation for in situ water removal *Chem. Eng. Process. - Process Intensif.* p 108475
[24] Mustafa A Karmali A and Abdelmoez W, 2016 Optimisation and economic assessment of lipase-catalysed production of monoesters using Rhizomucor miehei lipase in a solvent-free system *J. Clean. Prod.* **137** p 953–964
[25] Derawi D Azman N A Z and Jumadi M F 2017 Preliminary study on production of monoacylglycerol and diacylglycerol of virgin coconut oil via enzymatic glycerolysis using lipase Candida antarctica (Novozyme 435) (Kajian Awal Penghasilan Monoasilgliserol dan D *Malaysian J. Anal. Sci.* **21** p 37–45
[26] Nitbani F O Jumina Siswanta D Sholikhah E N and Fitriastuti D 2018 Synthesis and antibacterial activity of 1-monolaurin *Orient. J. Chem.* **34** p 863–867
[27] Koblitz M G B and Pastore G M 2005 Contribution of response surface design to the synthesis of monoacylglycerols catalyzed by Rhizopus sp. lipase *J. Food Sci.* **70** p c503–c505
[28] Widiyarti G Hanafi M and Soewarso W P 2010 Study on the synthesis of monolaurin as antibacterial agent againts Staphylococcus aureus *Indones. J. Chem.* **9** 1 p 99–106
[29] Sangadah K Handayani S Setiasih S and Hudiyono S 2018 Enzymatic synthesis of glycerol ester hydrolyzed coconut oil fatty acid and lauric acid as emulsifier and antimicrobial compound *AIP Conf. Proc.* **2023**

IOP Conf. Series: Materials Science and Engineering 1192 (2021) 012019 doi:10.1088/1757-899X/1192/1/012019

Modeling the effect of dammar additions on the mass loss response of soy wax/beeswax/dammar blends via response surface methodology

S I S Shaharuddin[1,4], N B A Aziz[1], N Bacho[1], N K M Khairussaleh[1], A Tumian[2], I A Zakaria[3]

[1]Department of Manufacturing and Material Engineering International Islamic University Malaysia Gombak, Selangor
[2]Facilities of the Future, Petronas Research Sdn. Bhd., Putrajaya, Wilayah Persekutuan
[3]Faculty of Mechanical Engineering, Universiti Teknologi MARA, Selangor

[4]Corresponding Author's E-mail: shaimihezri@iium.edu.my

Abstract. The wax used in the batik industry plays a significant role as its composition dictates the ease of wax rendering, dye layering, and wax removal process. This study aims to evaluate the effect of dammar additions on soy wax/beeswax blends based on the mass loss (%) of the wax-covered cotton fabric in 100°C water. Central composite design (CCD), a subset of response surface methodology (RSM) was used to develop a response model (Y: mass loss %) for three independent variables (X_1:beeswax, X_2:soy wax, X_3:dammar). The final quadratic response model obtained (F value =5.43, lack of fit F value = 4.70, adequate precision = 7.65) was proposed in this study. ANOVA analysis showed that the standard error of design was relatively small, ranging between 0.43 to 1.18 for the design space. It was deduced from the response model, that increasing the dammar content in the soy wax/beeswax blends increases the mass loss (%), possibly due to the compositional inhomogeneity of the blends. The result of this study shows great potentials in formulating new soy wax-based compositions that produce varying degrees of ease of wax removal for the batik industry.

1. Introduction

Batik is a process that applies resist-based materials such as wax to create batik motifs on fabrics. Once the batik motifs have been completely outlined using wax, the fabric is given soda ash (Na_2CO_3) pre-soak before being dyed. Alternatively, soda ash can also be mixed with the dye and immediately applied to the fabric. Soda ash increases the alkalinity of the solution and a pH reading in the range of 10 to 11.5 is desirable to encourage a stronger chemical reaction between the cellulosic fiber anion and the dye. This will result in batik cloth having better color vibrancy and steadfastness. Once the coloring phase has been completed, the final stage of batik production involves the removal of the batik waxes to reveal the batik motif outline via immersion in hot water. Traditional batik wax composition in Malaysia is composed of a blend of paraffin, microcrystalline wax, beeswax, dammar, and vegetable oil. The use of paraffin and microcrystalline wax has raised environmental concerns as these synthetically produced materials do not biodegrade easily in water. Recently, there is an overwhelming supply of vegetable-based oils, and soy wax, in particular, has been receiving great attention as a renewable and biodegradable replacement to paraffin wax [1].

Soy wax is derived from the hydrogenation process of soy oil by transforming full or partial unsaturated fatty acids into saturated form. Soy wax biodegrades better than paraffin and offers a cheaper option in comparison to beeswax [2]. Currently, the use of soy wax is not very popular amongst batik artisans. In comparison to the traditional batik wax blends, soy wax easily dissolves in water and therefore may not be suitable for multiple dye immersion techniques. In addition, soy wax is also hard and brittle or greasy [3]. To improve such shortcomings, Bowen [4] added beeswax to increase the water resistance and adhesive capacity of soy wax. Previously, dammar resin has also been added into the wax to impart stickiness or adhesive property [5]. The hydrophobic nature of soy wax, beeswax, and dammar has been studied to improve the hydrophilic properties of various materials [6–8].

Except for palm oil-based batik wax [9] and sago-based batik wax, it was realized that compositional study on naturally sourced batik wax alternatives has been quite limited [10]. Prior studies on dammar additions have mainly focused on chemical composition and properties [3,11,12] whilst their degradation reaction has been evaluated due to the aging effect from light and heat exposure. This study aims to investigate the effect of dammar additions on the mass loss response of soy wax/beeswax blends via response surface methodology (RSM). RSM is commonly adopted in physical and chemical experimental designs [13]. In this work, the RSM applies central composite design (CCD) to develop the design of the experiment (DOE) since it dictates a fewer number of tests and has been proven to effectively describe most steady-state process responses. CCD is also a useful approach to model various processes in comparison to the "one variable at a time" approach. The statistical analysis in this study was performed using ANOVA.

2. Methodology

2.1. Material preparation

The ratios of beeswax (Global Sdn Bhd, Malaysia), soy wax (Jargeous Sdn Bhd, Malaysia), and dammar resin (Rupert, Gibbon & Spider Inc.America) were weighed according to the weight ratios specified based on DOE using CCD in Figure 1. Each component of the batik wax compositions was then melted in a pan in the order of decreasing melting temperature as follows; dammar (melting point = 90°C), followed by beeswax (melting point = 70°C), and finally soy wax (melting point = 55°C). Upon complete melting, the blends were stirred and poured into metal pans covered with aluminum foil. The prepared batik wax blends were then left to solidify.

2.2. Response surface methodology

The RSM via CCD was represented by mass loss (Y, %) for three independent variables (k-3) consisting of beeswax (X_1, g) soy wax (X_2, g), and dammar (X_3, g). These independent variables were considered at two levels: low (30g) and high (100g). The total number of proposed experiments, used in this study was $N= 2^k + 2k + n_0 = 20$ where $2^k = 2^3 = 8$ factorial points, 2k = points located at "$\pm\sqrt{k}$" from the centre points = 6 axial points and n_0 = replicates = 6 central points. The experimental design in coded and uncoded values is listed and schematically shown in Table 1. Figure 1 shows the wax compositions (in gram) that represent the factorial points, axial points, and center points in a 3D design space. The ANOVA analysis and statistical values were determined using Design-Expert software (version 6.0.8). To evaluate how close the selected response model fits with the experimental values obtained, the parameters F-value, R^2, P-value, and lack of fit are used [13]. Insignificant lower-order terms in the initial response models were removed. Only insignificant lower-order terms that are included in significant higher-order terms were considered to form the final response model. ANOVA analysis was performed for the second time.

Table 1. CCD coded and uncoded values of the input variables for the experimental design.

Variables \ Codes levels	Lowest $-\sqrt{2}$	Low -1	Centre 0	High 1	Highest $\sqrt{2}$
(X_1) Beeswax (gram)	6	30	65	100	124
(X_2) Soy wax (gram)	6	30	65	100	124
(X_3) Dammar (gram)	6	30	65	100	124

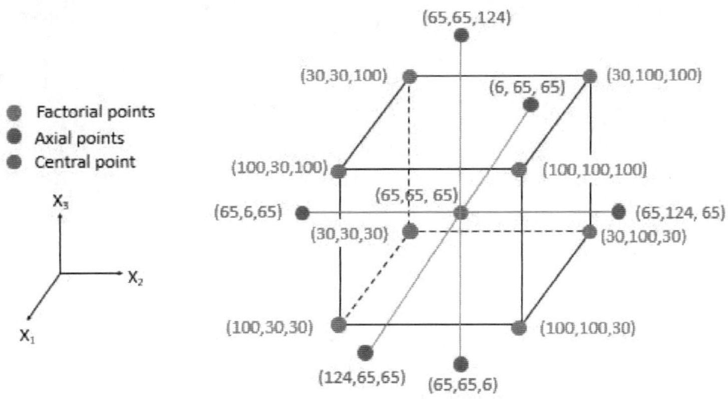

Figure 1. The 2^3 factorial design spaces representing the proportion of the independent variables. ($X1_{\text{g beeswax}}$, $X2_{\text{g soy wax}}$, $X3_{\text{g dammar}}$) used in this study.

2.3. Mass loss protocol

The wax blends were re-melted, and a small brush was used to apply the wax according to the sketched 8 x 8 cm guideline on the 10cm x10cm cotton fabric. The mass of the fabric with wax was initially weighed (M_i). Each waxed fabric sample was immersed in an alkaline solution for 1 hour. An alkaline solution of pH 10-11.5 was prepared by dissolving 0.62 grams of soda ash or sodium bicarbonate (Na_2CO_3) in 200 ml of water to increase the water's pH. After drying for 24 hours, each fabric was boiled in 100°C water for 1 hour. The sample fabric was then removed, dried and the final mass (M_f) was measured. The mass loss (%) was calculated according to Equation (1).

$$Mass\ loss\ (\%) = \frac{M_i - M_f}{M_i} \times 100\% \quad (1)$$

3. Results and Discussions

In this study, CCD was used to develop a polynomial model for the response (Y: mass loss %) based on a set of independent variables (X_1: beeswax (g), X_2: soy wax (g), X_3: dammar (g). A quadratic response model was then selected since the cubic model was aliased. An aliased model indicates the lack of insufficient unique design points and will result in unstable coefficient values and inaccurate graphs [13]. The standard error of design contour shown in Figure 2 at the central point of the DOE shows that the root means the square error is relatively small, ranging between 0.43 to 0.50. At the axial points of the DOE, the standard design error estimation only increases slightly in the range of 0.86 to 1.18 for the studied design space. It was concluded that near-uniform precision designs have been achieved since the error gradient is nearly uniform. Therefore, it was deduced that the response model (Y: mass loss %) will be able to deliver accurate predictions for the mass loss error for the intended range of batik waxes used in this study.

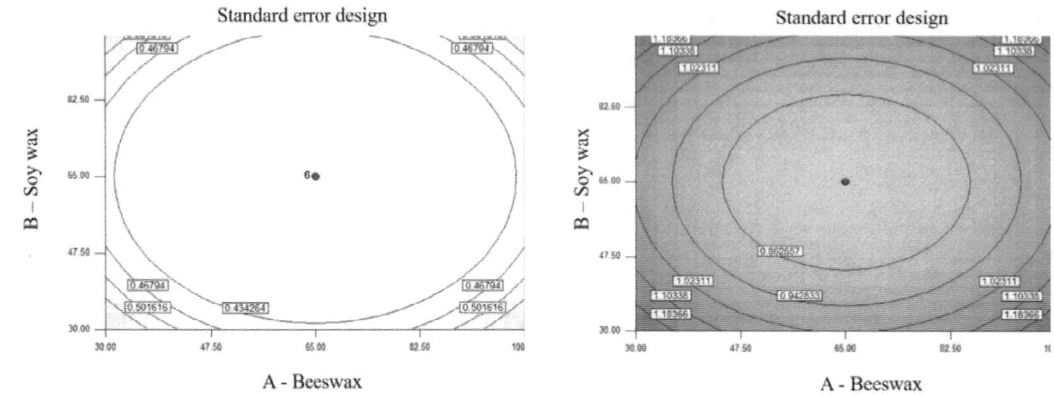

(a) Central points (b) Axial points

Figure 2. Standard design error estimation for the response based at (a) central points and (b) axial points for the design space.

The resulting mass loss (%) due to alkaline exposure and boiling water is presented in Table 2.

Table 2. Response (mass loss, %) for three-factor, two-level CCD matrices used in RSM in uncoded and coded values.

Run No.	Independent variables in uncoded value			Response
	X_1 (A) Beeswax (g)	X_2 (B) Soy wax (g)	X_3 (C) Dammar (g)	Y (Mass loss, %)
1	65	65	6	8.93
2	65	65	65	8.91
3	30	100	30	11.16
4	65	65	65	7.89
5	30	30	100	11.56
6	65	65	65	11.16
7	100	30	30	22.11
8	30	100	100	25.03
9	30	30	30	14.71
10	6	65	65	24.33
11	65	65	65	11.81
12	65	124	65	26.17
13	65	65	124	19.61
14	65	6	65	19.68
15	100	30	100	23.77
16	100	100	100	23.18
17	65	65	65	13.65
18	65	65	65	14.01
19	124	65	65	18.03
20	100	100	30	20.18

Table 3 shows the analysis of variance (ANOVA) for the final quadratic response model (Equation 2). The model has an F-value of 5.43. The p-value of 0.0055 of the response model indicates that there is only a 0.55% chance that the model F-value this large could occur due to noise. This implies the quadratic model is significant. The F value for Lack of Fit (4.70) has a corresponding p-value of 0.0516 which indicates that the lack of fit is not significant in comparison to the actual pure error. "Adequate Precision" measures the signal-to-noise ratio. A ratio value greater than 4 is highly desirable. The ratio value of 7.649 obtained in this study indicates an adequate signal, hence the final model can be used to navigate the design space. The R^2 value of 0.66 shows that there is a slightly high variability (mass loss (%) values) around the mean regression line. It is postulated that the R^2 value is affected by the method of wax deposition on the fabric. The manual brush stroke may have produced a slight inconsistent in the wax thickness layer. Wax that was spread thinly across the fabric can be easily degraded in high-temperature water compared to areas covered with a thicker layer of wax. However, important conclusions can still be drawn for the independent variables since the overall model is still statistically significant.

Table 3. Analysis of variance (ANOVA) for response surface quadratic model for mass loss (%).

Independent variables	F-value	P-value
Model	5.43	0.0055
Beeswax: X_1	1.15	0.3008
Soy wax: X_2	1.48	0.2442
Dammar: X_3	4.91	0.0437
X_1^2	8.80	0.0102
X_2^2	12.56	0.0032
Lack of fit	4.70	0.0516
Adequate precision (ratio>4) = 7.649		
R^2= 0.6599		

The proposed final regression model in coded value is shown in Equation (2).

$$Y = 12.09 + 1.19X_1 + 1.34X_2 + 2.45X_3 + 3.17{X_1}^2 + 3.79{X_2}^2 \quad (2)$$

The positive sign before each polynomial factor indicates that the response increases with the factor, while a negative sign indicates a reciprocal interaction [14]. As the P-value gets smaller, the significance of the coefficient increases [15]. Thus, the ranking of the first-order main effect that influences the response model is as follows: X_3 (dammar) > X_2 (soy wax) > X_1 (beeswax).

Figure 3 shows the normal probability and studentized residual plot for the mass loss (%) which indicates that the error terms were distributed normally and there are no extreme outliers. The interactions between RSM CCD design variables in 3D surface forms are shown in Figure 4. It was observed that further additions of dammar of more than 30g into the soy wax/beeswax increases the mass loss % of the blends. It was postulated that the increase in mass loss with the increase in dammar (wt%) was attributed to the loss in barrier efficiency. Soy wax is formed from partial or fully hydrogenated soybean oil [3] and is very sensitive to high temperatures. As the soy wax and beeswax were melt-blended approximately at 90 °C, the natural oil phase in soy wax may form and become separated from the wax. This explains the slightly greasy nature of the solidified blends. Yao et al.'s [2] study revealed that the addition of dammar into fully hydrogenated vegetable oil resulted in decreased compositional homogeneity. In general, the uniformity of hydrophobic multicomponent blends affects its barrier efficiency [16]. Therefore, it was postulated that the decrease in inhomogeneity due to higher dammar content (wt%) may have reduced the barrier efficiency of the blends.

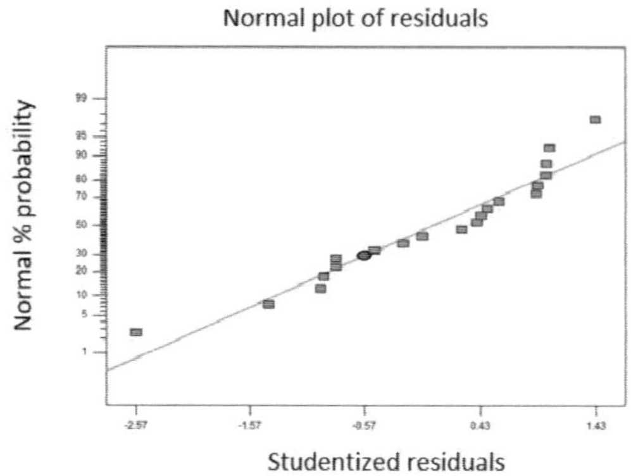

Figure 3. Normal residual plot for the batik waxes via ANOVA.

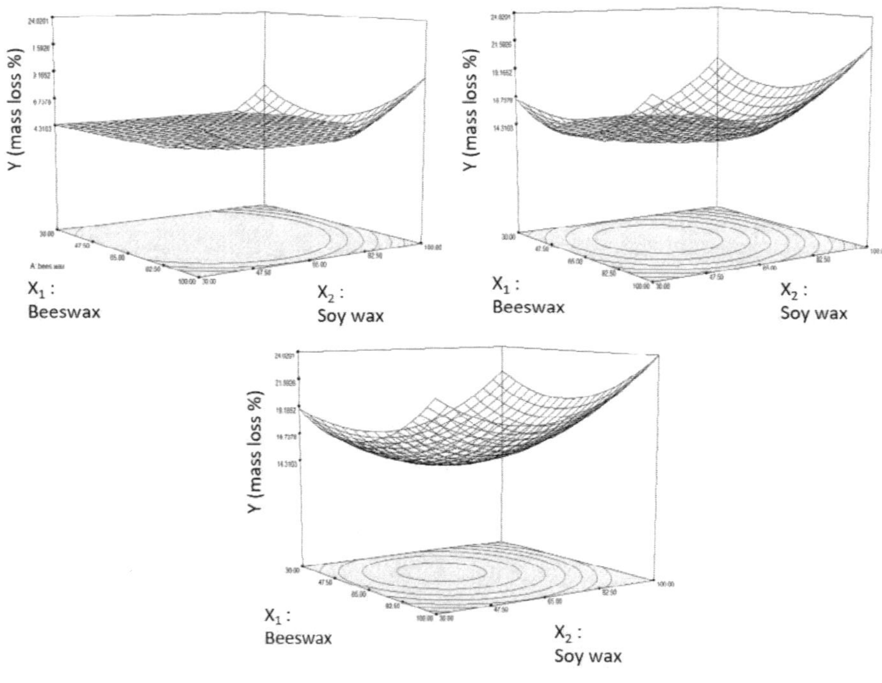

Figure 4. 3D surface interaction effect of X_1(beeswax) and X_2 (soy wax) and response, Y (mass loss%) at fixed values of (a) X_3 (dammar) = 30 g, (b) X_3 (dammar) = 65 g, (c) X_3 (dammar) = 100 g.

4. Conclusion

In this study, RSM via CCD was used to evaluate the effect of dammar additions based on mass loss % of the waxed covered fabric in 100°C water. The statistical ANOVA analysis for the obtained quadratic model is as follows; F-value of 5.43 with a corresponding p-value of 0.0055, 'lack of fit' at 4.70 with a corresponding p-value of 0.0516, 'adequate precision ratio value of 7.649 and R^2 value of 0.66. A statistically significant quadratic response model was then proposed in this study. The response model predicted that the addition of dammar increases the mass loss percentage attributed to the increase in inhomogeneity of the blend. The results of this study can be used to formulate new batik wax compositions with varying degrees of ease in wax removal during batik production. In addition, these compositions have the potentials to be used as an environmentally friendly batik wax composition for the batik industry.

Acknowledgement
This work was full supported by International Islamic University Malaysia and funded by the Ministry of Higher Education Malaysia (FRGS/1/2018/TK03/UIAM/02/3).

References

[1] Rezaei K, Wang T and Johnson L A 2002 Hydrogenated vegetable oils as candle wax *J. Am. Oil Chem. Soc.* **79** 1241–7
[2] Yao L and Wang T 2012 Textural and physical properties of biorenewable “waxes” containing partial acylglycerides *J. Am. Oil Chem. Soc.* **89** 155–66
[3] Yao L, Lio J, Wang T and Jarboe D H 2013 Synthesis and characterization of acetylated and stearylyzed soy wax *J. Am. Oil Chem. Soc.* **90** 1063–71
[4] Bowen D B 2005 Soy Wax : An Alternative Resist *Kuala Lumpur International Batik Convention* (Kuala Lumpur) pp 1–12
[5] Bomford D and Staniforth S 1981 Wax-resin lining and colour change: An evaluation *Natl. Gall. Tech. Bull.* **5** 1–10
[6] Han J H, Seo G H, Park I M, Kim G N and Lee D S 2006 Physical and mechanical properties of pea starch edible films containing beeswax emulsions *Food Eng. Phys. Prop.* **71** 290–6
[7] Fulbandhe V M, Jobanputra C R, Wadher K J, Umekar M J and Bhoyar G S 2012 Evaluation of release retarding property of gum damar and gum copal in combination with hydroxypropyl methylcellulose *Indian J. Pharm. Sci.* **74** 189–194
[8] Won S M, Koo J, Crawford K E, Mickle A D, Xue Y, Min S, Mcilvried L A, Yan Y, Kim S B, Lee S M, Kim B H, Jang H, Macewan M R, Huang Y, Iv R W G and Rogers J A 2018 Natural wax for transient electronics *Adv. Funct. Mater.* **1801819** 1–10
[9] Huei L W 2014 Palm based batik wax *Malaysian Palm Oil Board Inf. Ser.* 678–81
[10] Malik A, Rahmillah F I, Atmaja B D and Ihsan B F 2018 The effect of microwax composition on the staining quality of Klowong batik wax *MATEC Web Conf.* **154** 10–3
[11] Franz M H, Neda I, Maftei C V, Ciuca I, Bolcu D and Stănescu M M 2021 Studies of chemical and mechanical properties of hybrid composites based on natural resin Dammar formulated by epoxy resin *Polym. Bull.* **78** 2427–38
[12] Morkhade D M, Fulzele S V, Satturwar P M and Joshi S B 2006 Gum Copal and Gum Damar : Novel Matrix Forming Materials for Sustained Drug Delivery *Indian J. Pharm. Sci.* **68** 53–8
[13] Liu J, Wang J, Leung C and Gao F 2018 A multi-parameter optimization model for the evaluation of shale gas recovery enhancement *Energies* **11**
[14] Prasanna J L, Deepthi B and Rao N R 2012 Influence of formulation variables on the in vitro dissolution of gum damar based matrix tablet *Int. J. Pharm. Sci. Res.* **3** 4520–7
[15] Xiarchos I, Jaworska A and Zakrzewska-Trznadel G 2008 Response surface methodology for the modelling of copper removal from aqueous solutions using micellar-enhanced ultrafiltration *J. Memb. Sci.* **321** 222–31
[16] Debeaufort F, Martin-Polo M and Voilley A 1993 Polarity homogeneity and structure affect water vapor permeability of model edible films *J. Food Sci.* **58** 2–5

Simulation of process control for production of 5-HMF from old coconut water

N A Ahmad Joharee[1], J Mohd Ali[1], M N Mohd Fuad[1], and N Abd Rahman[1,2]

[1]Department of Chemical and Process Engineering, The National University of Malaysia, 43600 UKM, Selangor, Malaysia

[2]Corresponding Author's E-mail: norlizajkpp@ukm.edu.my

Abstract. 5-Hydroxymethyl furfural (5-HMF) is one of the intermediate chemicals produced from biomass through the process of dehydration of glucose and fructose in acidic conditions. Biomass resources are one of the sources that attract the attention of many users of unlimited biomass resources, in addition to being able to produce environmentally friendly products such as bio fuels and bio chemicals. Old coconut water and oil palm tree trunk sap was chosen in this study because of its abundance in Malaysia. It is an essential organic compound for the production of bioplastics, chemicals, building materials, fuels, textiles, cosmetics and food. Although the production of 5-HMF is relatively simple, in fact, the synthesis of 5-HMF is complicated by the occurrence of multiple reactions at one time, especially when using biomass as a raw material. In this report, process control of production of 5-HMF from biomass is presented using fuzzy logic controller for controlling temperature of a continuous stirred tank reactor (CSTR). The objective of this paper is to develop a control system for a non-isothermal temperature control of a CSTR with the most stable response curve. Through simulation results, fuzzy logic control performs better because there was no overshoot and successfully reaches the set point at faster settling time.

1. Introduction

Biomass is one of the most popular sources of environmentally friendly products such as bio fuels and bio chemicals [1]. According to [2], there are several biomasses suitable for raw materials for the manufacturing of 5-HMF such as corn stover, pine sawdust, grass, filter paper, corncob, red pine bark, maple wood, sugarcane bagasse and etc. By utilizing these biomass residues, we will be able to produce a valuable product while reducing agricultural waste and reducing environmental pollution. By using agricultural waste, the 5-HMF produced will be more reliable for use in the food industry because of its halal resources. Coconut, also known as Cocos Nuctfera Linn is one of the high demand products of over 90 countries. The volume of coconut water that is not used during the commercial process of coconut fruit can reach over 2.4 billion L a year. Coconut water is made from 96 percent water and 6 percent solid. The major components of solid in coconut water are soluble sugars, proteins, and minerals [3].

When it comes to chemical reactions and reactor design, there are some important parameters that need to be taken into account and controlled. Among the parameters that are often controlled in a process are temperature, pressure, flow rate, tank level, and pH [4]. One of the most common types of controls is fuzzy logic control. Fuzzy logic control is widely used in complex and non-linear systems [5]. Fuzzy logic is an extension of binary logic where the concept of partial truth is used as opposed to truth or falsehood. It has truth values in the range 0 to 1.0 where 0 indicates falsehood while 1 indicates absolute truth. Fuzzy logic controllers are able to provide automatic control over processes by systematically handling information. Simulation of process control on the continuous stirred tank reactor (CSTR) was performed by previous researchers by comparing the results of PI, PID and FLC control. Fuzzy logic control, FLC is said to be able to control non-isothermal systems better than PID controllers.

Studies on the direct conversion of biomass to 5-HMF have only been carried out in the past 10 years. Most studies are done using biomass from agricultural waste so that this waste can produce valuable products. Biomass is an unlimited source, making it an ideal source of feedstock, reducing energy consumption and carbon dioxide emissions. The objective of this paper is to develop mathematical model and block function using MATLAB to describe temperature control of CSTR for the production of 5-HMF from old coconut water. Simulation for process controller actions of PID and fuzzy logic was subsequently done using Simulink. The performance of both controllers in controlling the process will be evaluated through simulation study.

2. Methodology

This study is divided into two phases. The first phase is the collection of information and experimental data from previous researchers to identify the parameters to be controlled for a continuous stirred tank reactor (CSTR). Then, the simulation is carried out by developing the reactor model using the material balance and the energy balance of the reactor. Next, a mathematical model for describing temperature control for CSTR was also developed. The system will be modelled using MATLAB Simulink software with Equations (1)-(3) to represent component balance for old coconut water (reactant), reactor energy balance, and energy balance for heating jacket, respectively:

$$\frac{dz}{dt} = \frac{Fz_O}{V_R} - \frac{Fz}{V_R} - kz \tag{1}$$

$$\frac{dT_R}{dt} = \frac{FT_O}{V_R} - \frac{FT_R}{V_R} - \frac{\lambda kz}{Mc_p} - \frac{UA_J}{\rho c_P V_R}\left(T_R - T_J\right) \tag{2}$$

$$\frac{dT_J}{dt} = \left(\frac{F_J}{V_J}\right)\left(T_{JO} - T_J\right) + \frac{UA_J(T_R - T_J)}{\rho_J c_J V_J} \tag{3}$$

The reaction is first-order with respect to reactant concentration where the rate constant k is in Equation (4) [6]:

$$k = 7.2 \times 10^{10} e^{-8750/T_R} \tag{4}$$

The parameters of the model are listed in Table 1.

Table 1. Reactor model parameter values.

Parameter	Value
Working volume, V_R	1.35 m^2
Tank diameter, D	1.02 m
Tank height, H	2.05 m
Flow rate, F	2000 kg/hr
Reactor temperature, T_R	37 – 105 °C
Inlet temperature, T_0	37 °C
Jacket temperature, T_j	129 °C
Heat capacity of jacketed fluid, c_J	4.18 kJ/kg.K
Heat transfer coefficient	243.90 W/m^2 °C

2.1. PID controller
The PID control design is performed with a feedback control strategy where if an error is detected, a signal is sent back into the feed and performs adjustments to control the parameter according to its set point as shown in Figure 1.

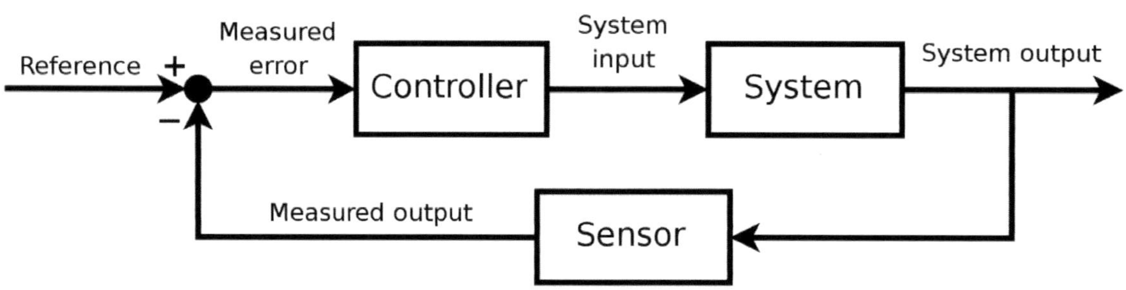

Figure 1. Block diagram of feedback control strategy.

PID controller is the most famous feedback controller strategy in the industry. The controller performance is subjected to the appropriate tuning of the three control elements which are proportional, integral, and derivative actions. The control action $u(t)$ of the controller is described by the following differential equation (Equation (5)):

$$u(t) = K_P e(t) + \frac{K_P}{T_I}\int_0^t e(\tau)\,d\tau + K_P T_D \frac{de(t)}{dt} \qquad (5)$$

where K_P, T_I, and T_D are the controller parameters that are associated with proportional gain, integral time, and derivative time, respectively. Best performance for PID control can be obtained for linear systems and mildly nonlinear systems if the corresponding parameters are tuned correctly.

2.2. Fuzzy logic controller
For temperature control of non-isothermal CSTR system, it is most critical to avoid overshoot and steady state error. This control problem can be handled appropriately by nonlinear control system such as fuzzy logic controller (FLC). FLC is based on fuzzy logic, which is an extension of binary logic where the concept of partial truth is used as opposed to complete truth or falsehood, thus can handle grey area pretty well. FLC is said to be able to control non-isothermal reactor systems better than PI and PID controller algorithms. The block diagram of the inference system used by typical FLC is shown

IOP Conf. Series: Materials Science and Engineering 1192 (2021) 012020 doi:10.1088/1757-899X/1192/1/012020

in Figure 2. In this system, the crisp numerical input is subjected to fuzzifier to transform it to fuzzy variable as subjected to a set of membership functions. The fuzzy variable is then sent to the inference system to determine the appropriate output, which is still a fuzzy variable before defuzzification turns it into a crisp numerical value, the transformation of which is accomplished by another set of membership functions.

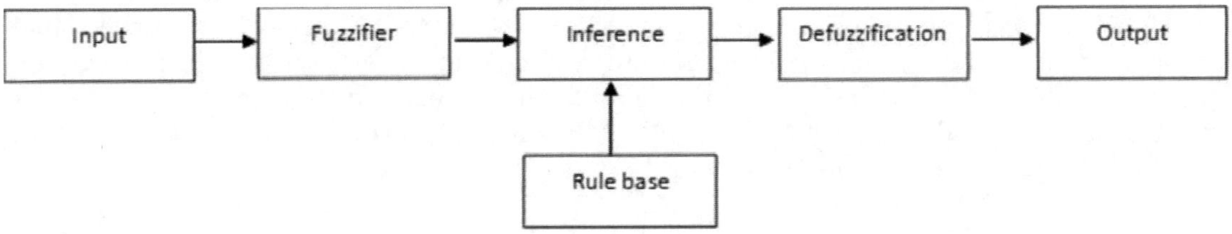

Figure 2. Fuzzy logic system block diagram.

3. Results and discussion

This study was conducted to study reactor control of 5-HMF production from old coconut water using PID control and FLC. The controlled process unit is a continuous reactor mixing tank (CSTR). Here, we consider setpoint tracking scenario whereby the reactor temperature will be controlled at 105 °C from the initial reactor temperature of 26 °C. The chosen setpoint temperature value, which is 105 °C is the optimum temperature to maximize 5-HMF production based on the experimental work of Maruf, 2018 [6]. Finally, the manipulated variable for the reactor temperature control is the heating jacket temperature.

3.1. *Simulation of PID controller*

First, the temperature control scenario where the reactor temperature is controlled by PID controller is studied. After tuning the controller using built-in PID tuner in Simulink, the following PID controller parameters without derivative action were obtained: K_P = 55.68 while T_I = 0.52 s. Figure 3 shows the response curve for this controller. From the corresponding figure, it can be seen that there is significant temperature overshoot occurred in the reactor temperature control by PID controller where the temperature value almost reached to 120 °C. Although this overshoot can be overcome by including derivative action in the PID control, however considering that the temperature measurement can be noisy, it might cause stability issue to the controller due to amplification of the measurement noise by the derivative action.

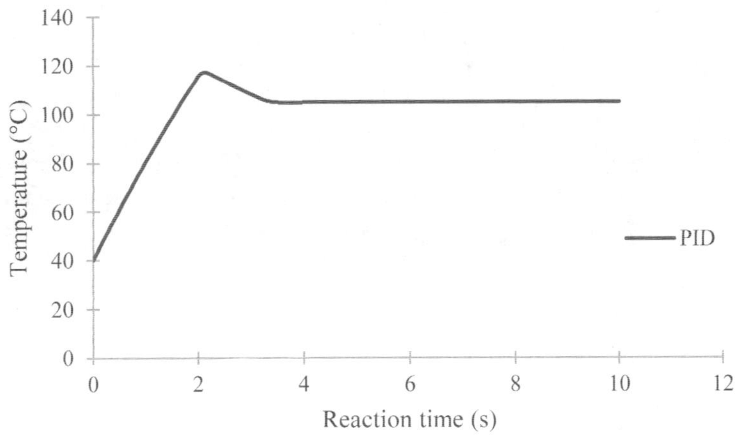

Figure 3. Response curve for reactor temperature control with PID controller.

3.2. Simulation of fuzzy logic controller

A FLC for the CSTR system was developed based on the Mamdani inference system. Among the reasons why the Mamdani system was chosen for this controller is because of its intuitive nature, suitable for human entry, basic rules that are easier to interpret, and suitable for various systems. As shown in Figure 4, the inputs to this Mamdani system are error in reactor temperature from the setpoint and the rate of change of this error with respect to time. The two inputs will be evaluated by the Mamdani inference system to determine appropriate course of action for the reactor jacket temperature value. For error and change in error, the values are subjected to three fuzzy variables which are NEGATIVE, ZERO, and POSITIVE. Meanwhile, for the output which represents the jacket temperature, the corresponding fuzzy variables are used: ZERO, SMALL, AVERAGE, and LARGE. The rules for this inference system are shown in Table 2.

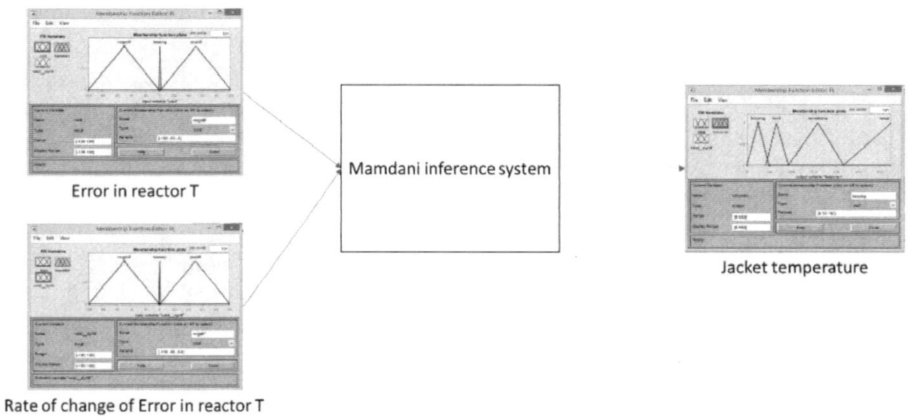

Figure 4. Mamdani inference system for fuzzy logic controller.

Table 2. Rules base for the Mamdani inference system.

Rule	Error	Change in error	Output
1	IF ZERO	AND ZERO	THEN ZERO
2	IF ZERO	AND NEGATIVE	THEN SMALL
3	IF ZERO	AND POSITIVE	THEN AVERAGE
4	IF NEGATIVE	AND ZERO	THEN ZERO
5	IF NEGATIVE	AND NEGATIVE	THEN SMALL
6	IF NEGATIVE	AND POSITIVE	THEN SMALL
7	IF POSITIVE	AND NEGATIVE	THEN AVERAGE
8	IF POSITIVE	AND ZERO	THEN SMALL
9	IF POSITIVE	AND POSITIVE	THEN AVERAGE

The results of the controller simulation using FLC for the CSTR system are shown in Figure 5. Through this graph, it can be seen that there is no temperature overshoot occurs in this FLC system, with the setpoint temperature successfully achieved stably and smoothly. Based on this result, we can surmise that FLC can control this non-isothermal reactor system better than PID controller.

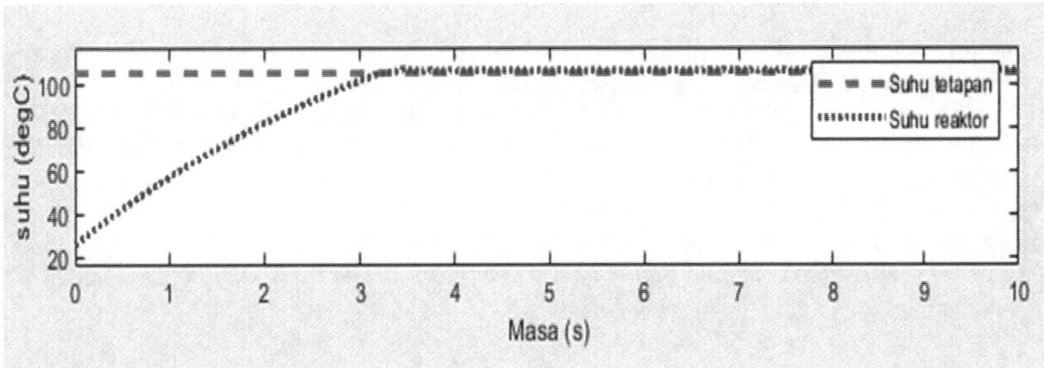

Figure 5. Response curve for reactor temperature control with fuzzy logic controller.

3.3. *Comparison of controller performance*

The comparison of PID and FLC performance on the reactor temperature control is shown in Figure 6 and further summarized in Table 3. It can be seen that both controllers show no steady-state offset error. However, it can be seen that FLC is much better controller for this system with no overshoot and faster process settling time.

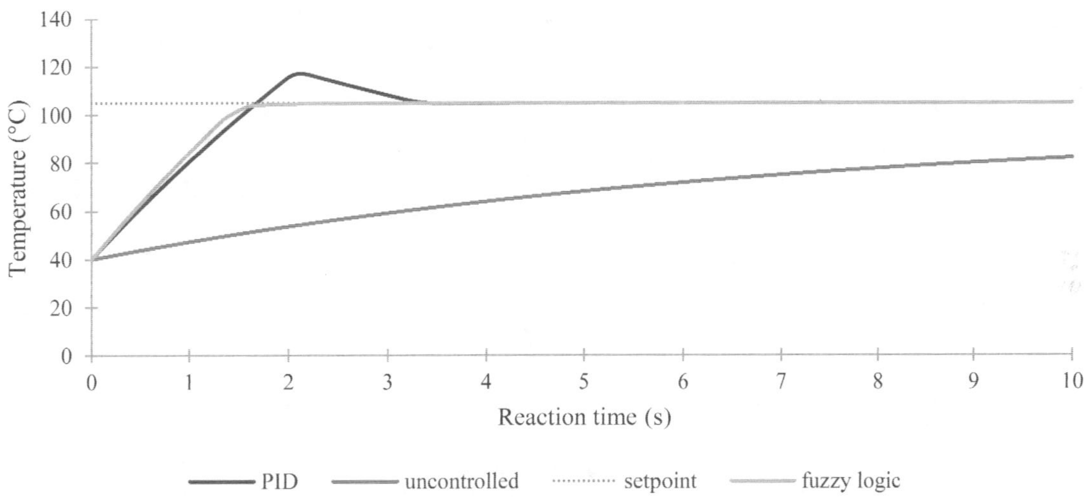

Figure 6. Comparison of response curve for uncontrolled system, and controlled system with PID and fuzzy logic controllers.

Table 3. Comparison of reactor temperature controllers.

Controller type	Peak temperature, °C	Settling time, s
PID	116.7	2.05
FLC	106.4	2.95

4. Conclusions

In conclusion, controller simulation study to assess performance of reactor temperature control using PID and FLC for 5-HMF synthesis reactor from old coconut water was successfully carried out. Simulation results indicate that FLC is superior to PID control for reactor temperature setpoint tracking that can avoid temperature overshoot and leads to faster process settling time. Next study will look into the corresponding controller performance for process disturbance rejection.

References

[1] Kougioumtzis M A, Marianou A, Atsonios K, Michailof C, Nikolopoulos N, Koukouzas N, Triantafyllidis K, Lappas A and Kakaras E 2018 *Waste Biomass Valori.* **9** 2433
[2] Menegazzo F, Ghedini E and Signoretto M 2018 *Molecules (Basel, Switzerland)* **23** 2201
[3] Gordon A and Jackson J 2017 *Food Safety and Quality Systems in Developing Countries* ed A Gordon (San Diego: Academic Press) p 185
[4] Luyben W L 2007 *Chemical Reactor Design and Control* (Wiley)
[5] Emami M S 2010 *J. Math. Comput. Sci.* **1** 339
[6] Maruf M K 2018 *Reka bentuk reaktor dalam penghasilan 5-hidroksimetilfurfural daripada air kelapa tua* (Universiti Kebangsaan Malaysia)

Effect of 45S5 bioactive glass on the sintering temperature of titanium-hydroxyapatite composite

D S Abu Bakar[1,2], **M G Kutty**[1] **and N A Yahya**[1]

[1] Department of Restorative Dentistry, Faculty of Dentistry, University of Malaya, 50603 Kuala Lumpur, Malaysia

[2]Corresponding Author's E-mail: dayana.abubakar@gmail.com

Abstract. Titanium (Ti) is widely known for its good mechanical properties and corrosion resistance. However, it has poor biocompatibility. Hydroxyapatite (HA) is a biocompatible material but has poor mechanical properties. Making Ti-HA composite creates a promising choice of biomaterial in dental and medical applications. However, creating a Ti-HA composite requires sintering at high temperatures which leads to oxidation of Ti. The aim of this study was to reduce the sintering temperature of Ti-HA composite by incorporating 45S5 Bioactive Glass (BG) without compromising the chemical, physical and mechanical properties of the composite. In this study, a Ti-HA-BG composite with wt% of 45-45-10 respectively was produced via powder metallurgy. This was compared with the control composite consisting of 50 wt% Ti- 50 wt% HA. Powders according to the above-mentioned ratio were milled at 200 rpm for 5 hours by using a planetary ball milling machine. Samples were then compacted into cylindrical pellets via uniaxial pressing at 1500 psi and sintered in an atmospheric furnace at 1000 °C, 1100 °C and 1200 °C for 4 hours. Ti-HA and Ti-HA-BG sample characteristics were examined and compared by using Fourier Transformed Infrared Spectroscopy (FTIR), Energy Dispersive X-Ray (EDX) and X-Ray Diffraction (XRD). The density and volumetric expansion of the composites were also measured and compared. Results from XRD data indicate the reduction of oxidation of Ti and decomposition of HA in Ti-HA-BG composite at lower temperature in comparison to Ti-HA composite. The density of Ti-HA-BG composites are higher compared to Ti-HA composite while the volumetric expansion of Ti-HA-BG composites is lesser than Ti-HA composite. Therefore, BG is a low melting point additive that acts as a good sintering aid to effectively lower the sintering temperature while maintaining the desired properties of initial components.

1. Introduction

Biocomposites for oral and dental tissue engineering have been specifically developed through the use of multiple material mixtures by utilizing their compatible properties [1-2]. Composite materials are fabricated using the combination of two or more materials, which often possess different properties. All materials work together to create the unique properties of the composite [3-4]. Notably, one of the most widely used composites in dental applications is the combination of metal-ceramics materials such as titanium (Ti) and hydroxyapatite (HA), and iron (Fe) and aluminium oxide (Al_2O_3) [5-6].

Ti has been widely used in dentistry, especially under load-bearing applications, mainly due to its good mechanical properties and anti-corrosion abilities. However, Ti is a bioinert material, which means that it may eventually result in loosening and malfunction of the implants as the interface between Ti and the host bone is a simple interlocking bonding [5, 7].

HA has been widely used as a bone substitute in restorative dental and orthopedic implants, due to its similar composition to the mineral component in bone and teeth structures. Nevertheless, HA is brittle and unable to resist high impact throughout the load-bearing applications. One of its key limitations to be used as a load-bearing implant is its poor mechanical properties. Hence, Ti-HA composite is said to be a promising approach in dental applications due to the good combination of the mechanical properties of Ti and bioactive properties of HA [5, 7].

However, the fabrication of Ti-HA composite can be a challenge because it involves a sintering process that requires very high temperature. It is important to note that sintering at a high temperature can cause the oxidation of Ti and decomposition of HA to occur. The undesirable interactions of these materials at a high temperature during the sintering process will change the microstructure and properties of the composite [5, 8-9].

A study done by Arifin et al. [8] has claimed that during the sintering of Ti-HA composite, tetra calcium phosphate (TTCP: $Ca_4(PO_4)2O$) and tricalcium phosphate (TCP: $Ca_3(PO_4)_2$) were produced after dehydroxylation and decomposition of pure HA. Dehydroxylation happened at approximately 900 °C in air, whereas in a water-free atmosphere, it occurred at about 850 °C. Despite that, Ti ions reacted with the dehydrated water of HA composite to produce titanium oxide (TiO_2). This caused the process of dehydroxylation and decomposition of HA to accelerate at about 800 °C. On the other hand, Ti was oxidized during the sintering of Ti-HA composites in air at 700 °C to 1200 °C. The observations stated that TiO_2 was produced at approximately at 800 °C, where only two crystalline phases, TiO_2 and calcium titanate ($CaTiO_3$) were produced [8]. Furthermore, the sintering products might include TCP, TTCP, $CaTiO_3$, titanium phosphides (Ti_xP_y), titanates (Ti_xO_y), calcium oxide (CaO), and amorphous phases in accordance with the experimental parameters [10]. All these reactions may contribute to the decrease in mechanical properties of the Ti-HA composite at high sintering temperatures.

One way to overcome this is to introduce a low melting point additive as a sintering aid to effectively lower the sintering temperature without compromising the desired mechanical properties. Several previous studies have proposed that bioactive glass (BG) can be a good candidate as a sintering aid in pure HA, which in results showed the enhancement of mechanical properties of HA as well as its biological properties [9, 11]. BG is a silicate-based material that can strongly bond to the tissues. It has high biocompatibility and can form a HA layer when implanted in the body or soaked in the simulated body fluid. BG is also an inorganic and synthetic material. These characteristics of BG have shown to be ideally used in medical and dental applications [12-14].

Therefore, this current study is concerned with the preparation and characterization of novel biocomposites containing Ti and HA where BG is incorporated into the system. This study evaluated the possible sintering temperatures effect on the chemical, physical and mechanical properties of the Ti-HA-BG composites that were obtained by the powder metallurgy method.

2. Materials and Methods

2.1. Sample preparation

A commercially available Ti powder with an average particle size of 45 μm was purchased from R&M Chemicals, while the HA powder with an average particle size of 5 μm and BG powder (type 45S5) with a particle size range of 45-75 μm were purchased from Sigma-Aldrich and XL Sci-Tech, Inc. respectively. Powders according to the wt% composition as shown in Table 1 were weighed accurately by using an electronic balance and were mixed and milled by using a planetary ball milling machine (Retch Germany) with zirconia balls and ceramic jars. During the ball milling process, a 10:1 of the ball:mass ratio was employed at 200 rpm for 5 hours. Then, 1.5 g of the milled powder was put into a stainless-steel die and was compacted by using a hydraulic hand press machine. The uniaxial pressing was done at 1500 psi for 5 minutes and produced cylindrical pellets with 20 mm diameter and 2 mm height. The pellets were then sintered in an atmospheric furnace (Elite Furnace, BRF 16/5, UK). Three groups of Ti-HA-BG composites were prepared according to the sintering temperatures at 1000 °C, 1100 °C and 1200 °C while the control group of Ti-HA composites was sintered at 1200 °C as shown in Table 2. The sintering process for each temperature group was done with 4 hours of holding time and a heating rate of 5 °C/min as shown in Figure 1.

Table 1. Weight composition of the prepared composite.

Samples	Composition (wt%)		
	Ti	HA	BG
Ti-HA	50	50	-
Ti-HA-BG	45	45	10

Table 2. Temperature group for the prepared composite.

	Sample Group			
	G1	G2	G3	G4
Samples	Ti-HA	Ti-HA-BG	Ti-HA-BG	Ti-HA-BG
Sintering Temperature (°C)	1200	1000	1100	1200

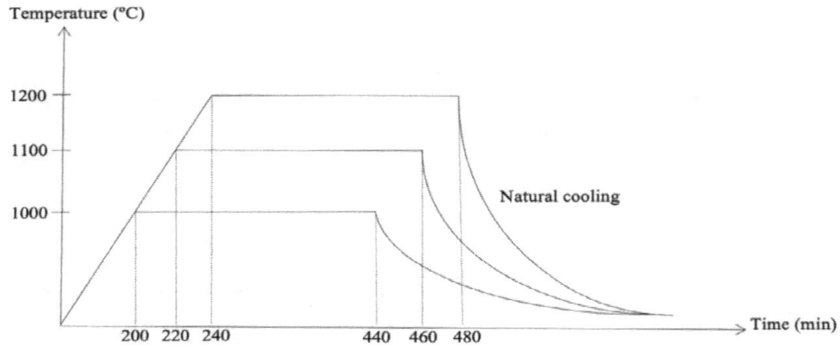

Figure 1. Sintering cycles of the samples.

2.2. Sample characterization
The identification of the crystalline phases of the prepared composites was determined by using x-ray diffractometry (XRD: Phillips PW1840), operating at 40 kV and 25 mA, using Cu-K radiation and Ni filter, scanned at 2-theta of 2 °/min. All data was analyzed by using X'Pert HighScore software (X'Pert Pro, Panalytical, Almelo, Netherlands) and the Joint Committee of Powder Diffraction Standard-International Center for Diffraction Data (JCPDS-ICCD) files were used as a reference data for the interpretation of XRD patterns.

The scanning electron microscope (SEM: FEI Quanta 250 FEGSEM) was used to observe the morphology of the composites. Composites from each group were viewed under SEM at 5000X magnification steps. The energy-dispersive x-ray spectroscopy (EDX: FEI Quanta 250) was used for elemental analysis purposes and to check the mean weight percentages for every element.

Fourier Transform Infrared Spectroscopy (FTIR-ATR: Thermoscientific Nicolet 6700) was used to support and verify the XRD findings as well as to identify the structural groups of the prepared composites.

The density (ρ) of the samples was determined by using the Archimedes' principle, where distilled water was used as the buoyant. All the weight measurements were done by using a densitometer (Sartorius AX224, Goettingen, Germany). Density of sample was obtained by employing Equation (1).

$$\rho = \frac{w_a}{w_a - w_b}\rho_b \tag{1}$$

Where w_a is the weight of sample in air, w_b is the weight of sample in buoyant and ρ_b is the density of the buoyant, gcm^{-3}.

The volumetric expansion was determined with a digital vernier caliper (Mitutoyo/Digimatic, Tokyo, Japan) by measuring the dimension and comparing the difference in the volume between green (V_g) and the sintered material (V_s) by using Equation (2).

$$Volumetric\ Expansion\ (\%) = \frac{V_s - V_g}{V_g} \times 100 \tag{2}$$

3. Results and Discussion

X-ray powder diffraction data of the prepared composites are shown in Figure 2. XRD pattern of G1 recorded some TiO_2, TTCP, TCP and $CaTiO_3$ that matched JCPDS card nos. 01-076-0319, 00-025-1137, 00-001-0941 and 22-153 respectively. No Ti and HA were recorded in G1. This shows that there were interactions between Ti and HA where the oxidation of Ti and decomposition of HA have occurred. These results were in agreement with previous studies that have claimed that Ti started to oxidize to TiO_2 in air at 700 °C to 1200 °C [5, 8]. On the other hand, some HA and Ti that matched JCPDS card no. 00-001-1008 and 00-44-1294 respectively were recorded in the XRD patterns of G2, G3 and G4. Although the results showed the existence of decomposition of HA and oxidation Ti, the undesired interactions between Ti and HA were minimized causing complete removal of $CaTiO_3$. Some $CaTi(SiO_5)$ that matched JCPDS card no. 01-073-2066 were also recorded which proves the interaction between TiO_2 and BG [15].

FTIR reflectance spectra in Figure 3 (a) shows clear sharp peaks at 969.52 and 1014.66 cm^{-1} that were attributed to v_1 and v_3 modes of phosphate in TTCP. Theoretically, phosphate ions (PO_4^{3-}) have four vibration modes; v_1, v_2, v_3 and v_4, all of which are infrared active [16]. On the other hand, as shown in Figure 3 (b), (c) and (d), the bands in the region of 900–1100 cm^{-1} almost vanished with increasing temperature. Peaks in this region are becoming weaker and broader as the sintering temperature increases, showing that most TTCP disappeared from the system [9, 16].

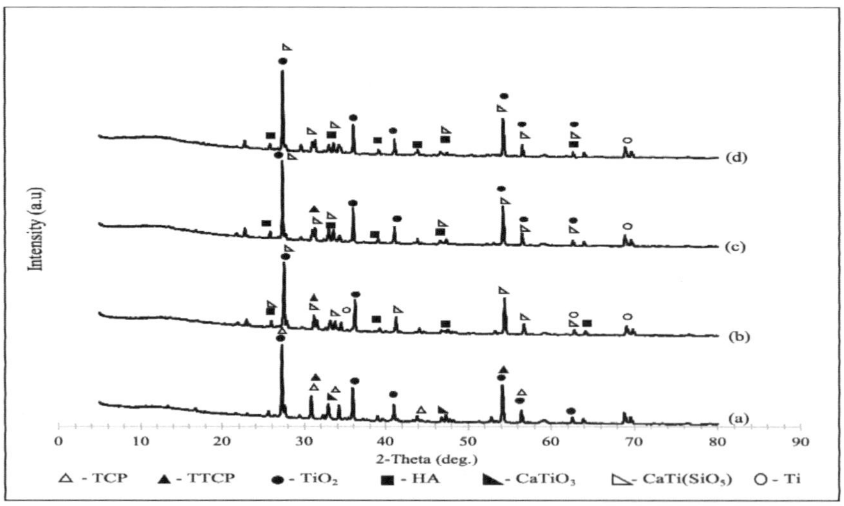

Figure 2. XRD pattern of sample (a) G1, (b) G2, (c) G3 and (d) G4.

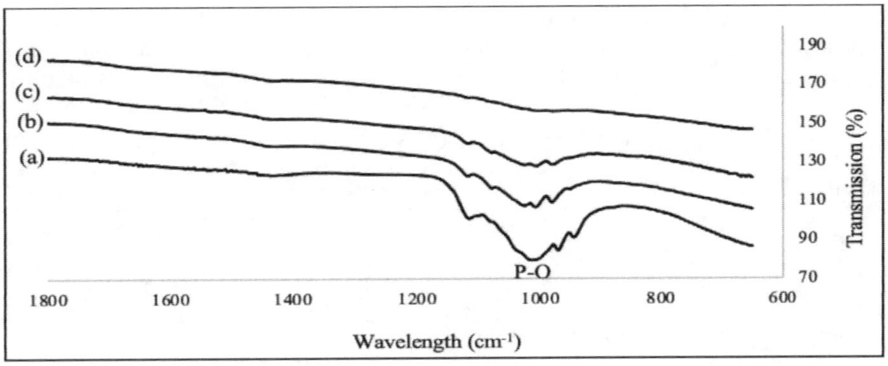

Figure 3. FTIR reflectance spectra of (a) G1, (b) G2, (c) G3 and (d) G4.

The surface morphology for each composite was determined by using SEM images as shown in Figure 4. Small pores and large pores are the two types of pores that were detected. Both types of pores can be observed near the grain boundaries. This study found that the microstructure of the porous composites demonstrates the presence of large defects when sintering temperature is raised from 1000 to 1200 °C. Generally, with an increase of sintering temperature, the porosity decreases and the strength decreases. The images showed three distinct morphological regions. As investigated by Weinand et al. [17], the dark regions were associated with high O concentrations with the probable formation of TiO_2. These regions also showed the presence of P and Ca. While in the gray regions, it is claimed to be predominated by Ti whereas O, P and Ca were detected at less ratio. Whereas the shining phase can be associated with the presence of high Ti concentration as can be seen in Figure 4 [17].

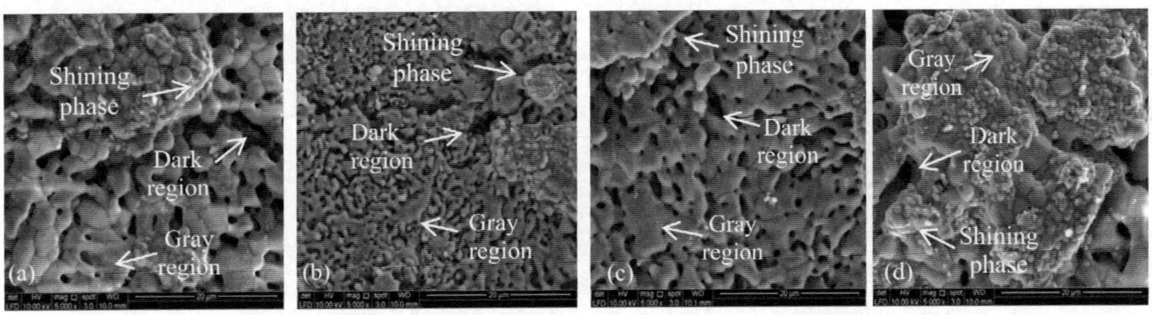

Figure 4. SEM images at 5000X magnification of sample (a) G1, (b) G2, (c) G3 and (d) G4.

Table 3 shows the result of the chemical composition as a function of sintering temperature. In Ti-HA composite (sample G1), no Si and Na were recorded as no BG was incorporated in its system. HA was indicated by Ca, O and P in all groups. However, in G3 shows no Si detected at that point of area which probably due to poor distribution of Ti, HA and BG components due to the processes during the composites' milling and sintering.

Figure 5 shows the density of each composite. Theoretically, density of Ti (4.5 gcm^{-3}) is higher than HA (3.156 gcm^{-3}) and BG (2.707 gcm^{-3}) [8]. Therefore, the theoretical density of Ti-HA and Ti-HA-BG composite is 3.828 gcm^{-3} and 3.7159 gcm^{-3} respectively. The experimental densities of Ti-HA-BG gradually decreased due to the sintering temperature. Results obtained in this study showed that the experimental densities were not much lower than the theoretical densities. The density of Ti-HA-BG is higher than the Ti-HA, with the lowest density value obtained is by the sample G4 and the highest value is given by the sample G2 which is 3.067 gcm^{-3} and 3.14 gcm^{-3}, respectively. Generally, densities can be affected by many factors such as powder characteristics and processing. One of the factors was believed that the decomposed and oxidized components were weaker than Ti and HA, making the density to be lowered [18].

Table 3. Chemical general analysis of each composite by EDX.

Sample Group	Composition (wt%)						
	Ti	Ca	P	O	C	Si	Na
G1	27.59	17.48	7.54	44.31	3.09	-	-
G2	12.11	23.85	12.38	46.56	-	1.60	3.50
G3	16.81	22.07	11.92	45.48	-	-	3.72
G4	10.46	6.36	13.20	41.74	5.00	0.68	2.56

Figure 6 shows the volumetric expansion of all composites. From this study, the volumetric expansion of Ti-HA-BG gradually decreased due to the sintering temperature. The Ti-HA-BG composites showed lesser volumetric expansion compared to the Ti-HA composite with highest value given by G2 sample (29.76%) and the lowest with G4 sample (23.97%). The expansion is most likely caused by a significant mismatch between the crystal lattice parameters of the initial and subsequent phases, which results in many defects such as vacancies, dislocations and stacking faults at particles interfaces, therefore resulting in unpacked and fragile matrix. The most essential parameters in powder metallurgy sintering are maximum sintering temperature, sintering time at maximum temperature and heating/cooling rates. Generally, low sintering temperatures and insufficient duration of time could provide in insufficient fusion of the particles and low densification of the composite. Improper cooling rates might be one of the contributing factors to rapid expansion of the composite [18].

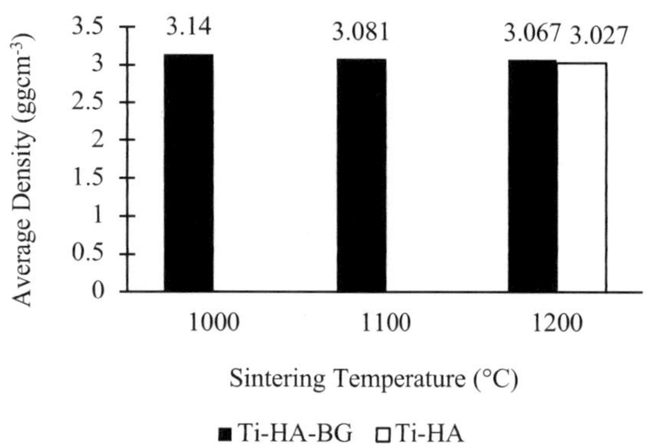

Figure 5. Average density of Ti-HA and Ti-HA-BG composites due to sintering temperature.

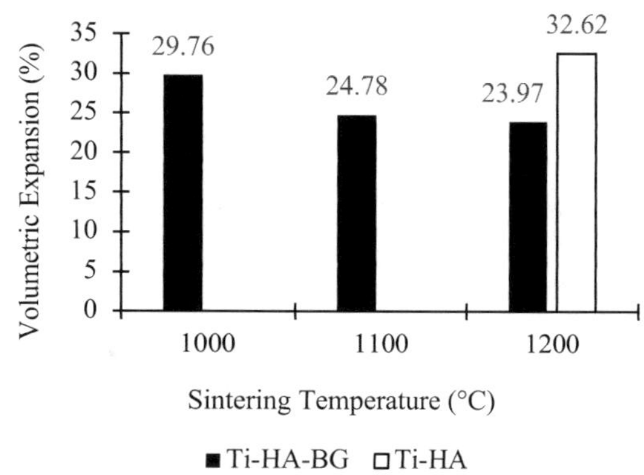

Figure 6. Volumetric expansion of Ti-HA and Ti-HA-BG composites due to sintering temperature.

4. Conclusions

In this study, the incorporation of BG into the Ti-HA system had successfully enhanced the chemical properties of the Ti-HA composite by minimizing the oxidation of Ti and decomposition of HA. Ti-HA-BG composites showed better physical properties with higher density and lower volumetric expansion compared to Ti-HA composite. Therefore, BG is a low melting point additive that acts as a good sintering aid to effectively lower the sintering temperature while maintaining the desired properties of initial components.

Acknowledgment
This research was supported by the FRGS No. FP019-2017A grant, from the Faculty of Dentistry, University of Malaya, Kuala Lumpur, Malaysia.

References

[1] Mohanty A K, Misra M, and Hinrichsen G 2000 Biofibres biodegradable polymers and biocomposites: An overview Summary of contents 1. Introduction 2. Reinforcing biofibres 2.1 Chemical constituents and structural aspects 2.2 Properties of biofibres 2.3 Degradation properties of biofibres 2.4 Cost asp *Macromol. Mater. Eng.* **276** 277 pp. 1–24
[2] Ciocca B E, Munhoz A L J, Cardoso G B C, Rodigues A A, Pattaro A F, Kaasi A, and Filho R M 2019 Viability Assays of PLLA Fibrous Membranes Produced by Rotary Jet Spinning for Application in Tissue Engineering *Brazilian Archives of Biology and Technology*, **62**
[3] Furtos G, Silaghi-Dumitrescu F L, Lewandowska K, Sionkowska and Pascuta P 2015 Biocomposites for orthopedic and dental application *Key Engineering Materials*, **672** pp. 261–275
[4] Yıldızhan Ş, Çalık A, Özcanlı M and Serin H 2018 Bio-composite materials: a short review of recent trends, mechanical and chemical properties, and applications *European Mechanical Science* **2** pp. 83–91
[5] Balbinotti P, Gemelli E, de Lima S A, de Jesus J, Camargo N H A, Henriques V A R, Soares G D A 2011 Microstructure development on sintered Ti/HA biocomposites produced by powder metallurgy *Materials Research* **14** pp. 384–393
[6] Eriksson M, Andersson M, Adolfsson E, and Carlström E 2006 Titanium - Hydroxyapatite composite biomaterial for dental implants *Powder Metallurgy* **49** pp. 70–77
[7] Ning C and Zhou Y 2008 Correlations between the in vitro and in vivo bioactivity of the Ti/HA composites fabricated by a powder metallurgy method *Acta Biomaterialia* **4** pp. 1944–1952
[8] Arifin A, Sulong A B, Muhamad N, Syarif J and Ramli M I 2014 Material processing of hydroxyapatite and titanium alloy (HA/Ti) composite as implant materials using powder metallurgy: A review *Materials and Design* **55** pp. 165–175
[9] Ning C Q and Zhou Y 2004 On the microstructure of biocomposites sintered from Ti, HA and bioactive glass *Biomaterials* **25** pp. 3379–3387
[10] Yang Y, Kim K H, Agrawal C M and Ong J L 2004 Interaction of hydroxyapatite-titanium at elevated temperature in vacuum environment *Biomaterials* **25** pp. 2927–2932
[11] Zarifah N A, Matori K A, Sidek H A A, Wahab Z A and Salleh M A M 2016 Effect of hydroxyapatite reinforced with 45S5 glass on physical, structural and mechanical properties *Procedia Chemistry* **19** pp. 30–37
[12] Abbasi Z, Bahroloolum M E, Shariat M H and Bagheri R 2015 Bioactive Glasses in Dentistry: A Review *Journal of Dental Biomaterial* **2** pp. 1–9, 2015
[13] Imran Z 2012 Bioactive Glass: A Material for the Future *World Journal of Dentistry*, **3** pp. 199–201
[14] Hench L L 2013 Chronology of Bioactive Glass Development and Clinical Applications *New Journal of Glass and Ceramics* **03** pp. 67–73
[15] Beherei H H, Mohamed K R and El-Bassyouni G T 2009 Fabrication and characterization of bioactive glass (45S5)/titania biocomposites *Ceramics International* **35** pp. 1991–1997
[16] Ye H, Liu X Y and Hong H 2009 Characterization of sintered titanium/hydroxyapatite biocomposite using FTIR spectroscopy *Journal of Materials Science: Materials in Medicine* **20** pp. 843–850
[17] Weinand W R, Gonçalves F F R and Lima W M 2006 Effect of Sintering Temperature in Physical-Mechanical Behaviour and in Titanium-Hydroxyapatite Composite Sinterability *Materials Science Forum* **530–531** pp. 249–254
[18] Wakily H, Dabbagh A, Abdullah H, Abdul Halim N F and Abu Kasim N H 2015 Improved thermal and mechanical properties in hydroxyapatite-titanium composites by incorporating silica-coated titanium *Materials Letters* **143** pp. 322–325

IOP Conf. Series: Materials Science and Engineering 1192 (2021) 012022 doi:10.1088/1757-899X/1192/1/012022

Evaluation of solid-state production of myco-coagulant using various lignocellulosic media to reduce water turbidity

M Fellah[1], M Z Alam[1,3], A Al-Mamun[2], N A Kabbashi[1] and N S B Engliman[1]

[1]Department of Biotechnology Engineering, Kulliyyah of Engineering, International Islamic University Malaysia, Gombak, 53100 Kuala Lumpur, Malaysia
[2]Department of Civil Engineering, Kulliyyah of Engineering, International Islamic University Malaysia, Gombak, 53100 Kuala Lumpur, Malaysia

[3]Corresponding Author's E-mail: zahangir@iium.edu.my

Abstract. High turbidity and suspended solids (SS) are among the significant issues that affect rivers due to the wastewater discharge, terrain condition, land cover, rainfall, agriculture, and other development activities. Chemical flocculants such as polyacrylamide and aluminum sulfate are widely employed for agro-industrial wastewater treatment. However, excessive use of chemical coagulants is harmful to human beings and the environment. Therefore, efficient and economically viable bio-coagulants from renewable biological sources is highly sought after. Myco-coagulant (My-coag) is an organic coagulant produced by fungi that are biodegradable and environmentally friendly. This research aimed to evaluate My-coag solid-state production from locally isolated fungal strains using various lignocellulosic media. The fungus was grown on different lignocellulosic substrates such as cocopeat, sawdust, rice bran, and palm kernel cake for 7 days with a pH of 7.0 at 30 °C. My-coag was extracted from the fungal culture using an aqueous buffer solution of pH 7. The fungal growth rate and dry mass were the highest on cocopeat supplement which was about 1.4 g of dry weight. My-coag extracted from the cocopeat showed good flocculating properties in kaolin suspension with the removal of 96.7% turbidity compared to other substrates such as sawdust, palm kernel cake, and rice bran with the removal of 53.7, 19.6, and 11%, respectively. It is expected that further optimization of this process parameters will lead to the efficient removal of turbidity and solids from water and wastewater to move forward in green technology for sustainable growth in the "Clean Water and Sanitation Sector" and protection of the environment too.

1. Introduction

Rivers are regarded as the primary source of water irrigation and water supply for agriculture, cities, and industries, critical for any country's long-term prosperity and development. Due to the rapid improvement of the economy and acceleration of urbanization, river pollution occurred continuously, which causes severe damage to river ecosystems [1]. High turbidity and suspended solids (SS) are significant problems affecting rivers due to the wastewater discharge, terrain condition, land cover, rainfall, soil type, agriculture, and other development activities.

Several treatment technologies are used to process raw water sources into drinking water and convert wastewater into treated effluent [2,3]. However, most water treatment processes consider coagulation and flocculation stage an essential part of the Treatment Processes, which cannot be separated. The coagulation and flocculation process are parts of a water treatment system that have the principal function of separating suspended particles in water and remove the high turbidity to produce clear and clean water [4].

The coagulation and flocculation process needs the use of composites known as coagulants and flocculants [5]. Chemical coagulants/flocculants have been traditionally used for water treatment because of their strong flocculating activity, cost-effectiveness, efficiency, and availability [6,7]. However, the use of such composites has been demonstrated to have significant side effects [8]. Several environmental issues associated with the long-term toxicity of chemical coagulants/flocculants are being addressed, especially for ecological observers worldwide [9,10]. In addition to these environmental effects, human health concerns also arise. Metal-based coagulants/flocculants are non-biodegradable or not biodegradable. Their residues in drinking water can directly affect human health if consumed and accumulate in somatic cells [10,11,12].

Many researchers are interested in biocoagulants/bioflocculants because of their biodegradability, benign qualities, and potential to replace traditional coagulants/flocculants. [10,13]. Biocoagulants are extracellular biopolymers secreted by microorganisms such as bacteria, yeasts, fungi, and algae [14]. Microbial coagulants mainly contain glycoproteins, polysaccharides, and proteins produced by microorganisms during active secretion and cell lysis [15].

Therefore, microbial coagulant has been extensively employed in removing pollutants (such as suspended solids, organic pollutants, and heavy metal ions) from wastewaters and sludge dewatering [16]. Furthermore, microbial coagulants are identified as a biotechnological tool in remediating contaminated water through the process of bio-coagulation, which is determined to be safer, economically toxic less, and environmentally friendly [17].

Conventional strategies are started using synthetic media as carbon and nitrogen source to produce biocoagulans/bioflocculants -producing microorganisms. New methods have been identified to produce microbial biocoagulant, using low-cost materials such as agricultural and industrial by-products. A variety of agricultural wastes are considered a source of carbon and nitrogen, and other nutrients that can sustain the growth of microorganisms [18]. However, bio-flocculants' commercial application and development are constrained by their high manufacturing costs, linked to the usage of expensive substrates. [19]. A suitable low-cost and abundant alternative, such as industrial wastewater and agricultural wastes, should be used to replace the synthetic substrates.

This study aimed to assess the solid-state production of myco-coagulant (My-coag) using various substrates, such as cocopeat, sawdust, palm kernel cake, and rice bran source of carbon and nitrogen instead of synthetic media. It also attempts to identify the best lignocellulosic media to produce an effective myco-coagulant that can increase the turbidity of synthetic turbid water using the Jar test.

2. Materials and method

2.1 Microorganisms
The microorganisms used in this study were fungi. The fungi were obtained from a moist area in the kitchen at Gombak, Malaysia. The fungal strain was maintained on 2% (w/v) of malt extract agar (MEA) and stored in the chiller at 4 °C until further use.

2.2 *Substate collection and preparation*
In this study, various substrates were used, including cocopeat, sawdust, palm kernel cake, and rice bran, collected from International Islamic University Malaysia, Kuala Lumpur. The substrates were ground and milled to a fine powder using a domestic blender. The fine particles that went through a 50-mesh sieve were maintained at room temperature until they were ready to be employed in solid-state fermentation (SSF).

2.3 Preparation of fungi growth media
Malt Extract Agar (MEA) was used in this study. The media was prepared according to the manufacturer s instructions. Thus, 20 g of malt extract and 15 g of agar-agar were suspended into 1 L distilled water and then sterilized at 121 °C for 15 minutes. The sterilized ME was then poured into Petri dishes. The ME plates were left almost open for solidification.

2.4 Inoculum preparation
Fungal strain obtained from the kitchen was inoculated and maintained on malt extract agar then sealed with parafilm to avoid contamination with undesirable microbes. The plate was then incubated at 30 °C for 12 days to complete the mycelial growth. This pre-culture was used as the standard inoculum preparation for further experiments.

2.5 Production of myco-coagulant
In the preparation of culture media, malt extract agar (MEA) was used and prepared according to the manufacturer's instructions. Thus, 20 g of MEA and 15 g of agar were dissolved in 1000 mL of distilled water and then autoclaved at 121 °C for 15 minutes. The sterilized MEA was then poured into petri plates contain 3 g of the lignocellulosic substrates and allowed the plates to solidify. Then a small incision of fungal culture was cut and transferred into the prepared plates. The plates were then incubated at 30 °C for 7 days to complete the mycelial growth to produce the desired coagulant.

2.6 Estimation of the fungal growth
In this study, two methods were used to estimate the fungal growth—first, continuous measurement of fungal colony diameter. Second, measure the weight of the colony (Mycelium Dry Weight). The methods were performed as follow:

2.6.1 Continuous measurement of colony diameter
Mycelial growth was measured using Image J software. Mycelial growth was first examined, with daily measurements taken for each substrate until mycelium reached the Petri dish's edge (90 mm diameter).

2.6.2 Measure the weight of the colony (Dry Weight of mycelium)
The colony and surrounding agar were removed from a petri dish and transferred to a 100 mL beaker of distilled water, which was then melted in a steam bath for 30 minutes. The mycelium that was still intact was cleaned in distilled water before being dried. The filter paper was weighed and placed on an aluminum tray, dried in an oven at 80 °C for 18 hours. After then, the filter paper was taken out of the oven. The filter sheets and mycelium were weighed after cooling to room temperature (30-20 °C), and the dry weight was calculated using the difference [20,21].

2.7 Extraction of Myco-coagulant from the lignocellulosic substrate
In this study, a fungal strain produced the myco-coagulants via solid-state fermentation (SSF). The extraction of the myco-coagulant from the biomass was conducted using an aqueous buffer solution pH 7. After 7 days of incubation, 27 mL of buffer solution of pH 7 were added to the plate and then was transferred into a flask. The flasks were shaken in the rotary shaker at 30°C and 250 rpm for 1 hour. The separation of the myco-coagulant from the biomass was done by centrifuging the culture broth at 10,000 rpm at 25 °C for 5 minutes. The clear supernatant was collected as the main myco-coagulant and used to determine their flocculating activity using the Jar apparatus method.

2.8 Evaluation of the Myco-coagulant using Jar test
The kaolin clay suspension method was used to determine the myco-flocculation coagulant's activity [22].

About 2 g of kaolin clay were suspended in 2.5 L of distilled water. Using a 2100Q HACH portable turbidimeter from the United States, the initial turbidity was measured at 736 NTU (standard technique 2130 B). Each jar contained 300 mL of kaolin suspension and 10 mL of the supernatant as My-coag was added in the jar test. The jar was set up for operation in three stages: rapid mixing at 250 rpm for 7 minutes, slow mixing at 90 rpm for 22 minutes, and finally settling for 1 hour.

The efficiency of the myco-coagulant was tested by collecting the top layer of water in each jar was. Then the flocculation activity was calculated based on the percentage removal of turbidity. All experiments were carried out in triplicate. The flocculation activity was calculated according to Equation (1).

$$\text{Turbidity removal efficiency (\%)} = \frac{\text{(A-B)}}{\text{A}} \times 100 \qquad (1)$$

whereby, A is the initial turbidity of kaolin suspension directly after preparation. Meanwhile, B is the final turbidity of kaolin suspension after the settling period, measured in the NTU unit.

3. Results and discussion

3.1 Pre-screening test for estimation of fungal growth

Estimating biomass is a critical step in a variety of microbial fermentation processes. Choosing the correct supplement can significantly increase the yield. On four distinct lignocellulosic substrates, a pre-screening test was performed in this research to monitor and compare the growth rate of fungal strains. As shown in Figure 1 on both cocopeat and sawdust, the fungal strain grew swiftly and created intact homogenous mycelium composite structures. On the other hand, fungal strains grew on palm kernel cake and rice bran slowly and had rigid mycelium structures.

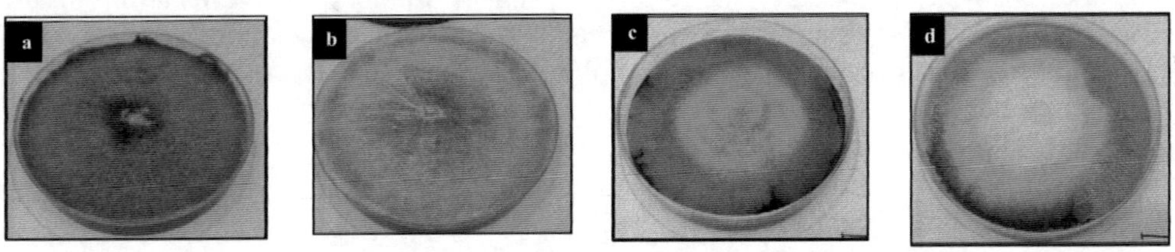

Figure 1. Culture of fungus on different substrates after seven days of incubation (a) culture of fungus on cocopeat; (b) culture of fungus on sawdust; (c) culture of fungus on PKC; (d) culture of fungus on rice bran.

3.2 Estimation of fungal growth

Biomass estimate is challenging in solid-state fermentation because the fungal mycelium frequently penetrates deep and remains attached to solid substrate particles; this research demonstrates the estimation of fungal growth. Figure 2(a) shows the growth rate of the fungus. In just seven days, mycelium in cocopeat substrate had grown to the edges of Petri dishes (90 mm). Cocopeat medium was the fastest-growing substrate in this study, growing at a rate of 1.24 cm/day. On the other side, the rice bran medium was the slowest growth rate (0.54 ± 0.01 cm/day).

The results showed the dry mass of fungus on different lignocellulosic media during the incubation period at 30 °C. which are presented in Figure 2(b) . The largest mean mycelial diameter was observed in cocopeat and sawdust media, with a mean of 9 cm and 8.8 cm of diameter and 1.4 g and 0.9 g of dry weight. On the other hand, palm kernel cake (PKC) and rice bran recorded the smallest mycelial diameter with a mean of 3.4 cm and 1 cm and 0.4 g; 1 cm and 0.1, respectively. Fungal mycelial growth is aided by the substrate, which contains modest concentrations of carbohydrates, lipids, proteins, inorganic substances, and water [23]. Physical characteristics of the substrate may reveal differences in mycelium growth [24, 25].

In addition, numerous combinations of fungal strains and substrates and various treatments used throughout the production procedures have been proposed to modify the properties of mycelium composites.

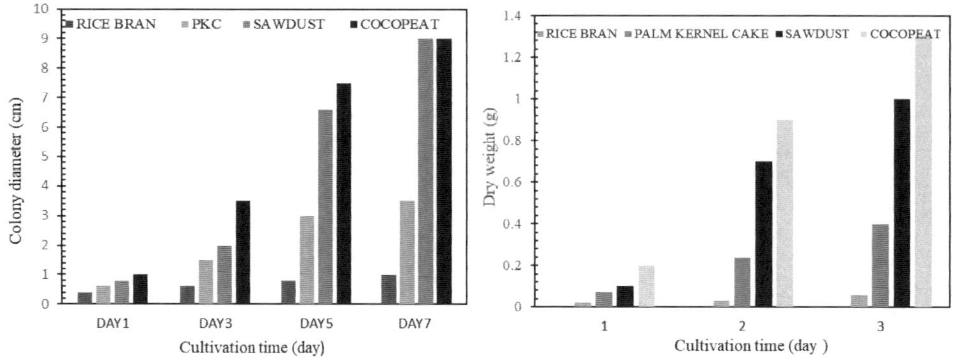

(a) (b)

Figure 2. Diameter (a) and dry weight (b) of mycelial growth of fungus on different lignocellulosic media.

3.3 *Evaluation of Myco-coagulant*

The myco-coagulant was evaluated to test its ability to remove turbidity by flocculating the particles' kaolin suspension using Jar apparatus. Table 1 shows the flocculation rate of colloid particles in terms of removal of turbidity of kaolin. In this experiment, the highest turbidity removal was indicated by the myco-coagulant extracted from cocopeat and flocculating rate of 96.1% for kaolin suspension, while the lowest turbidity removal was from rice bran of 11%. The other lignocellulosic media were 53% and 19% from sawdust and palm kernel cake, respectively. Based on the results, the fungal strain has proved its ability to produce an efficient myco-coagulant via solid-state using cocopeat as substrate in reducing turbidity removal. Luthfi, W.M. et al., (2020) reported that Aspergillus niger DWB showed a good flocculation activity rate at 60.5% from oil palm empty fruit bunch (OPEFB) fiber [26]. Among the substrates used, cocopeat provided the best carbon supply and produced a myco-coagulant with the highest flocculation activity (96.1%). The remaining substrate was removed from the remainder of the experiment.

Table 1. Removal of turbidity (%) from kaolin suspension.

Media	Removal of turbidity (%)
Cocopeat	96.1
Cocopeat control (without fungus)	16.4
Sawdust	53.2
Sawdust control (without fungus)	21.3
Palm Kernel Cake (PKC)	19.5
Palm Kernel Cake (PKC) control (without fungus)	19.2
Rice bran	11.0
Rice bran control (without fungus)	6.3

4. Conclusions

In conclusion, the fungus used in this research has demonstrated its potential to produce coagulants through solid-state fermentation. The findings of this study revealed that cocopeat was chosen as the best medium for producing an efficient myco-coagulant with the highest flocculation activity of 96.1% from kaolin suspension. In contrast, the other media sawdust, Palm Kernel Cake (PKC), and rice bran achieved 53.7%, 19.6%, and 11.0%, respectively. Thus, the results indicated a potential to produce myco-coagulants in the form of fungi using a solid-state bioprocessing system. Therefore, further optimization of the process parameters will help to reduce the turbidity of the water.

Acknowledgments
The ISDB Grant No SPI20-001-0001 provided financing for this research effort, which the authors gratefully acknowledge. The authors would also like to express their gratitude to the IIUM's Research Management Centre (RMC) for finance management and project progress monitoring and the IIUM's Biotechnology Engineering Department for scientific equipment and facilities.

References

[1] Wang J, Liu X D and Lu J 2012 Urban river pollution control and remediation. *Procedia Environmental Sciences* **13** pp 1856-1862
[2] Madaki Y S and Seng L 2013 Palm oil mill effluent (POME) from Malaysia palm oil mills: waste or resource *International Journal of Science Environment and Technology* **2** pp 1138-1155
[3] Rajasulochana P and Preethy V 2016 Comparison on efficiency of various techniques in treatment of waste and sewage water–A comprehensive review *Resource-Efficient Technologies* **2** pp 175-184
[4] Duan J and Gregory J 2003 Coagulation by hydrolyzing metal salts *Advances in colloid and interface science* **100** pp 475-502
[5] Koohestanian A, Hosseini M and Abbasian Z 2008 The separation method for removing of colloidal particles from raw water. American-Eurasian *J. Agric. & Environ. Sci* **4** pp 266-273
[6] Lapointe M and Barbeau B 2020 Understanding the roles and characterizing the intrinsic properties of synthetic vs. natural polymers to improve clarification through interparticle Bridging: A review *Separation and Purification Technology* **231** p 115893
[7] Moran S 2018 An Applied Guide to Water and Effluent Treatment Plant Design. *Butterworth-Heinemann*
[8] Barakwan R A, Trihadiningrum Y and Bagastyo A Y 2019 Characterization of alum sludge from surabaya water treatment plant, Indonesia *Journal of Ecological Engineering* **20**
[9] Imron M F, Kurniawan S B, Ismail N I and Abdullah S R S 2020 Future challenges in diesel biodegradation by bacteria isolates: a review *Journal of Cleaner Production* **251** p 119716
[10] Tomljenovic L 2011 Aluminum and Alzheimer's disease: after a century of controversy, is there a plausible link? *Journal of Alzheimer's Disease* **23** pp 567-598
[11] Kluczka J, Zołotajkin M, Ciba J and Staroń M 2017 Assessment of aluminum bioavailability in alum sludge for agricultural utilization *Environmental monitoring and assessment* **189** pp 1-8
[12] Mortula M, Bard S M, Walsh, M E and Gagnon G A 2009 Aluminum toxicity and ecological risk assessment of dried alum residual into surface water disposal *Canadian Journal of Civil Engineering* **36** pp127-136
[13] Al-Mutairi N Z 2006 Coagulant toxicity and effectiveness in a slaughterhouse wastewater treatment plant. *Ecotoxicology and environmental safety* **65** pp 74-83
[14] Aljuboori A H R, Idris A, Abdullah N and Mohamad R 2013 Production and characterization of a bioflocculant produced by Aspergillus flavus *Bioresource technology* **127** pp 489-493
[15] Aljuboori A H R, Uemura Y, Osman N B and Yusup S 2014 Production of a bioflocculant from Aspergillus niger using palm oil mill effluent as carbon source *Bioresource technology* **171** pp 66-70
[16] Gao B Y, Wang Y, Yue Q Y, Wei J C and Li Q 2007 Color removal from simulated dye water and actual textile wastewater using a composite coagulant prepared by ployferric chloride and polydimethyldiallylammonium chloride *Separation and Purification Technology* **54** pp 157-163
[17] Mu J, Cui X, Shao M, Wang Y, Yang Q, Yang G and Zheng L 2019 Microbial origin of bioflocculation components within a promising natural bioflocculant resource of Ruditapes philippinarum conglutination mud from an aquaculture farm in Zhoushan, China *PloS one* **14** p e0217679

[18] Siddeeg S M, Tahoon M A and Ben Rebah F 2020 Agro-industrial waste materials and wastewater as growth media for microbial bioflocculants production: a review *Materials Research Express* **7** p 012001

[19] Zhao G, Ma F, Wei L, and Chua H 2012 Using rice straw fermentation liquor to produce bioflocculants during an anaerobic dry fermentation process *Bioresource technology* **113** pp 83-88

[20] Paster N, Lisker N and Chet I 1983 Ochratoxin A production by Aspergillus ochraceus Wilhelm grown under controlled atmospheres *Applied and Environmental Microbiology* **45** pp 1136-1139

[21] Zill N and Schoenborn C A 1990 *Developmental, learning, and emotional problems: Health of our nation's children, United States, 1988* (No. 190) US Department of Health and Human Services, Public Health Service, Centers for Disease Control, National Center for Health Statistics

[22] Gomaa E Z 2012 Production and characteristics of a heavy metals removing bioflocculant produced by Pseudomonas aeruginosa *Pol. J. Microbiol* **61** pp 281-9

[23] Krishna, S. Quantum 2005 dots-in-a-well infrared photodetectors *Journal of Physics D: Applied Physics* **38.13** 2142

[24] Karimi S, Mahboobi Soofiani N, Mahboubi A and Taherzadeh M J 2018 Use of organic wastes and industrial by-products to produce filamentous fungi with potential as aqua-feed ingredients *Sustainability* **10** p 3296

[25] Mazur R 2015 Mechanical properties of sheets comprised of mycelium *A paper engineering perspective*

[26] Luthfi W M, Alias H, Tay G S and Lee C K 2020 Production and characterization of bioflocculant via solid state fermentation process using oil palm empty fruit bunch fibers as substrate *Biocatalysis and Agricultural Biotechnology* **23** p 101454

IOP Conf. Series: Materials Science and Engineering 1192 (2021) 012023 doi:10.1088/1757-899X/1192/1/012023

Simulation of biomass and municipal solid waste pellet gasification using Aspen Plus

A D D Diallo[1], M F R Alkhatib[1,2], M Z Alam[1], M Mel[1]

[1]Bioenvironmental Engineering Research Centre (BERC), Department of Biotechnology Engineering, Faculty of Engineering, International Islamic University Malaysia, Jalan Gombak, 53100 Kuala Lumpur, Malaysia.

[2]Corresponding Author's E-mail: maan@iium.edu.my

Abstract. The work deals with the simulation of biomass and municipal solid waste pellet gasification using Aspen Plus software. The effects of key parameters on the composition of the emitted gas are discussed, including gasification temperature, moisture content, and equivalence ratio. The sensitivity analysis was studied with the Aspen Plus Software, which includes FORTRAN modules. The simulation is validated using experimental results, which revealed that it was roughly correct. Using air as the gasification agent, the sensitivity analysis findings confirm higher temperatures promote syngas production with increased hydrogen and energy content. The simulation results demonstrated that CO_2 concentration (3.95%) increases from 450°C to 600°C and then decreased drastically near 0.225kmol/hr. at 900°C. As the gasification temperature rises from 450°C to 900°C, the CO concentration rises and the H_2: CO ratio falls. At 900°C, increasing the gasification temperature results in a product gas with more H_2 (65%) and CO (12.43%), resulting in a higher calorific value, whereas the contents of CH_4, CO_2, and H_2O followed an inverse correlation. CH_4 decreased with temperature because of the formation of exothermic methane reactions. When the gasification process reaches 800°C, all components except CO_2 become steady, and gasification reactions were achieved. The equivalence ratio (ER) ranged from 0.2 to 0.3. The gas produced by a gasifier is highly dependent on the ER value. The ER determines the gas quality, and it must be less than 1 to ensure that it gasifies the fuel rather than burnt. Moisture content was 10wt. %, this is an essential parameter for the optimum conditions during the gasification process. Moisture content determines the gas characteristics at the exit phase. The model predictions and calculated values are in good agreement.

1. Introduction

The world's growing energy demands, as well as the production of a large amount of biomass waste, have fueled the development of biomass gasification technology [1]. Furthermore, the lack of fossil fuels and environmental problems such as greenhouse gas emissions, deforestation, and other health-related issues, is driving the hunt for alternative energy sources.

To replace fossil fuel, it is necessary to find alternative energy sources that are abundant, inexpensive, and have a low environmental impact. Biomass was suggested as a solution to these problems because it is green, easy to find, and long-lasting. The amount of CO_2 absorbed from the atmosphere during photosynthesis of green biomass is nearly equal to the CO_2 produced during biomass combustion [2].

Gasification of municipal solid waste (MSW) and biomass waste offers significant opportunities for energy security and climate change mitigation. To avoid pollution of the environment due to waste accumulation, some developed countries encouraged the conversion of municipal solid waste into energy [3]. Thermochemical conversions such as combustion, pyrolysis, and gasification, as well as anaerobic digestion, were used to convert MSW to energy. The appropriate gasification operating parameters are also investigated, including gasification temperature, moisture content, and equivalence ratio. For this objective, a simulation model is developed, and the results are compared to published experimental data [4,5].

This study presents the gasification of the mix of MSW and empty fruit bunch (EFB) using the Aspen Plus simulator. The aim of this study was to use Aspen plus to predict the impact of key variables on gasification, such as temperature, the equivalence ratio, moisture content, and syngas caloric value).

2. Methodology

2.1. Development of a Gasification Model

The Aspen Plus gasification model was used to transform a mixture of municipal solid waste and biomass waste feedstock into refuse-derived fuel pellets using a solid-based simulation. The following steps are involved in the development of a model in the simulator: The definition of the process flow-sheet, the description of the stream class, the choice of the property approach, the configuration of software packages (via a database), and the identification of conventional and non-conventional components (via unit operation blocks and linking material and energy flows) [6].

The gasification process includes drying (lowering the moisture content of fuel to 10% prior to supplying it to the reactor), pyrolysis (the raw material was decomposed into volatile compounds and char), and then gasification occurs.

2.2. Model assumptions

The following are the model's fundamental assumptions:

1. Free kinetic model with a steady-state.
2. The (MSW+EFB) particles are a uniform size and temperature.
3. The model contains only ideal gases. To reduce the hydrodynamics complexity the influence of tar and other heavy products is not considered.
4. All sulfur is converted to H_2S.
5. There are no nitrogen oxides produced, only NH3, [1, 7].

Table 1 depicts the models of ASPEN Plus unit operation.

Table 1. ASPEN Plus unit operation features.

ASPEN PLUS ID	ID of the block	Description
RStoic	DRIER	Reduces the fuel's moisture content
RYield	DECOMP	Yield reactor- uses FORTRAN statements to convert non-conventional biomass into conventional components.
RGibbs	GASIFIER	Gibbs free energy reactor: By reducing Gibbs free energy, this reactor deals with three-phase equilibrium and estimates synthesis gas mixture.
Sep	SEPARATOR	Split fractions are used to remove gases from ash.

2.3. Physical property method

The PR-BM property approach was used for the Peng-Robinson (PR) equations. The Boston-Mathias (BM) alpha formula is used in the PR-BM property method to measure physical properties of the conventional components in the gasification.

The temperature-dependent variable alpha is included in this property package. Hydrocarbons and light gases like carbon dioxide, hydrogen sulphide, and hydrogen are examples. This property approach is very useful in high-temperature, high-pressure environments, such as hydrocarbon processing or supercritical extractions. HCOALGEN and DCOALIGT were used as the enthalpy and density models for the non-conventional components (MSW, biomass, and ash).

2.4. Description of the Model

The gasification process simulation model was created on Aspen Plus software using three (3) main reactor types: Stoichiometric Reactor (RStoich), the Yield Reactor (Ryield), and the RGibbs Reactor. The process steps are illustrated in Figure 1. The mixture (Biomass and MSW) is developed in the simulation model utilizing the proximal and ultimate analyses and is specified as a non-conventional component in ASPEN Plus.

Table 2 contains the data used to describe the feedstock. The 'RStoic' block was used to simulate the drying process (MW and biomass). Evaporation is used in this phase to reduce the moisture content of the feedstock Furthermore, the drying process has been regulated by asserting in a calculator block and describing the raw material drying process using FORTRAN modules. To all non-conventional elements, Aspen assigns a molecular weight of 1. The entire production of volatiles is assumed to be equal to the parent fuel's volatile content, whereas the average production of carbonaceous materials is proportional to the carbon content and ash concentrations calculated through proximate analysis.

Biomass particles are heated to 100-150 °C in a gasifier, dried, and the moisture is removed as steam. Because gasification is an endothermic reaction, heat is delivered to the particle, which is heated to 500 °C, knocking off all volatiles. During this process, approximately 70-80% of the dry biomass weight is lost. Char is the solid biomass that is left over after the combustion process. In the presence of heat and a medium agent, this char gasifies, releasing the second collection of gases. After the drying stage, the decomposition process model (Ryield block) begins, converting non-conventional feed components, in this case, a mixture of (MSW) and biomass, into conventional compounds (C, H, N, O, S, and ash). This is accomplished by employing the RYield model and performing calculations in the calculator block based on part output stipulations (FORTRAN). Because it takes less time and releases around 80-90% of the mass as steam, devolatilization is a significant step in gasification. Unlike pyrolysis, gasification focuses on boosting gas output. As a result, the operating conditions for gasification have been improved to allow the release of volatiles from biomass [8].

The converted elements are ready for gasification after decomposition. Furthermore, in this block, the air stream introduces air or steam as a medium. The RGibbs reactor model, which is rigorous and can measure chemical and phase equilibrium by minimizing the system's Gibbs free energy, is used to simulate this gasification step. The emitted gas passes through another separator block to extract the ash as slag from the GASIFIED stream as it exits the reactor (SEP2).

The RGibbs reactor receives the decomposed biomass and air, which undergoes partial oxidation and gasification reactions. H_2S is formed when all of the sulfur in the feedstock reacts with H_2. Because of the low sulphur content of fuel, the discrepancies of this simplification are negligible.

Other researchers [9-12], have assumed that just NH_3 is formed and no nitrogen oxides are generated. The unit activity model Sep was used to simulate the separation of ash from syngas. The RGibbs reactor's outlet stream enters the Sep block, which uses split fractions to separate gases from ash. Figure 1 depicts the Aspen Plus process flow for municipal and biomass mixture gasification.

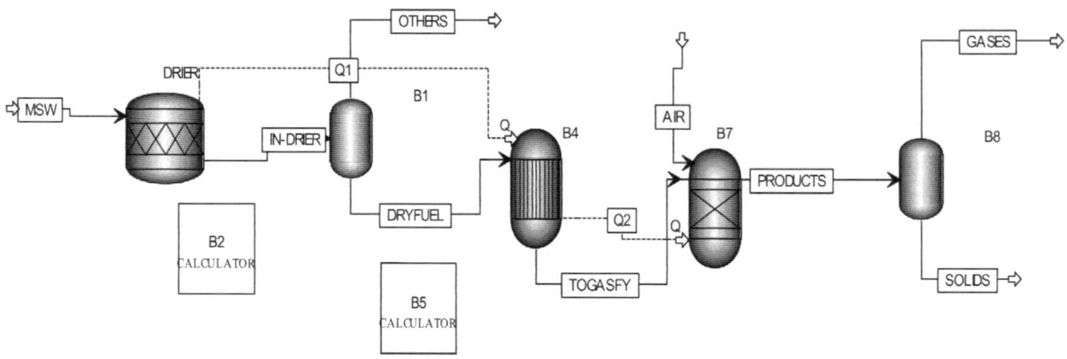

Figure 1. Aspen Plus process flow for municipal and biomass mixture gasification.

The analysis of feedstock conditions at the ultimate and proximate levels as well as the gasification process operating parameters are shown in Tables 2 and 3.

Table 2. Analysis of Feedstock Conditions at the Ultimate and Proximate Levels.

Municipal solid waste + Empty fruit bunch			
Proximate analysis (mass %)		Ultimate analysis (mass %)	
FC	85.02	C	52.50
VM	2.53	H	7.95
MC	8.23	O	34.91
ASH	4.22	N	0.33
		S	0.07

Table 3. Gasification Process Operating Parameters.

Parameter	Value	Unit
Feed rate	1000	Kg/h
Steam/ biomass ratio	0.5-1	
Feed Temperature	25	°C
Feed Pressure	1	atm
Air Pressure	1	atm
Air Temperature	25	°C
Gasifier Temperature	450-900	°C
Gasifier Pressure	1	atm

2.5. Validation of the model
Three sets of experimental results from pilot-scale fluidized bed gasification processes published in the literature were used to validate the simulation results. The design can forecast gasifier efficiency under a variety of conditions, such as temperature, and equivalence ratio, among others.

3. Results and Discussions
In general, there are two types of gasification processes. Basic pyrolysis is the early stages, which involves dissolution of municipal solid waste to generate gases, ash, and char. In the second phase, homogeneous and heterogeneous reactions take place in a reducing atmosphere [13].

As a function of gasification temperature, Figure 2 shows the generated gas molar fraction. The actual content of the output will be influenced by the temperature of the gasifier. Le Chatelier's principle states that higher temperatures favor endothermic reaction products. Endothermic reactions occur in a number of significant reactions within the gasifier. Moreover, as shown in Figure 2, increasing the combustion temperature results in the production of produced gas with increased H2 and CO concentrations, and hence a calorific value. The contents of CH_4, CO_2, and H_2O follow the opposite result, with CH_4 decreasing with the temperature due to the exothermic nature of methane reactions. Furthermore, at 800 °C as the gasification temperature, all components except CO_2 become stable, and the gasification reaction tends to complete. The findings show that as the temperature rises from 450 °C to 650 °C for H_2, the contents of Hydrogen and CO enhance. As a result, two reactions (R1) and (R2) increase H2, while (R3) and (R4) increase CO. The research results are similar to those reported in [14, 15].

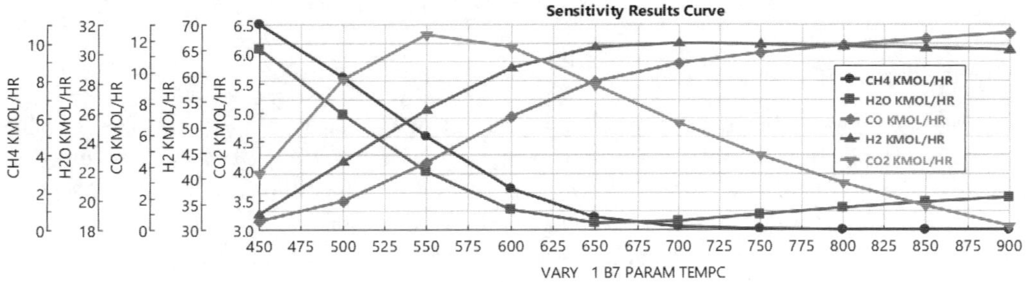

Figure 2. The influence of temperature on syngas composition.

Higher temperatures promoted hydrogen production and gas yield in general, but not always gas heating value, as too high a temperature decreased gas heating value. The quality of syngas degrades as moisture content rises, and gasifier temperatures have been found to have a greater impact on syngas composition [16]. Endothermic reactions, such as the Boudouard reaction, the water gas reaction, and steam methane reforming, prefer forward reactions as the gasification temperature rises.

$C + 0.5O_2 \leftrightarrow CO - 111$ MJ/kmol	Partial oxidation	R1
$C + H_2O \leftrightarrow CO + H_2 + 131$MJ/kmol	Water gas reaction	R2
$CO + H_2O \leftrightarrow CO_2 - 41$MJ/kmol	water gas shift	R3
$C + CO_2 \leftrightarrow 2CO_2 + 172$MJ/kmol	Boudouard	R4
$C + 2H_2 \leftrightarrow CH4 - 74$MJ/kmol	methanation	R5

Figure 3 demonstrates the effect of temperature on hydrogen and CO ratio.

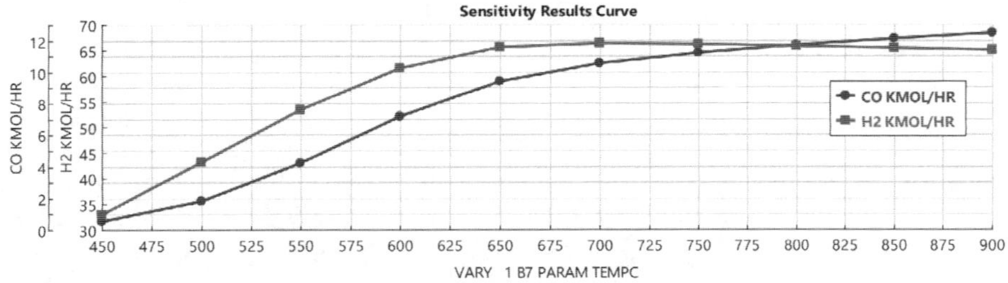

Figure 3. Effects of temperature on hydrogen and CO ratio.

As the gasification temperature increases, the H_2: CO ratio drops (refer to Figure 3). According to [17], the steam to biomass ratio (SBR) is an important process factor in the gasification of steam. The fundamental benefit of using steam as an agent is that it enriches the molecular fraction of H_2 and increases the concentration of H_2O, resulting in more water shift reactions.

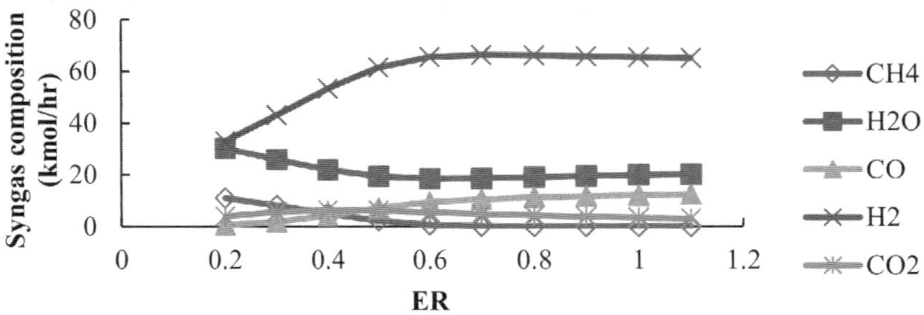

Figure 4. Effects of ER on the composition of syngas.

Suitable temperature range for a maximum amount of H_2 and CO is 650 to 900 °C. There is a similar variation between CO_2 and H_2O. Moreover, the equivalence ratio (ER) and steam biomass ratio (SBR) rise with the CO_2 and H_2O increase. CO_2 increases with temperature to a maximum value at 550 °C and later starts decreasing. While H_2O decreases, with an increase in temperature, reaches its minimum value at 650 °C, and next starts increasing. CO_2 and H_2O reach their peak at the same temperature interval. Increased ER speeds up the oxidation reaction and lowers the quality of product gas. A high ER, on the other hand, results in low H_2 and CO concentrations in the product gas, as well as a high CO_2 content. [18, 19] investigated the effect of ER on biomass air gasification in a fluidized bed gasifier from 0.15 to 0.35, and from 0.1 to 0.9 [1]. They discovered that as ER increased, H_2 and CO increased at first and then decreased, whilst CO_2, CH_4, and other hydrocarbons' content was gradually reduced, which is similar to our findings. Table 4 depicts a comparison of modeling and experimental results.

Table 4. A comparison of modeling and experimental results.

Temperatures	550 °C[18]		650 °C[18]		750 °C [5]		900 °C [6]	
Gas composition	vol%	Error %	vol%	Error %	vol%	Error %	vol%	Error %
H_2 Simulation	53.34	1.94	65.47	4.14	66.14	1.75	65.07	1.21
H_2 Experiment	54.4		68.3		65		65.87	
CO Simulation	4.19	7.43	9.37	14.8	11.19	15	12.43	12.8
CO Experiment	3.9		11		14		14.26	
CO_2 Simulation	6.31	7.86	5.46	16.51	4.26	19.3	3.06	4
CO_2 Experiment	5.85		6.54		5.28		3.2	
CH_4 Simulation	4.99	9.27	0.66	20	0.004	20	0.001	50
CH_4 Experiment	5.5		0.55		0.005		0.002	

By comparing the composition of the product gas generated from this work's constructed model to the data published in the literature by [5-6,18] the gasification model was validated. At the gasification temperatures of 550, 650, 750 and 900 °C, the mole fractions of gas product were compared. Our simulation results, as shown in Table 4, are comparable to literature values, and the main component errors are low, except CH_4. This implies that the chemical equilibrium model established in this study is reliable and can be used to simulate the gasification of MSW and EFB mixtures.

4. Conclusion

Variations in gasifier temperature, ER, and moisture content were tested to see how they affected syngas composition, HHV, and hydrogen generation. As the temperature of the gasifier rises, so does the generation of carbon monoxide and hydrogen. Higher moisture content affects the HHV and degrades gasifier performance. The simulation results were validated using three sets of experimental data from pilot-scale fluidized bed gasification systems described in literature. In the current investigation, the resultant syngas has a substantially larger mole fraction of H_2 than CO, and the operating conditions are very comparable to the results being refered. The major changes are found in the parameters of the raw material, specifically the moisture content (10 wt.%) and (12-48 wt %) in the other studies. It also has very similar working conditions to those from previous studies. The model predictions and the calculated values were in good agreement.

References

[1] Ramzan N, Ashraf A, Naveed S and Malik A 2011 *Biomass and Bioenergy* **35** 3962–3969
[2] Yaman S 2017 *Prediction of Calorific Value of Biomass from Proximate Analysis* **107** 130–136
[3] Ding S, Li Z, Zhou H, Hu Y and Liu L 2019 *Energy Sources, Part A: Recovery, Utilization and Environmental Effects* **41** 1640–1648
[4] Ding G and He B 2020 *Applied Sciences (Switzerland)* **10**
[5] Mahmut C and Acar Y E B 2018 *14th International Combustion Symposium* 25–27
[6] Pala L P R, Wang Q, Kolb G and Hessel V 2017 *Renewable Energy* **101** 484–492
[7] Hlaba A 2018 *Process Simulation of Municipal Solid Waste Derived Pellet Gasification for Fuel Production* IEEA **18** 59–64
[8] Kaushal P and Tyagi R 2017 *Renewable Energy* **101** 629–636
[9] Rupesh S, Muraleedharan C and Arun P 2016 *Resource-Efficient Technologies* **2** 94–103
[10] Formica M, Frigo S and Gabbrielli R 2016 *Energy Conversion and Management* **120** 358– 369
[11] Niu M, Huang Y, Jin B and Wang X 2013 *Industrial and Engineering Chemistry Research* **52** 14768–14775
[12] Dhanavath K N, Shah K, Bhargava S K, Bankupalli S and Parthasarathy R 2018 *Journal of Environmental Chemical Engineering* **6** 3061–3069
[13] Dong J, Chi Y, Tang Y, Ni M, Nzihou A and Weiss-hortala E 2018 *Energy Sources* **41**
[14] Li J, Liao S, Dan W, Jia K and Zhou X 2012 *Biomass and Bioenergy* **46** 174– 180
[15] Dong J, Chi Y, Tang Y, Ni M, Nzihou A, Weiss-Hortala E and Huang Q (2016) *Energy and Fuels* **30** 3994–4001
[16] Shehzad A, Bashir M J K and Sethupathi S 2016 *Renewable and Sustainable Energy, Reviews* **60** 1302–1311
[17] Tavares R, Monteiro E, Tabet F and Rouboa A 2020 *Renewable Energy* **146** 1309–1314
[18] Ahmad A A, Zawawi N A, Kasim F H, Inayat A and Khasri A 2016 *Renewable and Sustainable Energy Reviews*, ***53***, 1333–1347.
[19] Mohammed M A A, Salmiaton A, Wan Azlina W A K G, Mohammad Amran M S and Fakhru'L-Razi A 2011 *Energy Conversion and Management* **52** 1555-1561

Adsorption of Carbon Dioxide by Metal Organic Framework for Indoor Air Quality Enhancement

K.C. Chong[1], **S.S. Lee**[1], **S.O. Lai**[1], **H.S. Thiam**[1], **P.S. Ho**[1], **W.J. Lau**[2]

[1]Lee Kong Chian Faculty of Engineering and Science, Universiti Tunku Abdul Rahman, Jalan Sungai Long, Bandar Sungai Long, 43000 Kajang, Malaysia.
[2]Advanced Membrane Technology Research Centre (AMTEC), Universiti Teknologi Malaysia, 81310 Skudai, Johor, Malaysia.

Abstract. Air pollution has become a severe environmental issue among millions of people around the globe. However, the risk of exposure to indoor air pollution is much higher than outdoor air pollution. The most effective way to improve indoor air quality (IAQ) by reducing the indoor CO_2 content is by capturing and storing. There are several types of adsorbents used to capture CO_2, namely physical adsorbents and chemical adsorbents. Metal-Organic Framework (MOF) is one of the recent interests arising physical adsorbents which possesses high adsorption capability. In this study, MOFs fabricated with different metals and organic ligands were used to evaluate their performance in CO_2 adsorption under an enclosed office space. Magnesium, chromium, and copper metals were used as the main element in the MOF fabrication coupled with trimesic acid as an organic ligand. The MOFs' morphologies generally illustrated that magnesium MOF exhibited a dispersed nanorod flask crystal, chromium MOF showed agglomeration crystal, whereas fine crystal rod was observed in copper MOF. The elemental analysis from EDX and XRD confirmed that the metals were successfully embedded with the organic ligand, which is similar to the literature studies. The CO_2 gas adsorption study suggested that magnesium MOF fabricated with trimesic acid possess superior CO_2 adsorption capability as the recorded CO_2 concentration reduced from 960 ± 73 ppm to 895 ± 57 under 2 hours continuous sampling time. The CO_2 adsorption study reveals that the magnesium MOF with trimesic acid ligand yields a promising result on indoor CO_2 concentration reduction. This result suggested that the MOF possesses a great potential to be applied in the indoor air quality enhancement with the integration to the existing air purification and/or filtration system.

1. Introduction

Air pollution has become a serious environmental issue among millions of people around the globe. Modern people spend about 80% of their time indoors on average such as work in the office, school, residential, or even sport inside a building. However, the risk of exposure to danger for indoor air pollution is much higher than outdoor air pollution [1]. Sick building syndrome occurs when there is a poor level of indoor air quality (IAQ) [2]. The symptoms of sick building syndrome are watery eyes, headaches, dizziness, coughing, sneezing, fatigue and upper respiratory congestion. This might bring adverse effects to the respiratory or human health effects such as lung disease, worsening asthma, pharyngitis, epistaxis, nasal congestion, fever, vomiting, myalgia, hearing loss, conjunctival irritation and even affect the working performance and increase of absenteeism [3]. Therefore, IAQ play a vital role in both productivity and human health

One of the indoor air pollutants that will lead to poor IAQ and human health is CO_2. The increment of CO_2 concentration in the indoor air will lead to dizziness and difficulty to breathe. Additionally, a high concentration of CO_2 also will lead to a sharp and acidic odour in the room. Because of this, the National Institute of Occupational Safety and Health Malaysia (NIOSH) recommends the threshold CO_2 concentration in a room is below 1,000 ppm and the effect of CO_2 indoor levels for better IAQ [4]. The effect of the CO_2 mentioned had lead to servious health effect such as sick building syndrome and respiratory related disease which will led to absentism to work and school.

CO_2 could be reduced by the solid adsorbents, namely zeolite, activated carbon, and MOF. Zeolites are crystalline materials that are built up of silicon, aluminum, and oxygen in an ordered manner. The atoms formed a rigid framework and pore between them, such as a honeycomb-like structure. The ratio of silicon to aluminum can be adjusted to produce different zeolites for specific applications [5]. Next, activated carbons are amorphous solids made from carbonaceous materials. They have a high surface area and the capacity to serve as an adsorbent. Due to their non-polarity surface, the non-polar organic compounds tend to attach to their surface or pores [6]. MOFs are newly developed adsorbent which is a crystalline solids that have high porosity and tunable, which received significant attention from the industry [7] . The comparison of three solid adsorbents shows that the surface area of MOFs (500 to 10,000 m^2/g) is significantly higher than the surface area of zeolite (720 m^2/g) and activated carbon (214 m^2/g) [8 – 10]. The surface area of the adsorbent is proportional to the amount of adsorbate that will be adsorbed. This enables MOF to adsorb many more CO2. Besides, MOFs also have greater capacity and pore size (1.5 to 33.5 mmol/g, 2.5 to 11.1 Å) respectively relative to zeolite (0.5 to 8.0 mmol/g, 3.0 to 10.0 Å) and activated carbon (0.3 to 6.3 mmol/g, 3.0 to 100.0 Å) for CO2 adsorption [8 – 10]. In this study, three different types of MOF made from magnesium nitrate hexahydrate, chromium (VI) oxide and copper (II) nitrate trihydrate with trimesic acid were fabricated in a laboratory scale and test for its carbon dioxide gas adsorption.

2. Methodology

2.1. Materials

The magnesium, chromium and copper metal organic framework was made by magnesium nitrate hexahydrate, chromium (VI) oxide and copper (II) nitrate trihydrate, respectively. The organic ligand utilized in this study was trimsestic acid (TMA) and co-solvent used was ethanol. All the chemicals used in this study were purchase from Sigma Aldrich.

2.2. Fabrication of the metal organic framework

The metal organic framework in this study was carried out by the solvothermal method. The metal oxide (either magnesium, chromium or copper) was first dissolved in the distilled water, subsequently, the organic ligand (trimesic acid) and ethanol as co-solvent. The molar ratio of the mixture for the metal oxide and organic ligand is at the ratio of one to one in this work. The solution was stirred by an overhead mechanical stirrer to obtain a homogenous solution. The mixture is then placed in the Teflon line autoclave and heated in the vacuum oven at 80°C for three days. Further, the mixture was taken out from the vacuum oven and cooled at room temperature to obtain a crystal metal organic framework. The dried crystal shall keep in a vacuum cabinet to maintain its dryness before the gas adsorption test. The magnesium, chromium and copper metal organic framework used in this study are labeled as Mg-TMA, Cr-TMA and Cu-TMA for ease of reference.

2.3. Characterization

In this study, various characterization was applied to study the physical and chemical properties of the fabricated MOF. The surface morphology and chemical composition of the fabricated MOF were analyzed by scanning electron microscopy (SEM) and energy dispersive X-ray (EDX). SEM is a material characterization technique that will produce a clear and high-resolution image of sample size and surface morphology. The SEM is carried out under an accelerated voltage of 12kV. A small amount of powdered MOF samples on the sample holder and the samples are fixed on the holder by using carbon tape. The surface of samples is required to coat with a thin layer of gold by a sputter-coated machine to increase the conductivity of the samples and the small grain size of gold enables a clearer image. Energy dispersive x-ray (EDX) is an x-ray technique used to identify the MOFs elemental composition. The signals produced from SEM are the characteristics of x-rays. X-rays generate in SEM begin from the energy produced when the electron beam hits the surface of samples and so the atom electrons use the energy to jump to an energy shell. Therefore, the electron leaves and leads to a positive charge hole which will then attract the negative charge of electrons from a higher energy shell. Hence, the energy difference between two shells (the hole with low energy shell fills by high energy shell) is produced in the form of an x-ray which can be measure by EDX.

2.4. Carbon dioxide gas adsorption study

An experiment is set up as shown in Figure 1 to measure the CO_2 adsorption performance of the fabricated MOFs. Before the gas adsorption experiment, the mass of the empty teabag and the initial mass of MOF sample was weighed and recorded. Furthers, weighed and recorded the mass of the teabag with MOF again after the weighted MOF sample was placed into an empty tea bag. Then, the teabag with MOF sample was placed in a hollow stainless-steel cylinder tube which both ends of the tube were sealed tightly. This is to prevent the leaking of CO_2 gas when the CO_2 gas flows into the cylinder tube via a pipe for approximately two hours. Later, the sample MOF has sufficient time to adsorb the supplied CO_2 in a closed system. The inlet and outlet concentration of CO_2 were recorded every 5 minutes to get the difference which then can be used to calculate the CO_2 adsorption capacity of the MOF. The best performance MOF from the CO_2 adsorption study will be then used to evaluate the CO_2 reduction in an indoor enclosed volume of approximately 28.3m^3. The initial CO_2 concentration was measure by a CO_2 meter and the indoor air will be drawn into the stainless-steel vessel containing MOF sample. The CO_2 concentration outlet of the stainless steel vessel will be measured and the difference in initial and outlet CO_2 concentration will be the rate of CO_2 adsorbed (Figure 2).

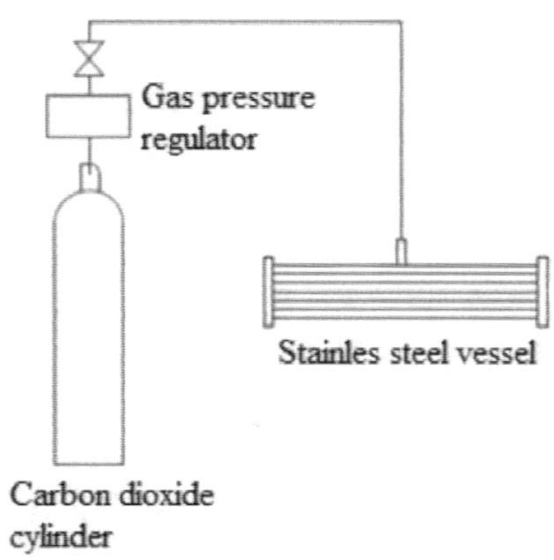

Figure 1. Carbon dioxide gas adsorption setup

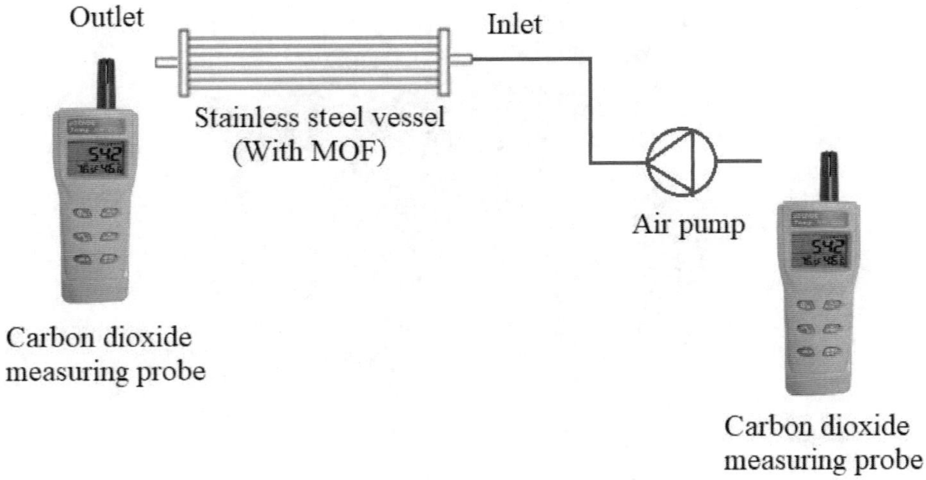

Figure 2. Indoor carbon dioxide gas adsorption setup

3. Results and Discussions

3.1. Morphologies of Metal Organic Framework

Mg-TMA possesses dispersed nanorod flask crystal as exhibited in Figure 3(a). These structures formed due to simultaneous nucleation and crystal growth under solvothermal method [11]. Tahmasian, Morsali and Joo (2013) reported the size of these structures is dependent on the concentration that produces by initial reagents. In the case of a higher concentration of initial reagents, uniform morphology and smaller size of inhomogeneous nanorods can be obtained [12]. Cr-MOFs exhibited a large amount of agglomerate crystal structure and the size of crystal structures of Cr-MOFs was relatively small which were estimated in the range of 100 micrometers to 5 micrometers (Figure 3(b)). Lower crystallinity of Cr-TPA can be seen when the molar ratio of the organic ligands increases. The surface area of the MOFs to volume ratio becomes lower when there is large agglomeration of the crystal structure which may hinder the good gas adsorption capabilities [13, 14]. Cu-TMA exhibit coarse crystal rod of Cu-TMA have been observed (Figure 3(c)). The formation of this coarse crystal structure was attributed to the interaction and the overall stoichiometry of Cu-MOFs might be affected by few factors such as the component concentration and the purity, counter ions size and the reaction conditions such as temperatures, solvent and mixing rates [15, 16].

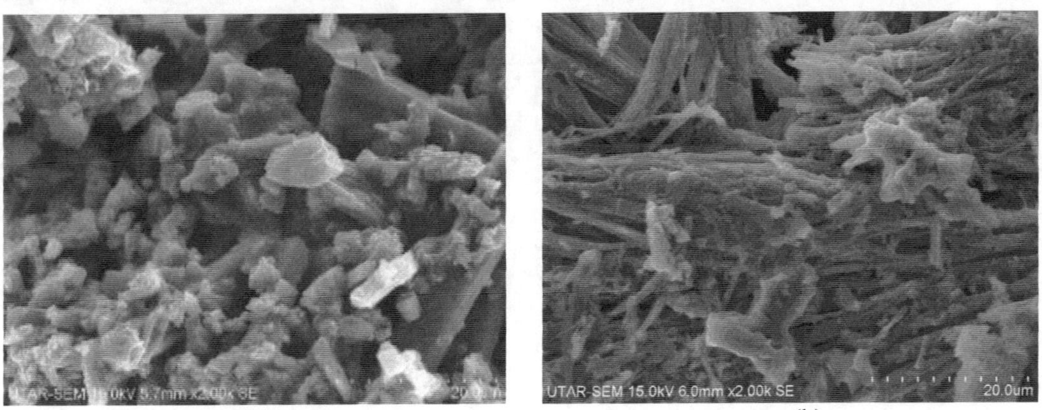

(a) (b)

(c)

Figure 3. SEM image of (a) Mg-TMA, (b) Cr-TMA and (c) Cu-TMA

3.2. EDX Analysis

The elemental content of the fabricated MOF was analyzed by the EDX techniques (Table 1). Given the irregularity of the samples, the measurement of the weight percentage was calculated by the average of 5 sampling spots from samples to ensure reproducibility. The fabricated MOF were suggested to be successful based on the result from the EDX analysis. The element of magnesium, chromium and copper were detected in the Mg-TMA, Cr-TMA and Cu-TMA, respectively, suggesting the presence of the metal in the fabricated MOF. Additionally, the occurrence of the carbon and oxygen in all the MOF suggested that the organic ligands were successfully linked to the metal oxide to form MOF.

Table 1. EDX analysis

MOFs	Carbon	Oxygen	Magnesium	Chromium	Copper
Mg-TMA	34.59 ± 1.73	53.67 ± 3.48	11.74 ± 0.93	Not detected	Not detected
Cr-TMA	49.54 ± 2.91	48.97 ± 2.87	Not detected	11.49 ± 0.83	Not detected
Cu-TMA	39.95 ± 2.43	43.39 ± 3.02	Not detected	Not detected	16.66 ± 0.71

3.3. Carbon Dioxide Gas Adsorption Study

Mg-TMA sample shows the highest CO_2 adsorption capacity recorded at 4.6 mmol/g followed by Cu-TMA and Cr-TMA in this study (Table 2). The good performance recorded by Mg-TMA might be attributed to the number of open metal sites in MOF framework greatly affect CO_2 adsorption capacity which means that bonding between Mg metal sites in the MOF with CO2 molecule that acts as adsorbate molecule is stronger than the other metal such as Cu and Cr. This is because the bonding of the ionic character of the Mg-O bond can trap the CO_2 into MOF easily. Different types of metal and organic ligands use might form different crystal structures of MOF. On the other hand, the lowest performance of CO_2 adsorption by Cu-TMA suggested the coarse crystal size led to a lower surface area to volume ratio, therefore, reducing the gas adsorption ability of this MOF. Subsequently, Mg-TMA was used to study the CO_2 adsorption in the indoor air condition. The initial enclosed space CO_2 concentration was recorded at 960 ± 73 ppm. The CO_2 concentration was then recorded at 895 ± 57 ppm after two hours of sampling time, indicating a reduction of 6.77% of CO_2 concentration.

Table 2. Carbon dioxide gas adsorption capacity of fabricated MOFs

MOFs	**CO_2 adsorption capacity ($\times 10^{-3}$ mmol/g)**
Mg-TMA	46 ± 4.72
Cr-TMA	39 ± 4.18
Cu-TMA	18 ± 3.27

4. Conclusions

In a nutshell, the main cause of the IAQ problem is indoor air pollution that releases gases or particles especially CO_2 molecules. Inadequate ventilation indoor may lead to an increase in the CO_2 level which can then cause sick building syndrome. According to NIOSH, the CO_2 concentration in a room must below 1,000 ppm else consider an inadequate ventilation room. Hence, the right ventilation can help to prevent the IAQ problem, and the researchers must develop some effective methods or technology to capture CO_2. Strong action and study are needed in the IAQ topic. In this paper, MOF samples have been chosen to capture CO_2 due to several outstanding features such as the topologies surface of MOFs which is more adaptable than the other existing porous materials. There are three types of MOF namely, Mg-TMA, Cr-TMA and Cu-TMA were successfully fabricated, and its gas adsorption performance was studied. The SEM images suggested that Mg-MOF exhibited the highest surface area attributed to its fine nanorod flask crystal. The EDX analysis suggested that the metal oxides were well attached to the organic linker. The gas adsorption study shows that Mg-TMA possesses the highest gas adsorption capacity. Subsequent study shows that the Mg-TMA able to reduce the 960 ± 73 ppm to 895 ± 57 ppm under 2 hours continuous sampling time suggested this MOF possess the potential to be applied for CO2 reduction for indoor air quality improvement.

References

[1] Chen S., Guo C., Huang X. 2018 *J. Environ. Econ. Manag.* **92** 465.
[2] Khamal R., Isa Z., Sutan R., Razif N., Ghazi N. H., 2019. *Ann. Glob. Health.* **85** 1.
[3] Liu H., Salvo A., 2018. *J. Environ. Econ. Manag.* **92** 300.
[4] Department of Occupational Safety and Health, Malaysia. 2010. *Industry Code of Practice on Indoor Air Quality. Kuala Lumpur.* Department of Occupational Safety and Health, Malaysia.
[5] Abd A.A., Naji S.Z., Hashim A.S., Othman M.R., 2020, *J. Environ. Chem. Eng.* **8** 104142.
[6] Li H., Wang K., Sun Y., Lollar C., Li J., Zhou H.C., 2018. *Mater. Today.* **21** 108.
[7] Wen Y., Zhang J., Xu Q., Wu X.T., Zhu Q.L., 2018. *Coord. Chem. Rev.* **376** 248.
[8] Younas, M., Sohail, M., Kong, L., Bashir, K., Sethupathi, S., 2015. *Int. J. Environ. Sci. Technol.* **13** 1839.
[9] Li, P., Chen, J., Zhang, J. and Wang, X., 2015. *Sep. Purif. Rev.* **44** 19.
[10] Hauchhum, L. and Mahanta, P., 2014. *Int. J. Energy Environ. Eng.* **5** 349.
[11] Jhung S. H., Jin T., Hwang Y.K., Chang J., 2007. *Chem. Eur. J.* **13** 4410.
[12] Tahmasian A., Morsali A., Joo S.W., 2013. *J. Nanomater.* **10** 1.
[13] Gargiulo N., Peluso A., Aprea P., Hua Y., Filipović D., Caputo D., Eić M. 2014. *RSC Adv.* **4** 90.
[14] Liang J., Huang Y., Cao R., 2019. *Coord. Chem. Rev.* **378** 32.
[15] Sahiner, N., Sel, K., Ozturk O.F., Demirci S., Terzi G., 2014. *Appl. Surf. Sci.* **314** 664.
[16] Liu Y., Ghimire P., Jaroniec M., 2019. *J. Colloid Interface Sci.* **353** 122.

IOP Conf. Series: Materials Science and Engineering 1192 (2021) 012025 doi:10.1088/1757-899X/1192/1/012025

Virtual Screening and *In Silico* Interactions Studies for Potential Antivirals and Diagnostics against the Spike protein from the Novel Coronavirus SARS-Cov-2

F I Che Abd Aziz[1], F A Ahmad Fuad[1,2], S Tanbin[1]

[1]Department of Biotechnology Engineering, Kulliyyah of Engineering, International Islamic University Malaysia, Jalan Gombak, 53100 Kuala Lumpur, Malaysia

[2]Corresponding author's email: fazia_adyani@iium.edu.my

Abstract. COVID-19 is a newly-emerged respiratory disease that is caused by the SARS-CoV-2, the seventh known Coronaviruses strain that has struck a global pandemic. The sharp increase in the number of positive cases worldwide necessitates highly-sensitive diagnostics kits and effective antiviral drugs to be developed for the populations. One of the antigens that is targeted for antibody neutralisation is the coronavirus Spike protein that consists of the S1 and S2 subunits, which mediated the entry pathway into the host's cell. Thus, the Spike protein has been suggested as a potential target for Covid-19 diagnostics and drug design. This study aims to evaluate the interactions between the SARS-CoV-2 Spike protein and the known monoclonal antibodies from Coronaviruses and to screen for potential Spike protein inhibitors. Virtual screening was conducted based on two compounds, N-acetyl-D-glucosamine (NAG) and Hesperetin, which is a small molecule that binds to the SARS-CoV-2 Spike protein structure and a natural compound that has prophylactic agents against SARS-CoV-2 infection as it binds to Spike protein, respectively. Protein-protein interaction studies were conducted by using the STRING webserver, prior to performing rigid docking using SWISSDOCK and visualised using USCF Chimera. Meanwhile, ligand-based screening was conducted through Ultrafast Shape Recognition Virtual Screening Database (USR-VS), and structure-based screening was performed *via* AutoDock4 software. The toxicity of the compounds was predicted using ProTox-II database. Possible interactions have been observed between the known monoclonal antibodies with the SARS-CoV-2 Spike protein, where M396 monoclonal antibody has shown the strongest interaction with a binding energy of -8.50 kcal/mol. Meanwhile, virtual screening has yielded several compounds that indicate the possibility to inhibit the SARS-CoV-2 Spike protein, where Tamarixetin has shown the strongest binding energy of -7.93 kcal/mol. These findings have potentials to be further evaluated in the future for the development of improved diagnostic kits and potential therapeutic drugs that specifically target the Spike protein of SARS-CoV-2.

1. Introduction

Coronavirus originates from the *Coronaviridae* family and a subfamily of *Coronavirinae,* which is divided into four genera, α-coronavirus, β-coronavirus, γ-coronavirus and δ-coronavirus [1]. It is a single-stranded, positive sense of RNA genome (+ssRNA) with 5'-cap and 3'-poly-A tail that is approximately 26-32 kilobases in size [2]. As of today, there are seven types of HCoV identified until early 2020, which are 229E, OC43, NL63, HKU1, SARS (Severe Acute Respiratory Syndrome), MERS (Middle East Respiratory Syndrome) and SARS-CoV-2, which is the newly-emerged novel coronavirus that caused the Covid-19 pandemic. The virus was firstly detected in Wuhan, Hubei Province of China from the Huanan local traditional market in December 2019. The infected patients showed similar symptoms to previous diseases known as SARS-CoV and MERS-CoV, which were declared epidemics in the year 2002 and 2012, respectively. The resemblance is due to SARS-CoV-2 carrying similar pathogenic characteristics with SARS-CoV and MERS-CoV with 79% and 50% similarity, accordingly [1].

SARS-CoV-2 contains several structural proteins that are crucial for fusion and assembly of the virus, including the Spike protein, Membrane protein, Envelope protein and Nucleocapsid protein. The Spike protein has been proposed as a potential drug target for disease treatment, as it consists of the S1 and S2 subunits, where S1 contains the receptor binding domain (RBD) that binds with the host cell receptor, while S2 facilitates the fusion and entrance process into the host's cell targeting the angiontensin-converting enzyme 2 (ACE2) to lower the host's immune response [3]. A ligand, 2-acetamido-2-deoxy-beta-D-glucopyranose (NAG) is one of the small molecules that bind between the receptor-binding-domain (RBD) of the SARS-CoV-2 Spike protein and the host cell receptor angiotensin-converting-enzyme 2 (ACE2). This ligand was added to stabilize the crystal structure of the SARS-CoV-2 Spike protein, due to the presence of the chemical bond between the carbohydrate moieties and the asparagine side chain [4]. Meanwhile, Hesperetin is a compound that can be found in *Citrus sinensis* (sweet orange), which is one of the flavonoids that contain flavanones, a group of phenolic natural chemical compound. It is rich in antioxidant as well as anti-microbial function that can inhibit the key proteins that are involved in the coronavirus infective cycle [5]. These compounds will be further analysed using *in silico* methods against the Spike protein, which is the basic requirement for the development of biomaterials involving substances that have been engineered, which subsequently can be used to directly control the interactions in the course of any therapeutic or diagnostic procedure.

In this study, protein-protein interactions (PPIs) between the Spike protein (antigen) and the known monoclonal antibodies were analysed, towards the development of antibody-based diagnostics that is designed to detect the presence of SARS-CoV-2 virus that caused the devastating Covid-19. The known monoclonal antibodies chosen are CR3022, M396 and 4C2, which are known antibodies for SARS-CoV-2, SARS-CoV and MERS-CoV, respectively [6][7]. For drug design, ligand-based and structure-based approaches were conducted to determine the potential inhibitors of Spike protein, by using NAG and Hesperetin as the query molecules to search for similar molecules that can bind to Spike protein with high binding affinity. Subsequently, the toxicity of these compounds were predicted using the ProTox-II webserver (https://tox-new.charite.de/protox_II/).

2. Materials and Methods

2.1 Programmes and databases

Table 1 lists the programmes and databases used during the *in silico* screening process.

Table 1: Programmes and databases used during the *in silico* screening process.

Programme / Database	Function
USR-VS	To screen for similar compounds or analogues of the query molecules.
AutoDock 4	To perform docking between the SARS-CoV-2 Spike protein with the analogues.
SWISSDOCK	To perform docking between the SARS-CoV-2 Spike protein with monoclonal antibodies.
USCF Chimera	To visualise the docking of monoclonal antibodies onto the SARS-CoV-2 Spike RBD-ACE2 protein.
ProToxII	To predict the toxicity of the compounds.
Protein Data Bank (PDB)	To obtain the PDB file of the structure of the SARS-CoV-2 Spike protein.
ZINC15	To obtain the structures of the selected ligands that will be used for docking.
PubChem	To obtain the SD file of the structure of the SARS-CoV-2 Spike protein and the analogues.
STRING	To study protein-protein interactions between the Spike protein and the monoclonal antibodies.

2.2 Protein-protein Interactions

The known monoclonal antibodies specific to Coronaviruses, which are M396 for SARS-CoV and 4C2 for MERS-CoV, as well as CR3022 for SARS-CoV-2 were evaluated by using the STRING webserver.

IOP Conf. Series: Materials Science and Engineering 1192 (2021) 012025 doi:10.1088/1757-899X/1192/1/012025

The STRING database features the largest number of 5090 organisms and 24.6 million proteins, which can be the benchmarked data sources, while providing intuitive and fast viewer for online use [8]. Chain H in M396 antibody, chain B in CR3022 antibody and Chain A in 4C2 antibody were found to have the highest combined score, which was determined prior to rigid molecular docking by using the SWISSDOCK webserver, in order to determine the binding energy between the SARS-CoV-2 Spike-ACE2 protein and the monoclonal antibodies.

2.3 Virtual Screening Analyses

The ligand-based method was conducted through USR-VS webserver by using NAG and Hesperetin as the query molecules. This program was able to screen 93.9 million 3D conformers of 23 million molecules using the ZINC database [9]. The similarity score was ranked based on the similarity of the screened compounds with NAG and Hesperetin and the top five compounds from each query molecule were chosen to be docked with the SARS-CoV-2 Spike protein. The structure-based method was performed with the top five compounds, where it was docked using the AutoDock4 software. The docking process was conducted to determine the binding energy between the Spike protein-compounds complexes. Finally, a toxicity test was performed for all selected compounds by using the ProTox-II webserver.

3. Results and Discussion

3.1 The Interactions between the SARS-COV-2 Spike protein and the Monoclonal Antibodies

The currently available rapid test kits are antigen and antibody-based detection systems, where these can be used for initial detection of viral infection and after the incubation period of 14 days post-infection of SARS-CoV-2, respectively. The common method to design a diagnostic kit is by using the enzyme-linked immunosorbent assay (ELISA) and some examples of the current rapid test kit developed are Sofia SARS Antigen from QUIDEL company and SARS-CoV-2 LgG Antibody from Beckman Coulter company, apart from real-time reverse transcription–polymerase chain reaction (RT-PCR) diagnostics kits [10]. On the other hand, this study was conducted to determine the known monoclonal antibodies from previous coronaviruses that are able to neutralize the SARS-CoV-2 Spike protein using *in silico* approach by evaluating the binding energy form between the known monoclonal antibodies with the Spike protein.

Based on the interactive network analysed by the STRING database, it was shown that the top three highest combined score of protein interactions among the antibodies are chain H in M396 antibody, chain B in CR3022 antibody and chain A in 4C2 antibody, as tabulated in Table 2. The combined scores are calculated based on the functional association, which links between the known monoclonal antibodies contributed due to the specific biological function [8]. Thus, the binding energy was generated between the SARS-CoV-2 Spike protein's RBD-ACE2 with the antibody, which shows that the M396 antibody exhibits the highest binding affinity towards the Spike protein, as shown in Figure 1, indicating the formation of stronger complexes that is ideal for antigen-antibody interactions. Therefore, the M396 antibody may well have potential to be used as the capture antibody in the diagnostics kit to selectively detect the SARS-CoV-2 Spike protein. However, the specificity of M396 antibody can be improved *via* antibody engineering for future diagnostics development for detection of Covid-19.

Table 2: Three antibodies with the highest combined scores and binding energy.

Monoclonal antibody	Chain	Combined Score	Binding Energy (kcal/mol)
CR3022	B	0.934	-7.90
M396	H	0.998	-8.50
4C2	A	0.903	-7.67

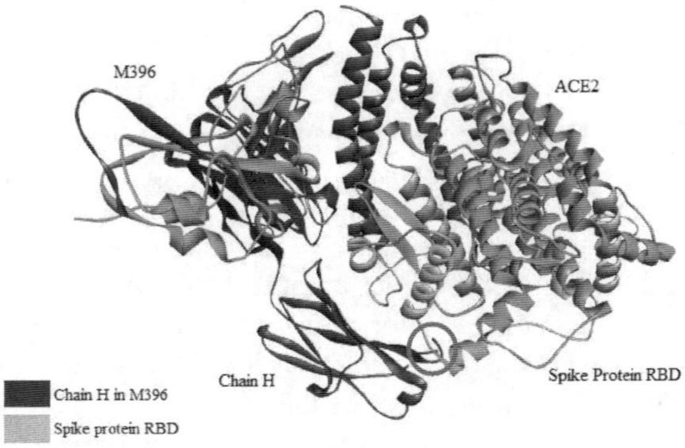

Figure 1: The interactions between the M396 antibody with the RBD-ACE2 region of the SARS-CoV-2 Spike protein.

3.2 Virtual Screening of Compounds against the SARS-CoV-2 Spike Protein

Since there are no clinically proven treatment for Covid-19, only supportive symptomatic treatments such as oxygen therapy, steroids on a per case basis, fluid management and broad-spectrum antibiotics against the co or secondary bacterial infection were provided for patients [11]. Other than that, similar to any viral diseases, Covid-19 can be aided by therapeutic intervention, which can be extended to either direct acting antivirals (DAAs) or immunomodulatory adjuvants and substances [12]. Several drugs have been tested for repurposing to treat Covid-19, which can be divided into two categories; those that targeted the viral replication cycle and the ones that aim to control the symptoms of the diseases. One of the drugs tested is known as chloroquine, which were classically used as anti-malarial medication, meanwhile in Covid-19, it was found to be an effective antiviral agent that can inhibit SARS-CoV-2 infection [13]. Moreover, there are several clinical trials that have reported that the use of chloroquine can alter terminal glycosylation of the angiotensin-converting enzyme 2 (ACE2) receptor by suppressing the SARS-CoV-2 Spike protein binding, which notably reduces viral replication by blocking the virus fusion process [14].

Through virtual screening analyses, two potential inhibitors, which are ZINC 6614489 and Tamarixetin were shown to possess lower binding energy compared to the query molecules; NAG and Hesperetin respectively, as shown in Table 3. Each compound fulfilled the Lipinski's Rule of Five, which is the rule of thumb in designing oral drugs. It was shown that each compound must consist of no more than 5 hydrogen bond donors, have no more than 10 hydrogen bond acceptors, molecular mass less than 500 Dalton and partition coefficient (log P) must not be greater than 5 [15]. The analyses showed that the binding energy for ZINC 6614489 is close to NAG, while Tamarixetin's binding energy is lower than Hesperetin, which indicates the possibility of these potential inhibitors to bind strongly to SARS-CoV-2 Spike protein. The dockposes of both ZINC 6614489-Spike protein RBD and Tamarixetin-Spike protein RBD protein-ligand complexes are shown in

IOP Conf. Series: Materials Science and Engineering 1192 (2021) 012025 doi:10.1088/1757-899X/1192/1/012025

Figure 2. The first potential inhibitor, ZINC 6614489 is a compound similar to NAG that possesses the binding energy of -4.92 kcal/mol, which is very close to the binding energy of NAG. Meanwhile, Tamarixetin is an analogue of Hesperetin and exhibited the binding energy of -7.93 kcal/mol, which is lower compared to Hesperetin (-7.43 kcal/mol). The well-established and potent inhibitor of Spike protein, Chloroquine was shown to have a binding energy of -7.53 kcal/mol, which means that Tamarixetin has higher potential to inhibit the SARS-CoV-2 Spike protein, compared to both Hesperetin and Chloroquine. Meanwhile, the toxicity prediction value (LD_{50}) is required for drug design evaluation in order to determine whether the compounds are safe for human use. The toxicity classes are categorised into six classes, where the lowest is Class 1, indicating that the compound is fatal if swallowed when the LD50 value is less and equal to 5, while the highest is Class 6 that is non-toxic when the LD50 value is above 5000. In this study, the toxicity prediction value (LD_{50}) for each compound falls under the category of class 5 within the range of ($2000 < LD50 \leq 5000$), which indicated that the compounds maybe harmful for consumption. Therefore, further analysis is required to find the right amount of dose of the compounds that have potentials as drugs that are safe to be consumed. Hence in this study, virtual screening has successfully indicated that ZINC 6614489 and Tamarixetin could be the potential inhibitors of the SARS-CoV-2 Spike protein that can be further developed in the drug design pipeline for Covid-19 treatment.

Table 3: Potential inhibitors (ZINC 6614489 and Tamarixetin) obtained from virtual screening analyses along with the query molecules, NAG and Hesperetin, respectively. Also shown in the table is a repurposed drug that is currently used to treat Covid-19, Chloroquine.

Compound Name	**Chemical Structure**	**Binding Energy (kcal/mol)**	**ProTox-II Prediction (LD_{50}, mg/kg)**	**H-bond donor (<5)**	**H-bond acceptor (<10)**	**MW (<500Da)**	**XlogP (<5)**
Chloroquine	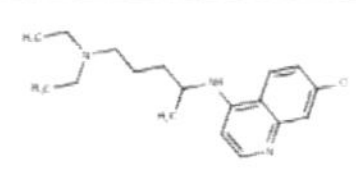 $C_{18}H_{26}ClN_3$	-7.53	-	1	2	319.87	4.63
NAG	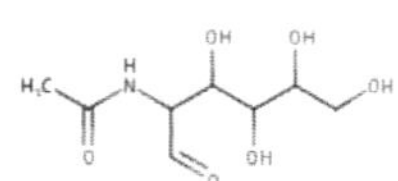 $C_8H_{15}NO_6$	-5.17	-	5	6	221.21	-1.72
ZINC 6614489	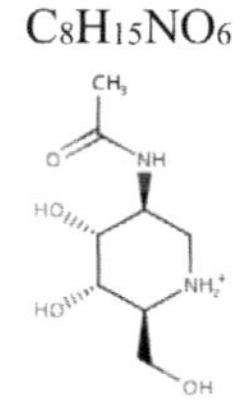 $C_8H_{17}N_2O_4$	-4.92	5000 Category 5	5	7	221.209	-3.03
Hesperetin	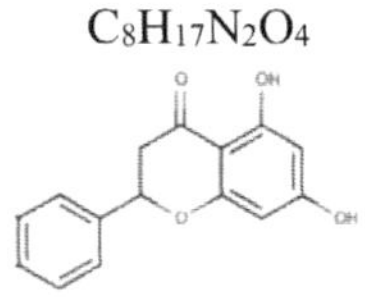 $C_{16}H_{14}O_6$	-7.43	-	3	6	302.28	2.60

Tamarixetin	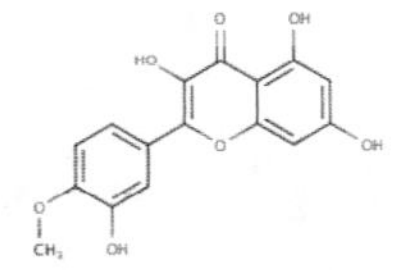 $C_{16}H_{12}O_7$	-7.93	5000 Category 5	4	7	316.265	1.99

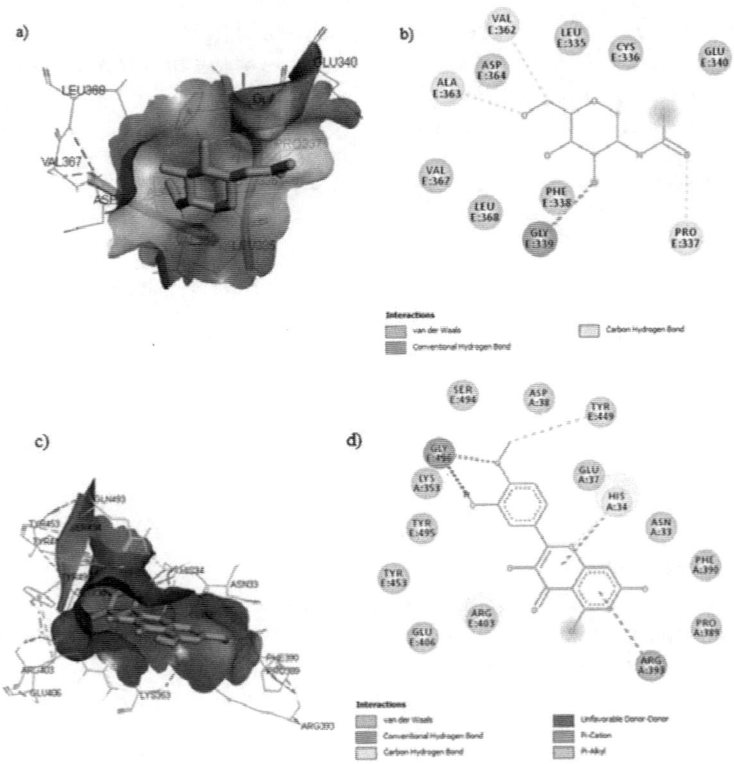

Figure 2: a) The docked ZINC 6614489 ligand with SARS-CoV-2 Spike-RBD. b) The type of interactions formed during the docking process with the interacting amino acid residues. c) The docked Tamarixetin ligand with SARS-CoV-2 Spike-RBD. d) The type of interactions formed during the docking process with the interacting amino acid residues.

4. Conclusion

Protein-protein interactions between the SARS-CoV-2 Spike protein with M396 antibody has shown the strongest binding energy of -8.50 kcal/mol, suggesting that this antibody can be further improved in future antibody engineering and design studies, to be developed as a specific probe for rapid detection kit. Meanwhile, the virtual screening analyses, which were conducted to search for potential inhibitors of SARS-CoV-2 Spike protein had shown that Tamarixetin has the strongest binding energy compared to other screened compounds. Therefore, it is suggested that this compound has potentials to be designed as a future therapeutic drug. Overall, this study has computationally shown potential monoclonal antibody-Spike protein interactions, as well as potential candidates for rational drug design against the SARS-CoV-2 Spike protein, which have paved the path towards the control of Covid-19 transmission and future medication.

Acknowledgement

The authors would like to acknowledge the Department of Biotechnology Engineering, Kulliyyah of Engineering, IIUM for the support to conduct this study.

References

[1] Fani M, Teimoori A, and Ghafari S 2020. Comparison of the COVID-2019 (SARS-CoV-2)

pathogenesis with SARS-CoV and MERS-CoV infections. *Future Virology* https://doi.org/10.2217/fvl-2020-0050.

[2] Weiss S and Navas-Martin S 2005. Coronavirus Pathogenesis and the Emerging Pathogen Severe Acute Respiratory Syndrome Coronavirus. *Microbiology and Molecular Biology Reviews*, vol **69** pp 635–664. https://doi.org/10.1128/mmbr.69.4.635-664.2005.

[3] Bharath B, Damle H, Ganju S, and Damle L 2020. In silico screening of known small molecules to bind ACE2 specific RBD on Spike glycoprotein of SARS-CoV-2 for repurposing against COVID-19. *F1000Research*, https://doi.org/10.12688/f1000research.24143.1

[4] Wlodawer A, Dauter Z, Shabalin I and *et al* 2020. *Ligand-centered assessment of SARS-CoV-2 drug target models in the Protein Data Bank FEBS J* vol **287** pp 3703-3718.

[5] Bellavite P, and Donzelli A 2020. Hesperidin and SARS-CoV-2: New light on the healthy function of citrus fruits. *Antioxidants* vol **9** pp 1–18.

[6] Jiang S, Hillyer C, and Du L 2020. Neutralizing Antibodies against SARS-CoV-2 and Other Human Coronaviruses. Trends Immunol , vol **41** p 355. https://doi.org/10.1016/j.it.2020.03.007.

[7] Choudhry H, Bakhrebah M, Abdulaal W and *et al* 2019. "Middle East respiratory syndrome: Pathogenesis and therapeutic developments," *Future Virol*, vol **14** pp 237–246, doi:10.2217/fvl-2018-0201.

[8] Szklarczyk D, Gable A, Lyon D and *et al* 2019. STRING v11: Protein-protein association networks with increased coverage, supporting functional discovery in genome-wide experimental datasets. *Nucleic Acids Research*, vol **47** pp 607–613. https://doi.org/10.1093/nar/gky1131

[9] Schreyer A M and Blundell T 2012. USRCAT: Real-time ultrafast shape recognition with pharmacophoric constraints. *Journal of Cheminformatics*, vol **4** pp 1–12. https://doi.org/10.1186/1758-2946-4-27.

[10] Khalaf K, Papp N, Chou J, Hana D, Mackiewicz A, and Kaczmarek M 2020. SARS-CoV-2: Pathogenesis, and Advancements in Diagnostics and Treatment. In *Frontiers in Immunology*, vol **11** (Frontiers Media S.A) p 570927. https://doi.org/10.3389/fimmu.2020.570927

[11] Libster R, Pérez M, Wappner D and *et al* 2021. Early High-Titer Plasma Therapy to Prevent Severe Covid-19 in Older Adults. *New England Journal of Medicine*, vol **384** pp 610–618. https://doi.org/10.1056/nejmoa2033700.

[12] Nitulescu G, Paunescu H, Moschos S and *et al* 2020. Comprehensive analysis of drugs to treat SARS-CoV-2 infection: Mechanistic insights into current COVID-19 therapies (Review). In *International Journal of Molecular Medicine* vol **46** (Spandidos Publications) pp 467–488. https://doi.org/10.3892/ijmm.2020.4608.

[13] Singh B, Ryan H, Kredo T, Chaplin M, and Fletcher T 2020. Chloroquine or hydroxychloroquine for prevention and treatment of COVID-19. Cochrane Database Syst Rev 2021. https://doi.org/10.1002/14651858.CD013587.pub2.

[14] Bolarin J, Oluwatoyosi M, Orege J and *et al* 2021. Therapeutic drugs for SARS-CoV-2 treatment: Current state and perspective. In *International Immunopharmacology* vol **90** (Elsevier B.V) p 107228. https://doi.org/10.1016/j.intimp.2020.107228.

[15] Banegas-Luna A J, Cerón-Carasso J P and Pérez-Sánchez H 2018. A review of ligand based virtual screening web tools and screening algorithms in large molecular databases in the age of big data. *Future Medicinal Chemistry*, vol **10** pp 2641–2658. https://doi.org/10.4155/fmc 2018-0076.

IOP Conf. Series: Materials Science and Engineering 1192 (2021) 012026 doi:10.1088/1757-899X/1192/1/012026

Activated carbon assisted electrocoagulation process for treating biotreated palm oil mill effluent

A Tahreen[1], M S Jami[1,2] and F Ali[1]

[1]Department of Biotechnology Engineering, Faculty of Engineering, International Islamic University Malaysia, 50728 Kuala Lumpur, Malaysia

[2]Corresponding Author's e-mail: saedi@iium.edu.my

Abstract. Electrocoagulation (EC) is a sustainable wastewater treatment alternative that is widely studied because of its environmentally friendly nature, versatility, and simplicity in setup and operation. However, EC alone cannot treat wastewater up to reusable standards and requires integration with other processes, mostly by adding highly hazardous oxidants. This work aims to investigate the combination of powdered activated carbon (AC) with biotreated palm oil mill effluent (BPOME) as wastewater sample, in the EC reactor, and to optimize its concentration for maximum pollutant removal efficiency. Ranging from 0.5-1.5 wt. % concentration of AC mixed with EC reactor, EC was carried out with its critical parameters set to a current of 1.75 A (i.e., 160 mA/cm^2 current density) and initial pH 6 and 10 mm interelectrode distance with aluminum electrodes. The EC treated wastewater was sampled from 5 minutes to 60 minutes and the parameters monitored were total suspended solids (TSS), turbidity, color and chemical oxygen demand (COD). Turbidity, TSS and color were removed nearly to completion within 5 to 15 minutes of EC, whereas maximum COD removal was determined to be 84.6 % with 1 wt. % powdered AC combined with EC, which is an increase of about 14.6% compared to EC with no AC addition. The optimum concentration of AC for maximum removal efficiency on BPOME was 1 wt. % (2 g per 200 ml). Addition of AC in EC resulted in a faster pollutant removal rate, with enhanced process efficiency.

1. Introduction

It is crucial to develop an alternative treatment system to sustainably treat industrial effluents without requiring or producing hazardous chemicals, to lead towards a circular economy with environmental sustainability. Therefore, electrocoagulation (EC) is a topic of exhaustive study due to its simple and inexpensive operation and maintenance, eco-friendly nature, and the ability to treat a wide range of pollutants in various types of wastewater [1]. EC is performed by passing current in the target wastewater through metal electrodes, typically aluminum or iron electrodes. Consequently, metal hydroxides form in the EC reactor that destabilize the pollutants which come together as flocs, which are easily separated from the wastewater. However, EC alone cannot treat heavily polluted industrial wastewater, up to reusable standards. There is a need for an integration of a sustainable process to enhance EC. EC is often employed as a pre-treatment process for membrane based purification systems as the huge amount of pollutants removed by EC significantly contributes to membrane fouling mitigation [2]. The addition of hazardous oxidants to the wastewater, such as hydrogen peroxide through advanced oxidation processes, is widely explored to achieve a desirable outcome [3–5]. But it is not environmentally friendly and is a chemical intensive process, requiring toxic oxidants. An interesting approach is to add eco-friendly adsorbents, such as activated carbon (AC) and to investigate the synergistic effect of EC with AC adsorption to effectively treat severely contaminated industrial effluents. Sher et al. (2021) achieved promising results by coupling powdered AC with EC for general wastewater treatment [6]. This study explores the addition of commercially available powdered AC to combine with EC to treat heavily polluted biotreated palm oil mill effluent (BPOME). BPOME is the final form of the effluent after biological treatment, ready for environmental discharge from the palm oil industries. The characterization of BPOME is presented in Table 1.

Table 1. Characterization of BPOME.

Parameter	Average value
Chemical Oxygen Demand (COD), (mg/L)	1981
Total Suspended Solids (TSS), (mg/L)	192
Color, (PtCo)	2882
Salinity, (ppt)	8.2
Total Dissolved Solids (TDS), (g/L)	7.91
Conductivity, (mS/cm)	14.1
pH	7.9
Turbidity, (NTU)	332

2. Methodology

A 250 ml beaker was used as the EC reactor. Aluminum electrodes with dimensions 1.55 mm x 20 mm x 100 mm and HACH COD vials High Range (HR) were purchased from Jecre Trading, Malaysia. Alligator clips and direct current (D.C) supply unit, Twintex (TP-2303K, Taiwan) were used to pass current through the electrodes to form the electrochemical cell. Figure 1 shows the schematic diagram of the EC setup.

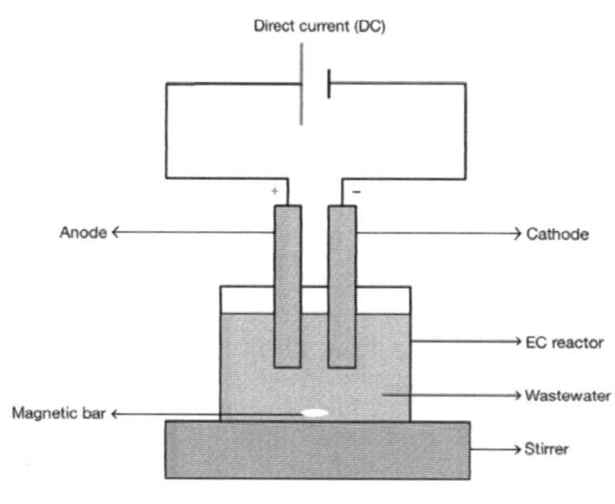

Figure 1. Schematic diagram of EC setup [7].

BPOME was collected from the final discharge pond released at 23 tons per hour capacity-based discharge, in Negeri Sembilan based palm oil mill, Malaysia. The collected effluent was stored at 4°C temperature, to prevent degradation due to biological activity.

Commercially available powdered AC was added to the EC reactor and EC process was run with the parameters, current density 160 mA/cm^2 (1.75 A), initial pH 6 and 10 mm of interelectrode distance. The parameters namely COD, turbidity, TSS and color were monitored from 5 to 60 minutes of EC operation. To control the initial pH of the samples, 5% HCl and 0.1M NaOH solutions and pH meter (Mettler Toledo, MP220 model, USA) were used. With 250-300 rpm stirring speed, and 1 g of AC, the EC-AC coupled process was carried out like the work of [8]. To optimize the AC dosage for optimum EC performance and COD removal, the EC coupled adsorption process was run with AC ranging from 1 g to 3 g (0.5 wt % to 1.5 wt %), as shown in Table 2.

Table 2. Wt.% of powdered AC and its equivalent mass per working volume.

AC (wt %)	AC mass (g) per 200 ml sample
0.5	1.0
1.0	2.0
1.5	3.0

The resulting removal efficiency (R_e) values were calculated with the following Equation (1).

$$R_e\,(\%) = \frac{X_i - X_f}{X_i} \times 100 \tag{1}$$

where, X_i and X_f are the initial and final concentrations of COD, TSS, turbidity and color.

The wastewater samples were measured for COD, color and total suspended solids (TSS) before and after EC using spectrophotometer (HACH DR 5000, USA) according to American Public Health Association (APHA) procedures [9]. Multi-meter (HACH sensION5, USA) was used to measure the total dissolved solids (TDS), conductivity, and salinity of the wastewater samples. Turbidity of the samples were monitored with turbidimeter (HACH 2100P, USA). Filter papers of 0.45 μm pore size were used to separate flocs from the EC treated solution for characterization. X-Ray Diffraction (XRD) was carried on D2 Phaser 2nd Gen, Bruker and operated with 0.25s of time/step. All experiments were carried out in room temperature.

3. Results and discussion

The optimized EC operational parameters, intial pH 6, and current density 160 mA/cm^2 (current 1.75 A) from previous work [7], were used for all the experiments from 0 to 60 minutes, and the treated wastewater was monitored for TSS, color, turbidity and mainly COD values. The outcome of this control study (with 0 wt. % AC) is summarized in Figure 2.

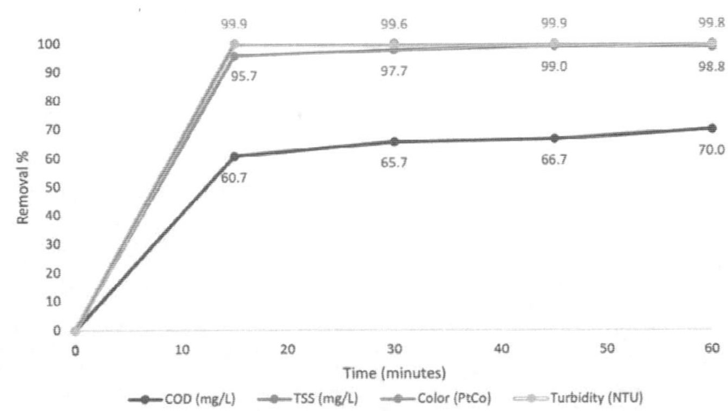

Figure 2. Removal efficiency of COD, TSS, color and turbidity % without AC in EC of BPOME.

Turbidity and TSS depicted a remarkably similar removal rate compared to the other parameters (color and COD), without addition of any AC. The colloidal destabilization by the aluminum hydroxide-based coagulants generated in situ worked effectively to remove all suspended solids from the BPOME. Besides, the quantity of suspended solids results in the turbidity of the wastewater, therefore, their removal rates depict a negligible difference between them. The same pattern was observed in EC coupled with powdered AC, but with a higher rate of removal, presented in Figure 3.

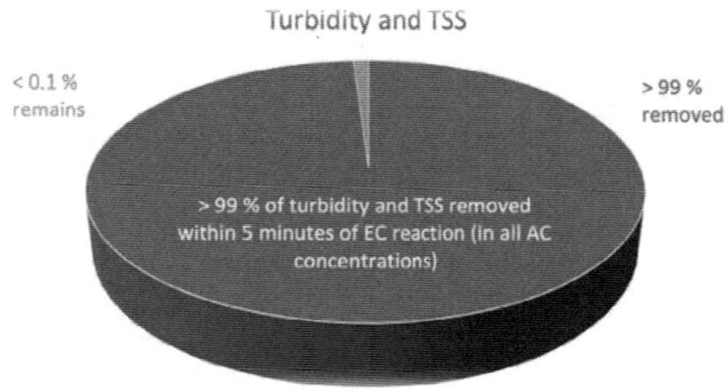

Figure 3. Turbidity and TSS removal % within 5 minutes of EC (combined with AC at 0.5 wt.% - 1.5 wt. %).

Table 3. Characterization of EC treated BPOME with 0.5 wt. % AC.

Parameters/EC time (minutes)	0	5	10	15	30	45	60
COD (mg/L)	1981	588	596	592	493	467	560
TSS (mg/L)	192	1	0	0	0	0	0
Color (mg/L)	2882	112	78	52	11	6	0
Turbidity (NTU)	332	0.89	0.4	0.5	0.45	0.78	0.47

Table 4. Characterization of EC treated BPOME with 1.0 wt. % AC.

Parameters/EC time (minutes)	0	5	10	15	30	45	60
COD (mg/L)	1981	435	306	316	452	436	490
TSS (mg/L)	192	0	0	0	0	0	0
Color (mg/L)	2882	16	16	14	7	0	1
Turbidity (NTU)	332	0.95	0.7	0.6	0.5	0.9	0.1

Table 5. Characterization of EC treated BPOME with 1.5 wt. % AC.

Parameters/EC time (minutes)	0	5	10	15	30	45	60
COD (mg/L)	1981	351	504	492	448	457	459
TSS (mg/L)	192	0	1	0	0	0	0
Color (mg/L)	2882	26	30	5.5	19	5	0
Turbidity (NTU)	332	1.9	0.5	0.4	2.0	1.2	0.4

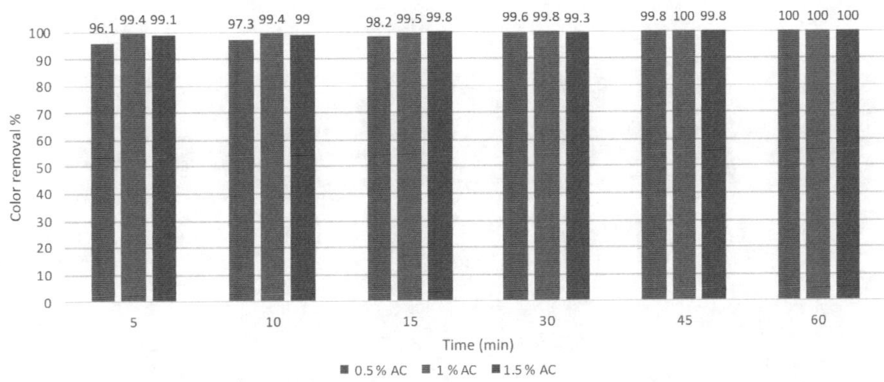

Figure 4. Bar graph representation of EC treatment efficiency (combined with AC) on BPOME in terms of color removal % vs time, with current density 160 mA/cm^2 and pH 6.

A significant reduction in color, as shown in Figure 4, was observed within 5 minutes of EC (coupled with AC) through a visible inspection, which was verified with chemical analyses. Tables 3-5 depict the sampling results in terms of COD, TSS, color and turbidity removal of the EC-AC treated BPOME from 5 to 60 minutes of treatment. It is evident from the data, that within 5 minutes of EC operation with addition of powdered AC, over 95% of turbidity, TSS and color were removed. This outcome is impressive as the process does not require harmful oxidants to break down the organic pollutants in the sample effluent. In maximum 15 minutes, all TSS and turbidity were removed from BPOME with EC alone with 0% AC (control). Using EC (set parameters: current density-160 mA/cm^2, initial pH- 6) on BPOME without the addition of powdered AC, resulted in a maximum of 70% removal of COD in 60 minutes of EC (Figure 2). In the range of 0 to 60 minutes, maximum color removal of 99 % was achieved at about 45 minutes of EC, and maximum COD removal was 70% in 60 minutes. However, with AC addition, the fastest removal % noted was of TSS from BPOME for all concentrations of powdered AC added, followed by turbidity, color and COD. The combined action of the adsorption of colloids by both AC along with the metal coagulants generated in the EC reactor, greatly enhanced the efficiency of the EC process, that contributed to a much higher pollutant removal rate. This conclusion also corresponds to the study reported by [8], where addition of granular AC in EC process, remarkably reduced the time required for humic acid removal from paper industry based wastewater.

No substantial difference in terms of TSS and turbidity removal rate was noted with AC addition in EC (Figure 3). The metal hydroxides generated by EC in situ was enough to completely destabilize the suspended solids and subsequently, turbidity. However, faster color and COD removal was observed with AC addition, relatively. The powdered AC addition enabled the breakdown of pigmented compounds (mainly phenols), present in the BPOME. Figure 4 shows that most of the color (>99%) is removed with 1 and 1.5 wt. % AC addition to EC, within the first 5 minutes of the process.

The primary parameter chosen to determine the best removal efficiency in this study is COD, as it denotes the total amount of oxygen required to degrade all the organic pollutants in the wastewater. Therefore, COD gives an estimation of the level of organic pollution of wastewater [10]. Besides, the COD values depicted a distinct difference while observing the EC efficiency with and without AC. With COD removal %, as the main criteria for measuring the EC effectiveness, the values are compared and visualized in Figure 5.

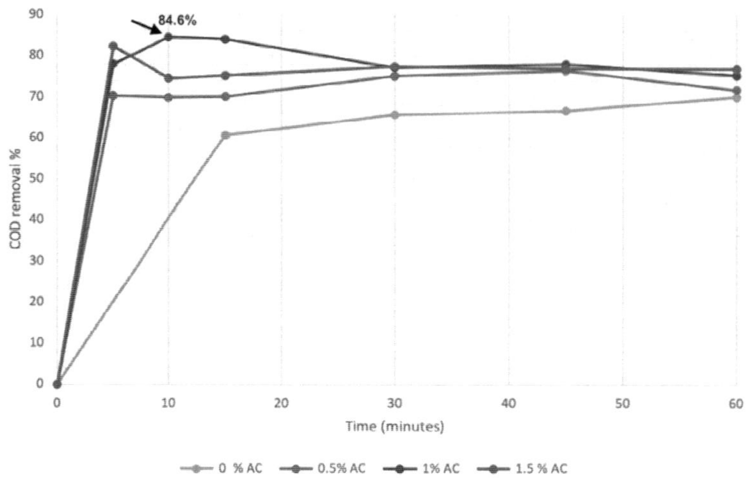

Figure 5. COD removal % with and without AC coupled with EC.

Powdered AC concentration of 1 wt. % (2g in 200 ml) is the optimum concentration for addition to EC process for best treatment efficiency in terms of overall COD removal (84.6%) for BPOME as observed in this study, in 10 minutes (Figure 5). With 0.5 wt. % AC, there were more space for pollutant adsorption from BPOME, while with 1.5 wt. %, in the upper limit, the AC quantity exceeding the desired active sites for the coagulant-colloid adsorption, could partly destabilize some of the flocs that have already formed complexes with the metal hydroxides and AC, due to flotation with H_2 release at the anode, or the constant stirring provided to maintain the homogeneity of the reactor constituents. With 1 wt. % AC added to the EC process yielded relatively the highest removal efficiency in treated BPOME in 10 minutes, the overall outcome of which is summarized in Figure 6. Besides, the visual observation of the color and turbidity removal after EC treatment with 1 wt. % AC, is presented in Figure 7.

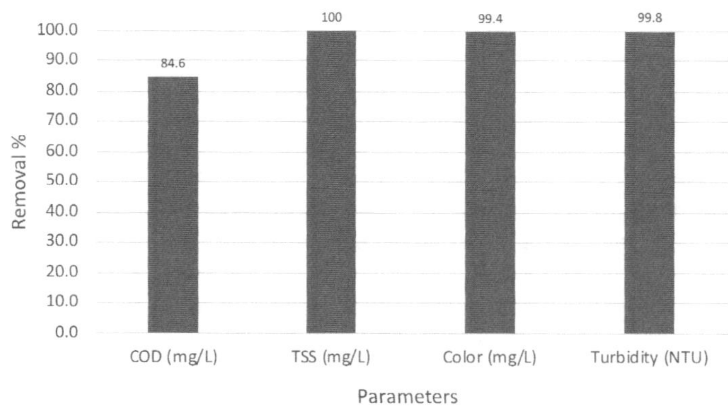

Figure 6. Summary of the overall treatment efficiency in terms of COD, TSS, color and turbidity removal % in 10 minutes of EC coupled with 1% AC.

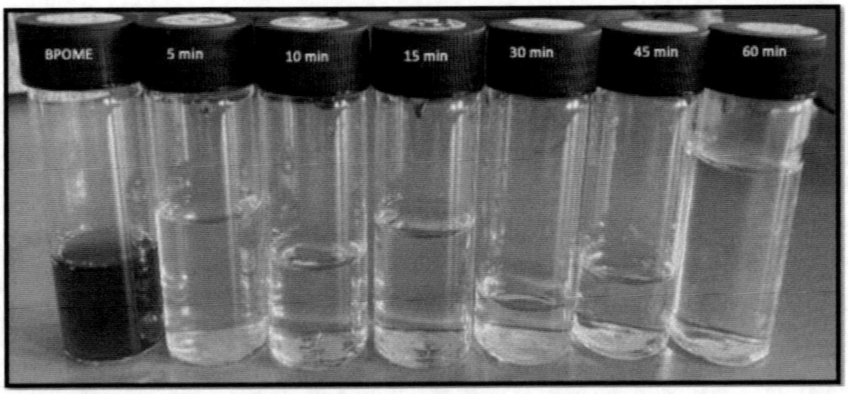

Figure 7. Visual observation of color removal on BPOME from 15 to 60 minutes of EC (left to right) 160 mA/cm^2 pH 6 with 1wt. % AC addition.

The powdered XRD spectra of AC and the sludge resulting from EC assisted with powdered AC on BPOME is presented in Figure 8. The powdered XRD analysis has been carried out with contact angles varying from 20° to 80° in the 2θ range. The powdered AC depicts two notable strong peaks (2θ) at around 25° and 43°, corresponding to the planes (002) and (101); same pattern of which was also observed by [11,12] for powdered AC. The graph represents the hexagonal structure of the carbon material in its graphitic form [13]. It is evident that the resulting sludge from EC is mostly powdered AC due to the similarities in the diffractogram pattern, with a visible shift in the peak intensity. The adsorption of organic pollutants combined with flocs of aluminum salts on the AC surface caused a distortion in its crystalline lattice structure, which is reflected by the variation in the peak intensity. Consequently, new sharp peaks at 28° and 29° were observed in the XRD spectrum of EC sludge, differing from AC's XRD spectrum. Priya and Sureshkumar (2020) made a similar conclusion suggesting the peak intensity variation resulting from AC adsorption [12]. As a large portion of the AC structure retains after EC, the reusability of the AC after EC process is worth exploring.

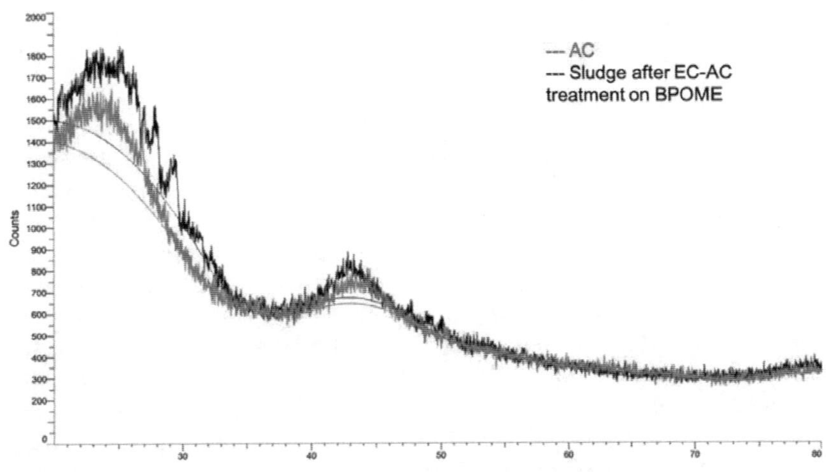

Figure 8. XRD spectra of AC before after EC treatment on BPOME (as sludge).

4. Conclusion

EC was investigated with AC addition to investigate its ability to enhance the overall treatment efficiency. Within 10 minutes of EC operation, coupled with 1 wt. % of AC, a transparent effluent with 100% TSS removal, 99% removal of both color and turbidity was observed. The maximum COD removal (84.6%) was achieved at this point with a final value of 306 mg/L. The EC treated wastewater was sampled from 5 minutes to 60 minutes and the parameters monitored were TSS, turbidity, color and

COD. Turbidity, TSS and color were removed nearly to completion within 5 to 15 minutes of EC, whereas maximum COD removal was determined to be 84.6 % with 1 wt. % powdered AC combined with EC. Therefore, the optimum concentration of AC for maximum removal efficiency on BPOME was 1 wt. % (2 g per 200 ml). A much faster treatment efficiency was noted after AC addition to EC, compared to no AC. Hence, the addition of AC in EC process shows a promising potential for a faster and more efficient treatment of heavily polluted BPOME.

Acknowledgement
The authors express thanks to Ministry of Education (MOE) Malaysia for granting a Fundamental Research Grant Scheme (FRGS), project no. FRGS-19-194-0803 and International Islamic University Malaysia, Faculty of Engineering, for the financial support under Tuition Fee Waiver (TFW) 2019 scheme for outstanding students awarded to AT, to support this work.

References
[1] Tahreen A, Jami M S and Ali F 2020 Role of electrocoagulation in wastewater treatment: A developmental review *J. Water Process Eng.* **37** 101440
[2] Tahreen A and Jami M S 2021 Advances in Antifouling Strategies in Membrane Ultrafiltration : a Brief Review *J. Adv. Res. Mater. Sci.* **76** 10–6
[3] Bahadur N and Bhargava N 2019 Novel pilot scale photocatalytic treatment of textile & dyeing industry wastewater to achieve process water quality and enabling zero liquid discharge *J. Water Process Eng.* **32** 100934
[4] Sharma S, Aygun A and Simsek H 2020 Electrochemical treatment of sunflower oil refinery wastewater and optimization of the parameters using response surface methodology *Chemosphere* **249** 126511
[5] Bashir M J, Lim J H, Abu Amr S S, Wong L P and Sim Y L 2019 Post treatment of palm oil mill effluent using electro-coagulation-peroxidation (ECP) technique *J. Clean. Prod.* **208** 716–27
[6] Sher F, Iqbal S Z, Rasheed T, Hanif K, Sulejmanović J, Zafar F and Lima E C 2021 Coupling of electrocoagulation and powder activated carbon for the treatment of sustainable wastewater *Environ. Sci. Pollut. Res.*
[7] Tahreen A, Jami M S, Ali F, Mohd N, Mat F and Ngabura M 2021 Promising Potential of Electro - Coagulation Process for Effective Treatment of Biotreated Palm Oil Mill Effluents *Pollution* **7** 617–32
[8] Barhoumi A, Ncib S, Chibani A, Brahmi K, Bouguerra W and Elaloui E 2019 High-rate humic acid removal from cellulose and paper industry wastewater by combining electrocoagulation process with adsorption onto granular activated carbon *Ind. Crops Prod.* **140** 111715
[9] APHA 2002 American Public Health Association; American Water Works Association; Water Environment Federation *Stand. Methods Exam. Water Wastewater*
[10] Islam M M M, Shafi S, Bandh S A and Shameem N 2019 *Impact of environmental changes and human activities on bacterial diversity of lakes* (Elsevier Inc.)
[11] Sonal S, Prakash P, Mishra B K and Nayak G C 2020 Synthesis, characterization and sorption studies of a zirconium(IV) impregnated highly functionalized mesoporous activated carbons *RSC Adv.* **10** 13783–98
[12] Sathya Priya D and Sureshkumar M V. 2020 Synthesis of Borassus flabellifer fruit husk activated carbon filter for phenol removal from wastewater *Int. J. Environ. Sci. Technol.* **17** 829–42
[13] Zuo P, Feng H, Xu Z, Zhang L, Zhang Y, Xia W and Zhang W 2013 Ultrastructural studies of rod photoreceptor nuclei from SCA7 mouse *Chem. Cent. J.* **7** 317–8

IOP Conf. Series: Materials Science and Engineering 1192 (2021) 012027 doi:10.1088/1757-899X/1192/1/012027

Control of Plant Growth by Monitoring Soil Moisture, Temperature and Humidity in Dry Climate

N K Madzhi[1,2] and M A Nor Akhsan[1]

[1]Department of Electrical Engineering, School of Engineering, Universiti Teknologi Mara Shah Alam Selangor, Malaysia

[2]Corresponding author's e-mail: ninak654@uitm.edu.my

Abstract. Monitoring of environment parameter such as soil moisture, temperature and humidity are important parts of plant growth. This paper focused on the development of an instrumentation system and analysis on the effect of the water volume to the soil moisture, effect rate of soil moisture, temperature and humidity for an indoor greenhouse. Data were collected through two experiment. First experiment focused on effect volume of water to soil moisture. Soil hygrometer sensor used to measure soil moisture in real time. Five bottles contained different volume of water poured into soil which the soil is fixed to 200gram.Three different rate of soil moisture applied to plant and the data were analysed to determined relationship between soil moisture to the plant growth. It can be concluded that the rate of soil moisture does have effect on the stem diameter and leaves length based on the observations of the plant growth for three weeks.

1. Introduction

Indoor Greenhouse System creates an artificial garden in the house. It can prevent disease, insect pest of breeding, reduce the dosage of pesticide, pollution and climate change [1-3]. According to Ibrahim Al-Adwan [4], the environment inside the greenhouse will affect the quality of the plants. Unlike outdoor greenhouse, monitoring environment parameter such as soil moisture, temperature and humidity inside indoor greenhouse are easier because of rate of change of these parameters are low.

The important parameter which needs to be monitored to produce a quality plan is soil moisture. Inefficient use of water will result in a waste of water supply [5]. If too little watering a plant, it will cause uneven water in soil. Prevention of overwatering is an aspect that can improve water usage efficiency and pest prevention. For example, a recent study from Slatyer reported that larvae of shore fly can be reduced if the soil moisture level is maintained around 30%. Similar approach can also reduce root zone diseases such as Pythium [6].

Delivering water to plants at the right time and right quantity is a key to effective and efficient irrigation. Knowing when and how much to water is two important aspects of irrigation. To do this, sensors is needed to determine when plants may need water. Water is absorbed by the root system and lost through transpiring leaves. Evaporation from the leaves is the driving force for transfer of water across the plant and only a small proportion of the uptake water is used for growth. It was calculated that the water lost per day by transpiration from some plants is equal to twice the weight of the plant. The rate of water lost depends on the condition of soil, air flow, relative humidity in air and the temperature of the environment. Loss of water from the soil by means of drainage is quite common during the dry season [5-6].

In this system, commercialized soil hygrometer sensor is used to monitored soil moisture. This sensor consists of two probes which are used to measure the volumetric content of water. The two probes allow the current to pass through the soil and then it gets the resistance value to measure the moisture value. When there is more water, the soil will conduct more electricity which means that there will be less resistance. Therefore, the moisture level will be higher. Dry soil conducts electricity poorly, so when there will be less water, then the soil will conduct less electricity which means that there will be more resistance. Therefore, the moisture level will be lower [7].

Another major factor influencing plant growth and healthy is the temperature [8], too cold or too hot temperature will result in unhealthy plants. Temperature influences most plant development process including photosynthesis, transpiration, absorption, respiration and flowering. In general, growth is promoted when the temperature rises and inhibited when temperature falls. The growth rate of a plant will not continue to increase with the increasing of temperature. Each species of plant has a different temperature range in which they can grow [8-9].

Humidity is important to plants because it partly controls the moisture loss from the plant. The leaves of plants have tiny pores, CO2 enters the plants through these pores, and oxygen and water leave through them. Transpiration rates decrease proportionally to the amount of humidity in the air. This is because water diffuses from areas of higher concentration to areas of lower concentration. Due to this phenomenon, plants growing in a dry room will most likely lose its moisture overtime. The damage can be even more severe when the difference in humidity is large. Plants stressed in this way frequently shed flower buds or flowers die soon after opening. High humidity can also affect the development of plant. Under very humid environments, fungal diseases most likely to spread, on top of that air becomes saturated with water vapour which ultimately restricts transpiration [10]. Monitoring these two parameters which is temperature and humidity need an accurate sensor. Study by Diaa Mehdi Faris suggested to use DHT11 temperature and humidity sensor where this component is cheap, reliable and easy to use. Each DHT11 element is strictly calibrated in the laboratory that is extremely accurate on humidity calibration. [11].

Monitoring of the crops exposure from excess cold or heat and unwanted pests nowadays can be done via an intelligent environment monitoring and control system based on Arduino UNO board for a small-scale agriculture namely greenhouse. The system user can monitor and control the greenhouse climate conditions remotely via web interface/mobile applications and GSM in a real-time manner [12-14]. The Arduino was said for interfacing all the sensors and integrates the results of each sensor. The sensors that connected to the Arduino should be programmed and the code was embedded in the Arduino [12].

To grow plants inside indoor greenhouse effectively, some parameters need to be monitor. Soil moisture, temperature and humidity are parameters that will affect a plant if too much or too low inside greenhouse. For example, as mention in introduction, too little watering a plant, it causes uneven water to soil and stunted growth of plant. But if too much water it causes root zone diseases such as Pythium and waste the water. Knowing when and quantity of the water needed to watering plant is also problem if soil moisture not to be monitor. By creating indoor greenhouse system equipped with soil moisture and DHT11 sensor will help to growth plant with a great quality.

In this work, two experiments were conducted to analyse the effect of water volume to soil moisture and relationship between rates of soil moisture to the plant growth. These experiments were done in AGROSEN lab at FKE, UiTM Shah Alam using Basil plant since it is the most suitable indoor plant to observe on the growth of its stem and leaves. Soil hygrometer and DHT11 temperature and humidity sensor were placed inside indoor greenhouse to senses change of parameters which is soil moisture, temperature and humidity.

The limitation of this study is the system cannot control the parameters which is temperature and humidity inside indoor greenhouse because the system did not have control element. If temperature and humidity are too low or too high it will affect plant growth.

2. Materials and Method

Figure 1 shows the basic block diagram of the system. The soil hygrometer sensor as shown in Figure 2 was inserted into the soil and the change of the moisture in the soil caused the resistance value varies and display the voltage output. The resistance measured decreased because of the increase of moisture and as resulted the voltage drop. DHT11 temperature and humidity sensor were placed inside indoor greenhouse to monitor the surrounding temperature and humidity. The processor used in this word was an Arduino UNO whereby it interpreted the data from these input sensors send the output to display in the serial monitor. Serial monitor became the output device that displayed soil moisture, temperature and humidity reading.

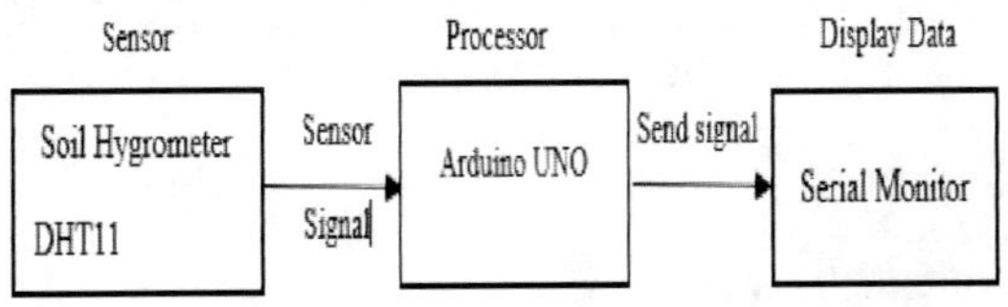

Figure 1. Block diagram of the system.

Figure 2. Soil hygrometer and DHT11 sensor.

2.1. Overall function of the system

Figure 3 shows the overall function of the system. The system starts by the sensors placement inside the indoor greenhouse. Soil hygrometer sensor was inserted into the soil and the DHT111 sensors mounted at the pillar of the greenhouse. These sensors measured the changes of parameters or values of soil moisture, temperature and humidity and the data displayed at the serial monitor.

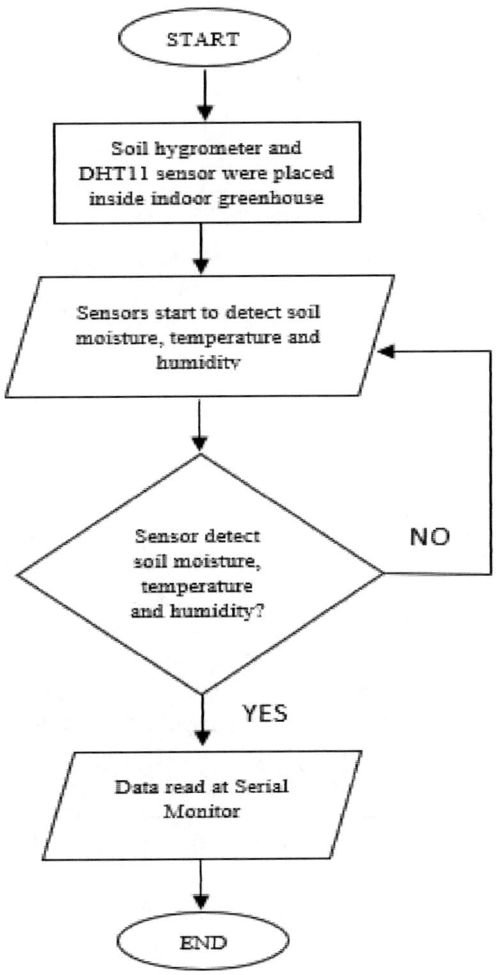

Figure 3. Flowchart of the system.

2.2. Data collection

The collection of data was performed in two stages. First was by observing the soil moisture at different conditions and secondly was by observing the relationship of soil moisture to plant growth. Figure 4 represents a beaker which contained Holland soil to observe the soil moisture. It is a type of loam soil made with a balance of the three main types of soil particularly sand, silt and clay soil which is perfect texture for plant growth.

Figure 4. 200 g of Holland soil.

2.2.1. Soil moisture experiment

This experiment has been run in lab is to analyze effect volume of water to the soil moisture. Five bottles placed inside greenhouse with 200 g of Holland soil each were poured by different volume of water. Soil moisture was monitored time by time and serial monitor displayed the result. Besides that, temperature and humidity were also recorded. Table 1 represents the bottles with specified volume of water poured into it.

Table 1. Bottles with different water volume.

Bottle No.	Water Volume (ml)
1	100
2	200
3	300
4	400
5	500

2.2.2. Relationship of soil moisture to plant growth

This experiment analyzed the relationship between rates of soil moisture to the plant growth. The soil moisture of the plant maintained at 30% for the first week. Second week, soil moisture increased to 60% and the final week, the soil moisture increased to 90%.

3. Result and Discussion

3.1. Indoor greenhouse monitoring system

Figure 5 shows the prototype of indoor greenhouse monitoring system equipped with sensors. The frames of greenhouse were made of Polyvinyl chloride (PVC) pipes. PVC pipes was chosen because of PVC's good mechanical strength and toughness are key technical advantages for its use in building and construction of greenhouse [15]. PVC pipes can be cut, shaped, welded and joined easily in a variety of styles. Its light weight reduces manual handling difficulties. This indoor greenhouse was design with ½ meter high, ½ meter width and ½ length. Transparent plastics were used to cover up the greenhouse to prevent insect pest. The soil hygrometer and DHT11 temperature and humidity sensors act as input to detect change of environment parameter inside this indoor greenhouse.

IOP Conf. Series: Materials Science and Engineering 1192 (2021) 012027 doi:10.1088/1757-899X/1192/1/012027

Figure 5. Prototype of indoor greenhouse monitoring system.

3.2. Soil moisture experimentation

Figure 6 represents the experimentation of five bottles which contained 200 g of Holland soil with different volume of water poured. The results are presented in Tables 2-6.

Figure 6. Five bottles with 200g of Holland soil.

Tables 2-6 represent the observation of soil moisture. Initial soil moisture reading of five bottles before adding the water were recorded. After that, water was poured into bottles with different volumes.

Table 2. Initial reading before adding the water to the soil.

	Input	Output			
Bottle	Water (ml)	Soil Moisture (%)	Temperature (°C)	Humidity (%)	Waste water (ml)
1	100	51	24	88	0
2	200	49	24	88	0
3	300	50	24	88	0
4	400	50	24	88	0
5	500	52	24	88	0

Table 3. Initial reading after adding water to the soil.

	Input	Output			
Bottle	Water (ml)	Soil Moisture (%)	Temperature (°C)	Humidity (%)	Waste water (ml)
1	100	81	24	89	0
2	200	100	24	87	10
3	300	100	25	85	50
4	400	100	25	88	100
5	500	100	24	88	180

IOP Conf. Series: Materials Science and Engineering 1192 (2021) 012027 doi:10.1088/1757-899X/1192/1/012027

Table 4. Reading after day one.

Bottle	Input	Output			
	Water (ml)	Soil Moisture (%)	Temperature (˚C)	Humidity (%)	Waste water (ml)
1	100	79	26	88	0
2	200	95	23	89	30
3	300	97	23	90	90
4	400	100	23	90	150
5	500	100	23	90	230

Tables 2 and 3 show the changes of soil moisture before and after adding the water according to the specified volumes. Only bottle 1 does not reach 100 % of soil moisture due to the limitation of water poured. This means that watering the plant already absorbed enough water. Waste water in Table 3 was recorded immediately after adding the water. Bottles 3, 4 and 5 have the most waste of water with more than 50 ml. Bottle 2 shows the best result with low in waste water but still maintained soil moisture of 100%. After day one of experiment, soil moisture in bottles 1, 2 and 3 started to decrease but not for bottles 4 and 5.

Table 5. Reading after day three.

Bottle	Input	Output			
	Water (ml)	Soil Moisture (%)	Temperature (˚C)	Humidity (%)	Waste water (ml)
1	100	73	27	88	0
2	200	85	26	89	30
3	300	88	27	89	120
4	400	90	27	88	150
5	500	95	27	88	230

Table 6. Reading after one week.

Bottle	Input	Output			
	Water (ml)	Soil Moisture (%)	Temperature (˚C)	Humidity (%)	Waste Water (ml)
1	100	58	25	89	0
2	200	72	25	89	30
3	300	78	25	89	120
4	400	83	25	92	150
5	500	88	25	90	230

Table 5-6 represents the results for each bottle soil moisture reduced gradually after one week. Besides that, temperature and humidity also recorded. The lowest temperature was recorded is 23°C and maximum reading is 27 °C. The average of humidity is 88%. There is no extreme climate change happened. So, it's suitable for a plant to growth.

3.3. Relationship of soil moisture to plant growth

Table 7 represents data collection during the second experiment. The data analyzed relationship between rates of soil moisture to the stems diameter. Four stems from plant type Basil were used to measured diameter. Photosynthesis is the process by which plants prepared their own food using carbon dioxide and water in the present of sunlight. In this process, functions of stem are to conduct water and mineral to leaves and to transport these products from the leaves to other parts of the plant including the roots. The stem conducts water and nutrient minerals from their site of absorption in the roots to the leaves. The higher soil moisture which means soil contains more water than bigger stem diameters [16-17]. Figure 7 represents the chart of four stems growth for three weeks. It can be observed that the rate of soil moisture effected the stems diameter.

Table 7. Data collection during Second experiment.

Week	Average Soil Moisture (%)	Stem Diameter (cm)			
		Stem 1	Stem 2	Stem 3	Stem 4
1	30	0.8	0.9	0.7	1
2	60	1.1	1	0.9	1.2
3	90	1.6	1.4	1.3	1.5

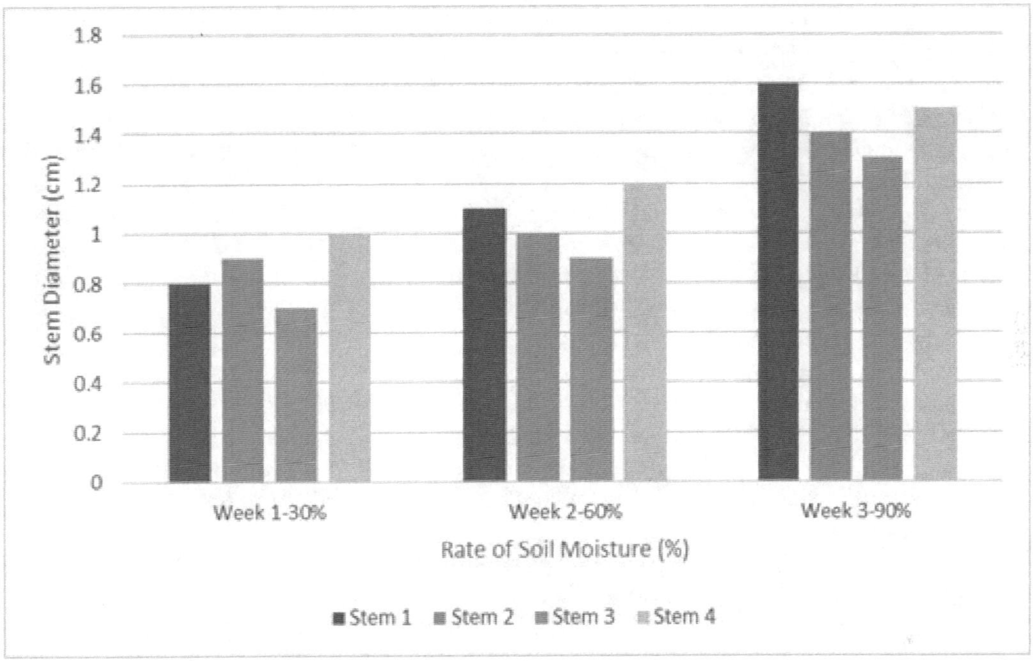

Figure 7. Stems diameter vs rate of soil moisture.

4. Conclusion

The prototype of indoor greenhouse monitoring system equipped with soil hygrometer and DHT11 temperature and humidity sensor were successful develop. From both experiment that have been run, result shows the effect volume of water to soil moisture and relationship between rates of soil moisture to the plant growth. In this study, soil moisture plays important role to growth plant. If too much watering plant, its effects on soil moisture. Thus, it's contributing to waste water. Besides that, there is no critical changed in temperature and humidity was recorded. For the recommendation in next project, this indoor greenhouse monitoring system should have control element to overcome the limitation. For example, temperature can be control by using fan.

Acknowledgments

Authors wishing to acknowledge assistance or encouragement from colleagues of Department of Electrical Engineering, School of Engineering Universiti Teknologi MARA Shah Alam with financial support from RMC Universiti Teknologi MARA Geran Penyelidikan Khas.

References

[1] Karimi M, Ahmadi A, Hashemi J, Abbasi A, Tavarini S, Guglielminetti L, et al. 2015 The effect of soil moisture depletion on stevia (*Stevia rebaudiana Bertoni*) grown in greenhouse conditions: Growth, steviol glycosides content, soluble sugars and total antioxidant capacity. *Sci Hortic*. **183** pp 93–99

[2] Pradeep Kumar S and Byregowda B K 2017 Greenhouse Monitoring and Automation System Using Microcontroller *International Journal of Engineering Trends and Technology (IJETT)* **45**
[3] Rosily A, Mini P K and Neenu M S 2014 Automated Greenhouse Monitoring System *International Journal of Engineering and Innovative Technology (IJEIT)* **3**
[4] Ibrahim Al-adwan, M S-D 2012 The Use of Zigbee Wireless Network for Monitoring and Controlling Greenhouse Climate *International Journal of Engineering and Advanced Technology (IJEAT)* **2** 35
[5] Barsoum N and Peter A 2015 GSM Greenhouse Monitoring and Control of Temperature and Soil Moisture *European International Journal of Science and Technology* **4** pp 1-10
[6] Slatyer R O. Plant-water relationship. *Academic Press*, New York
[7] Ling D P A review of Soil Moisture Sensors. Retrieved December 2014, from *Florinet Floriculture Newsletter- Moisture Sensors*
[8] Both A 2008 *Greenhouse Temperature Management* (New Jersey:Department of Plant Biology and Pathhology)
[9] Newton L A 2009 High-temperature inhibition of flowering of Phalaenopsis and Doritaenopsis orchids *HortScience* **44** pp 1271–1276
[10] Omid M and Shafaei A 2005 Temperature and relative humidity changes inside greenhouse *International Agrophysics* **19** pp 153-158
[11] Faris D M and Mahmood B 2014 Data Acquisition of Greenhouse Using Arduino *Journal of Babylon University/Pure and Applied Sciences* **22**
[12] Enokela J and Othoigbe T O 2015 An Automated Greenhouse Control System Using Arduino Prototyping Platform *Australian Journal of Engineering Research*
[13] Singh C B and Fielke J M 2017 Recent Developments in Stored Grain Sensors, Monitoring and Management Technology *IEEE Instrumentation & Measurement Magazine* pp 32-55
[14] Jiménez A, Jiménez S, Lozada P and Jiménez C 2012 Wireless Sensors Network in the Efficient Management of Greenhouse Crops *9th International Conference on Information Technology-New Generations, IEEE Computer Society*
[15] *Benefits of PVC over other materials*, available: http://www.pvcconstruct.org/en/p/material
[16] Stanhill G 1957 *The effect of differences in soil moisture status on plant growth* Soil Sei **84** pp 205-214
[17] Sarangi D, Irmak S, Lindquist J L, Knezevic S Z, Jhala A J 2016 Effect of water stress on the growth and fecundity of common waterhemp (*Amaranthus rudis*) *Weed Sci.* **64** pp 42–52

IOP Conf. Series: Materials Science and Engineering 1192 (2021) 012028 doi:10.1088/1757-899X/1192/1/012028

The potential of biomass waste in Malaysian palm oil industry: A case study of Boustead Plantation Berhad

N H S Jafri, D N Jimat, N F M Azmin, S Sulaiman, Y A Nor

Department of Biotechnology Engineering International Islamic University Malaysia Gombak, Selangor

Corresponding Author's E-mail: jnoraini@iium.edu.my

Abstract. The oil palm industry is known as the primary producer of biomass in Malaysia. The massive production of oil palm biomass has resulted in a significant waste disposal problem. This study aims to assess the potential of waste derived from oil palm plantations and processing mills to be transformed into value-added products for various applications. The study was conducted on three oil palm estates managed by Boustead Rimba Nilai Berhad using data on the generation rates of fresh fruit bunches (FFB), crude palm oil (CPO), and oil palm biomass waste from 2018 to 2020. The availability of FFB, CPO, empty fruit bunches (EFB) and excess fibers, and palm oil mill effluents (POME) in 2030 was predicted using a forecasting approach, with trend analysis as the tool of choice. From 2018 to 2020, the output of FFB, EFB and excess fibers, as well as POME grew by 16%, 14%, 23%, and 14% respectively. Based on trend analysis, the projection of FFB, CPO, EFB and excess fibers, and POME outputs in 2030 is expected to be 80%, 56%, 92%, and 56% respectively, which are greater than the figures recorded in 2020. The state of the palm oil industry in Malaysia was described in this study and also highlighted the potential of oil palm biomass in various industries and as a source of green energy. Oil palm biomass can be commercialized in a wide range of value-added products.

1. Introduction

Lignocellulose fibers have a great potential for use as important raw materials in a developing country such as Malaysia with high agricultural production. Oil palm is one of the most important crops grown in Malaysia, accounting for 34.56 %, followed by rubber (39.67 %), cocoa (6.75 percent), rice (12.68 %), and coconut (6.34 %). Oil palm is a high-yielding crop that produces one of the most affordable vegetable oils. Palm oil acts as a natural preservative, extending the shelf life of the product. Most of the products we used daily contain palm oil [1]. It is the most extensively traded vegetable oil globally, with 50 million tonnes produced every year. Since the 1970s, the Malaysian palm oil industry has thrived, with oil palm plantations spreading across rural and suburban areas. As of 2021, oil palm cultivation occupied 5.9 million hectares of land, producing around 19.86 million tonnes (MT) of palm oil and 2.32 MT of palm kernel oil [2].

The financial feasibility of palm oil is undeniable. Malaysia is one of the world's leading suppliers and exporters of palm oil, producing 8.4% of global oil and fat production and 19.1% of worldwide oil and fat export commerce [2]. Agriculture continues to be an important element of Malaysia's economy, with oil palm cultivation helping to revolutionize the sector. This development has a tremendous impact on the country's economic prosperity, increasing gross domestic product (GDP) and creating jobs to improve the economic well-being of the people [3]. The nation's export revenues from palm oil and palm oil products totalled RM 72.8 billion in 2020 [2]. Aside from the excellent monetary returns, the receiving environment is a crucial factor to consider. Despite the visible advantages of such industrial growth, the Malaysian palm oil industry has encountered significant challenges as the environmental impact of palm oil monoculture has become more widely recognized.

The majority of the oil palm tree (90%) is perceived as biomass, in which palm oil constitutes only 10% of the total biomass [4]. Oil palm biomass refers to agricultural by-products produced by the oil palm industry during replanting, pruning, and milling operations. Oil palm trunks (OPT) and oil palm fronds (OPF) are formed at plantations during replanting and pruning. In contrast, oil palm shell (OPS), palm kernel shell (PKS), empty fruit bunch (EFB), and palm oil mill effluent (POME) are generated during the milling process at the oil palm processing mills [5].

The oil palm sector is often associated with the environment because it is a land-intensive industry. The annual total dry matter (TDM) production rate is around 55 tonnes per hectares of land [6]. It is frustrating to learn that a significant portion of oil palm biomass waste is either disposed of by burning or dumped into the fields, posing a severe environmental hazard. The generation of massive amounts of biomass waste has heightened concerns regarding the sustainability and environmental implications of oil palm cultivation and processing, notably in terms of biodiversity loss, degradation of rainforest, and air pollution [1].

Palm oil is derived from the fruit of oil palm tree, which grows in tropical climates. However, edible oil production has a severe impact on the environment, leading to deforestation and loss of wildlife habitat. Deforestation and palm oil production go hand in hand. Palm oil producers damage biodiverse ecosystems by cutting down trees in tropical rainforests to construct oil palm plantations. Deforestation is the main contributor to climate change, causing global warming. A large portion of land in Malaysia is covered in tropical peat soils. Carbon is sequestered in this nutrient-dense soil. When the trees are lost, a large amount of carbon is emitted into the atmosphere, contributing to global warming [4]. Besides that, animals inhabiting biodiverse regions where oil palm is cultivated are at risk. Animals lose their habitats as a result of deforestation. Orangutans, Sumatran tigers, Sumatran rhinoceros, clouded leopards, and sun bears are among the wildlife species endangered by deforestation. Furthermore, the massive volumes of pesticides and herbicides required to sustain the oil palm plantation pollute the watercourse nearby [7].

Malaysia generates a substantial amount of biomass waste as a prominent player in the palm oil industry, which presents a great potential for this biomass to be commercialized in a wide range of biomass-based products. Several studies have been performed to transform this widely available and sustainable biomass into value-added products with various applications to minimize the challenges caused by improper disposal [5]. Oil palm biomass fibers have been reported to have exceptional properties when compared to other fibers. As a result, they can be utilized as reinforcement fillers to assist enzyme immobilization matrices and renewable sources of materials for the production of value-added products, like pulp and paper, bio-composite, and hybrid composites in the paper and furniture industries. In addition, oil palm biomass waste can be utilized for harnessing biomass energy in an eco-friendly and commercially viable manner [8].

Boustead Plantations Berhad (BPB) is one of Malaysia's most well-established upstream oil palm plantation companies with decades of experience. BPB is a subsidiary of Boustead Holdings Berhad (BHB), one of the oldest and biggest diversified conglomerates in Malaysia. Supported by BHB's involvement in the plantation sector, BPB has more than 50 years of experience managing oil palm plantation estate. BPB and its subsidiaries and associates managed the Group's plantation activities. They are primarily involved in the ownership and management of oil palm plantations, cultivation of oil palm and harvesting its fresh fruit bunches (FFB), and the processing and selling of crude palm oil (CPO) and palm kernel oil (PKO).

Boustead saw a 34.78% year-on-year increase to RM167.06 million in 2020 from RM123.95 million in the previous year, mainly due to higher pricing of palm goods, which boosted sales of CPO and PKO around the world [9]. BPB owns and manages 48 oil palm plantation estates in Malaysia, with 19 in Peninsular Malaysia, 20 in Sabah, and 9 in Sarawak. They also operate ten palm oil mills, with three in Peninsular Malaysia, five in Sabah, and two in Sarawak. One of its subsidiaries is

Boustead Rimba Nilai Sdn Bhd, which manages the group estates and mill on the East Coast of Malaysia. This review study was carried out on three oil palm plantation estates managed by Boustead Rimba Nilai Sdn Bhd using FFB processing data from 2018 to 2020.

Given the rising global demand for biomass, an overview of Malaysia's palm biomass industry is required to assess the country's biomass utilization potential. In addition to incorporating Malaysia into the current global trade pattern, the growth of Malaysia's palm biomass industry positions the country better for potential market challenges. Based on this notion, this review study aims to assess the potential of waste resulting from oil palm plantations and palm oil processing industries and provide information on the opportunity to utilize them further.

2. Data Collection and Analysis Techniques

This study relied on secondary data. The study was conducted using the information collected from the Sg Jernih Business Unit of Boustead Rimba Nilai Sdn Bhd. Three oil palm plantations located on the East Coast of Malaysia have participated. The oil palm plantations include Sg Jernih Estate (SJE), Bebar Estate (BBE), and Ladang Tabung Tentera Terengganu Estate (LTTTE). Among the participating estates, SJE is the only one with a palm oil mill, known as Sg Jernih Palm Oil Mill (SJPOM). Since the estates are close to each other and there are no other mills nearby, BBE and LTTE supply their fresh fruit bunches (FFB) to the SJPOM. The location of the Boustead Plantations estates and mills in Peninsular Malaysia is depicted in Figure 1.

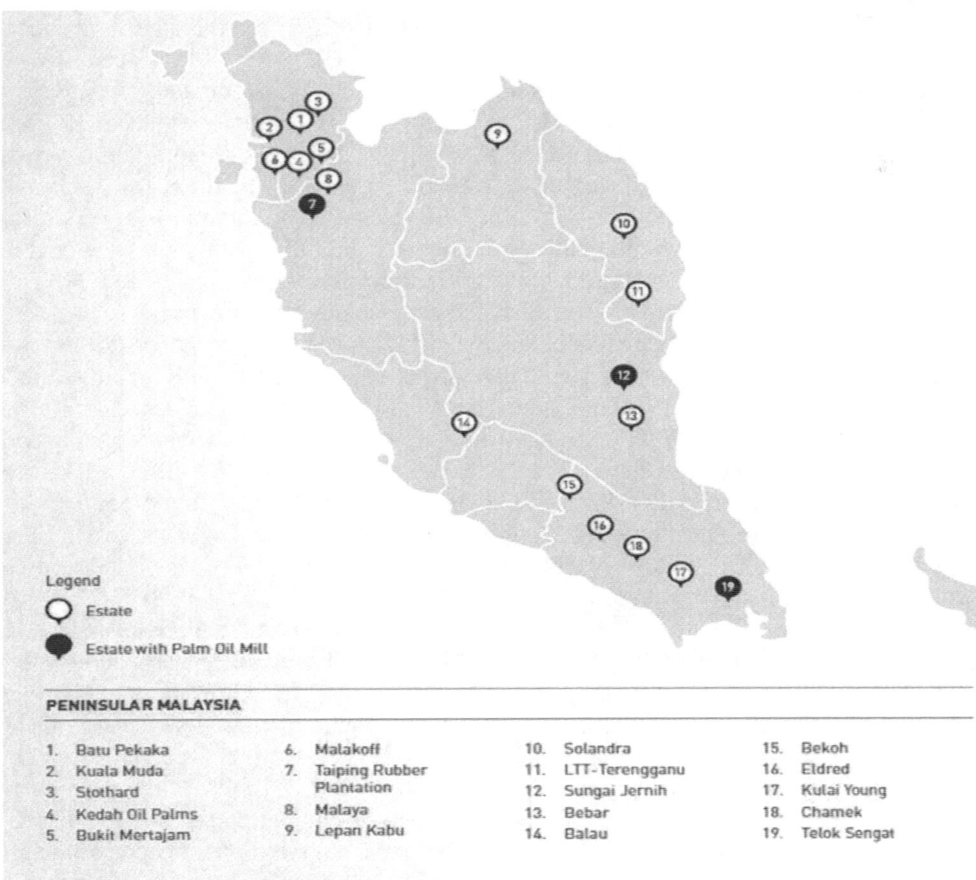

Figure 1. Location of Boustead Plantations estates and mills in Peninsular Malaysia.

The production and economic performances of SJPMO were investigated. As a result, the yield of FFB and FFB processed data was obtained from all participating estates from 2018 to 2020. The FFB data is critical for estimating the generation of oil palm by-products. The current condition and potential of oil palm production and the availability of oil palm waste were defined, explained, and characterized using descriptive analysis. The analysis was conducted using graphic and cross-tabulation techniques to provide explanations and detailed estimates, and statistical tools to process the data. The availability of fresh fruit bunches (FFB), crude palm oils (CPO), and oil palm wastes in 2030 was predicted using a forecasting approach, with trend analysis as the tool of choice. The data on the supply of FFB, CPO, and oil palm wastes in SJPMO from 2018 to 2020 was used to conduct the trend study.

2.1. Participating Estates

The study sites are located on the East Coast of Malaysia, where most of the existing plantations or estates involve oil palm cultivation. The oil palm plantations estates that participated in this study include Sg Jernih Estate (SJE), Bebar Estate (BBE), and Ladang Tabung Tentera Terengganu Estate (LTTTE).

Sg Jernih Estate (SJE) is situated along the Tun Razak Highway (Federal Route 12), which runs from Segamat in Johor to Kuantan in Pahang. The estate is located 1.5 kilometres southeast of the Pahang Peat Swamp Forest and 2.11 kilometres southwest of the Bukit Ibam Forest Reserve. Muadzam Sham is the closest town, which is about a 20-minute drive from the estate. The primary activities of the estate are oil palm cultivation (mature, young mature, replanting, and nursery). SJE is the only plantation with a palm oil mill located on the East Coast of Malaysia. The mill is known as Sg Jernih Palm Oil Mill (SJPOM), and it began operations in 1992. Therefore, the estate received a substantial supply of FFB from neighboring plantations such as BBE and LTTTE since it was the only one with mills. As of 2020, the mill has a processing capacity of 30 MT/h and consumes over 140,000 metric tonnes (MT) of FFB per year. The mill has two steam boilers that generate electricity from fiber and shell. POME is treated in eleven ponds (three of which are anaerobic) before being released and applied to plantation land within the SJE through channels and drains. The total land applications for the SJE comprise around 30 ha. At present, all of the EFB produced by the mill is applied directly to the estate. The electricity generated by the boiler was used for the mill operation, office, and housing complex of the SJPOM and the SJE. As of 2020, the mill produces approximately 32,000 MT of CPO, 119,000 MT of POME, and over 32,000 MT of EFB and excess fiber annually.

Ladang Tabung Tentera Terengganu Estate (LTTTE) is situated about 66 kilometres from Kuantan and 38 kilometres from Kemaman on the Jerangau-Jabo highway. Other plantations nearby include Felda Neram, Felda Sebarang Tayur Estate, Ketengah Perwira Estate, and Ladang Rakyat. The rivers of Sg Dadong and Sg Paloh flow along the estate's north and south borders, respectively, draining into Sg Kemaman and Sg Cherul. The estate generated 22,722.86 MT of FFB in the year 2010, which was all supplied to SJPOM. LTTTE is more hilly and undulating from a topographic perspective. Nonetheless, LTTTE introduced suitable agricultural practices such as vegetative ground cover maintenance, slope planting of Guatemala grass, proper frond stacking, and even gabion construction.

Bebar Estate (BBE) is situated approximately 65 kilometres from Gambang to Bandar Muadzam Shah on the left side of the Tun Razak Highway. The land is neighboring the Felda Merchong in the south, Lam Soon Estate in the north and northeast, Bukit Musuh State Land Forest in the mid-west, and Prosper Jawasa Estate in the east. Hutan Simpan Bukit Ibam marks the estate's northern boundary. A drain has been built through the estate, allowing surface water to flow from north to south and eventually into Sg Bebar. From 2005 to 2012, BBE is actively going through replanting. Ramets (tissue culture) and D x P seedlings were used in the plantings. Rhinoceros beetle damage on seedlings was common in the field, and pheromone traps were used to combat the problem. Both BBE and LTTTE supplies FFB to SJPOM.

IOP Conf. Series: Materials Science and Engineering 1192 (2021) 012028 doi:10.1088/1757-899X/1192/1/012028

3. Result and Discussion

3.1. Area coverage and Annual Production of FFB, CPO, and Oil Palm waste from 2018-2020
Oil palm plantations in the three estates (SJE, BBE, and LTTTE) remain constant from 2018 to 2020. Thus, the combined area for the three estates has been 6405.6 ha, with SJE accounting for 2571.6 ha, BBE for 2241.1 ha, and LTTT for 1592.9 ha. The plantation area for all estates is depicted in Figure 2.

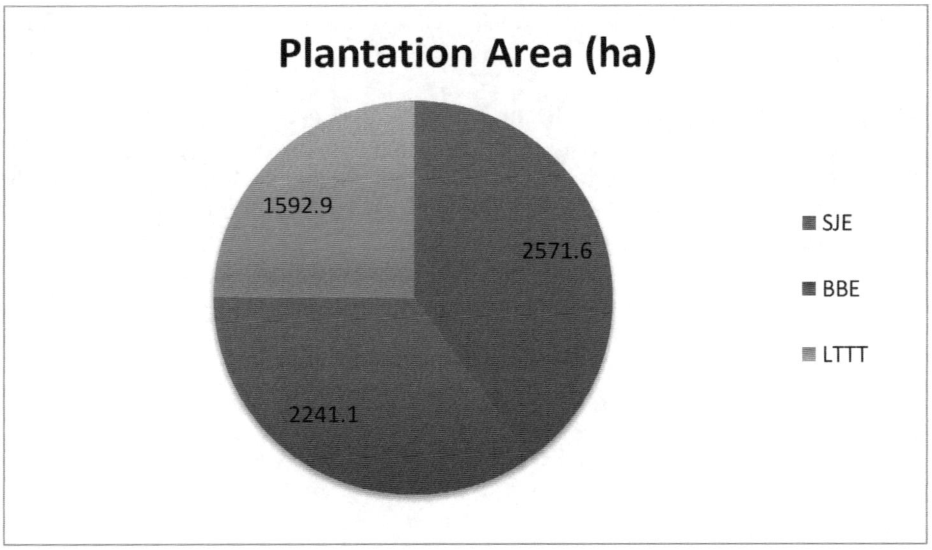

Figure 2. Plantation Areas (2018-2020).

The oil palm industry generates a large amount of oil palm biomass waste in the field and at oil palm mills. The wastes from the mill comprised of empty fruit bunches (EFB), oil palm shell (OPS), palm kernel shell (PKS), and palm oil mill effluent (POME). In contrast, oil palm fronds (OPF) and oil palm trunks (OPT) are wastes generated during replanting activities after reaching their economic life spans. During replanting, the increase in oil palm plantations has resulted in large quantities of wastes, particularly OPT and OPF. Approximately 24% of OPF is obtained from every oil palm tree over the course of a year of FFB harvesting in the field. Besides that, OPT was accounted for 70% of the replanting activities. Meanwhile, EFB is thought to be 22% of the FFB, and OPS are projected to be 5.5% of the FFB. The mill produces 0.67 tonnes of POME for every ton of FFB processed [1]. Based on the data obtained from Boustead Rimba Nilai Sdn Bhd, the participating estates (SJE, BBE, and LTTTE) directly supply all of their harvested FFB to the SJPOM for the production of CPO. Table 1 depicts the cumulative amount of FFB, CPO, and oil palm waste production obtained from all three estates from 2018 to 2020.

Table 1. Total FFB, CPO and oil palm waste production in 2018-2020

Year	Production (MT)			
	FFB	CPO	Oil Palm Waste (EFB, Excess Fiber and etc.)	POME
2018	118,315.21	27,756.00	25,041.19	101,978.00
2019	124,268.55	29,152.00	28,106.48	101,681.00
2020	140,930.28	32,054.00	32,579.48	119,586.00

Based on Table 1, the amount of FFB received from the three estates increased rapidly from 2018 to 2020, rising from 118,315.21 MT in 2018 to 140,930.28 MT in 2020 representing a 16% growth in FFB output from 2018 to 2020. Besides that, crude palm oil (CPO) production grew 14% from 27,756.00 MT in 2018 to 32,054.00 MT in 2020, indicating an upward trend as shown in table 1. The increase was primarily attributed to an increase in the amount of FFB yielded per hectare as a result of improved management and agricultural inputs. The volume of waste, which includes EFB and excess fibers, rises by 23% from 25,041.19 MT in 2018 to 32,579.48 MT in 2020. However, the amount of POME generated decreased marginally from 101,978.00 MT in 2018 to 101,681.00 MT in 2019, before steadily increasing to 119,586.00 MT in 2020. Between 2018 and 2020, the amount of POME generated increased by about 14%.

3.2 Projection of FFB, CPO, and Oil Palm Waste Production in 2030

A forecasting approach was used with trend analysis as the method of choice to predict the growth trend of FFB, CPO, and oil palm wastes output in the coming years. The data collected on the supply of FFB, CPO, and oil palm wastes in SJPMO from 2018 to 2020 was used to conduct the trend study. This estimate was used to measure the potential production of FFB, CPO, and oil palm wastes in 2030. As of 2020, the total oil palm harvested area for the three participating estates is about 6405.6 ha. Likely, the cultivated area does not grow significantly in the immediate future. This is a rational assumption given that Malaysian governments are limiting the expansion of oil palm plantations due to the rising concerns on the environmental impacts such as tropical deforestation, biodiversity loss, air pollution, and water pollution. Table 1 shows that between 2018 and 2020, the FFB output grew by 16%. Therefore, based on the trend analysis, the first forecast is that the FFB production will increase steadily to 80% by considering a 16% growth for every two years until 2030. On the other hand, the CPO sees a 14% increase in yield from 2018 to 2020. Therefore, this rising trend can forecast CPO production in 2030, resulting in a 56% increase in output. Furthermore, if the growth of 23% every two years is assumed until 2030, the yield of EFB and excess fibers is projected to increase to around 92%. Finally, based on a 14% increase in volume from 2018 to 2020, the volume of POME generated in 2030 is expected to be 56% higher than the value recorded in 2020.

4. The Scenario of Oil Palm in Malaysia

Malaysia is a tropical country with hot and humid weather throughout the year. This climate promotes oil palm growth and, subsequently, increased the oil palm cultivation in Malaysia. Oil palm *(Elaeis guineensis)* was originally an African plant that has been extensively cultivated in semi-wild orchards in tropical Africa. Even though oil palm is historically an African crop, South East Asia currently produces most global palm oil. It was first introduced to the region of South East Asia in the early 19th century. The oil palm was initially introduced in Malaysia in 1870 as an ornamental plant for planting in the Botanical Gardens in Singapore.

The first commercial oil palm estate in Malaysia was created in 1917 at Tennamaran Estate in Selangor. In the late 1950s, there was a massive shift from rubber to oil palm due to government funding for crop diversification. The oil palm has been developed fast as the most important commodity crop in Malaysia since 1960 due to the rapid growth of planting areas [7]. There is a surge in replanting across the country in the 1990s. By 1999, the land used for the cultivation of oil palm was reported to have grown to 3.31 million hectares, with 62 % of the total area in Peninsular Malaysia, while 28 % and 10 % are in Sabah and Sarawak [10]. Malaysia becomes the second-largest producer of palm oil in the world after Indonesia by the 2000s [7].

The seedlings for oil palm cultivation take about three months to germinate, after which they are cultivated in small plastic bags and kept in the so-called pre-nursery for a few months. They are then transplanted into larger plastic bags and nurtured in a nursery for several months before being planted in the field at around one year of age. The new enhanced crosses begin to blossom and bear their first bunches of fruit after less than a year of transplantation. Their leaves are taller than 2 meters in height and diameter at this age. In its early stages, the trunk grows at a rate of 35 to 75 cm per year and

IOP Conf. Series: Materials Science and Engineering 1192 (2021) 012028 doi:10.1088/1757-899X/1192/1/012028

produces alternate rows of leaves, depending on the gene. The number of leaves in oil palm crops increases from 30 to 40 every year at the age of 10-20 years, and the palm tree reaches a height of 7-13 meters. At the age of more than 20 years, the base of the old leaves covers the stem and starts to fall off. Growth and production had declined at this stage [1]. Leaf production decreased roughly 20 to 25 annually after that. Thus, the average economic lifespan of oil palm is between 25 years and 30 years [6].

Substantial growth in oil palm plantations started in the 1960s, with a peak in replanting in the1990s. During replanting, the oil palm reaches heights of 7 to 13 meters and has a width of 45 to 65 cm, 1.5 meters above the soil's surface. Each frond of the mature oil palm has about 41 leaves. The re-plantation method yields approximately 8.36 million tons of dried biomass in 2000, comprising 7.02 million tons of trunk and 1.34 million tons of leaves [6]. The freshly cut tree trunk cannot be burnt in the plantation because of the high moisture content of roughly 70% fresh weight. Leaving the old trunk for natural decomposition impedes the replanting process and attracts insects, harming the new trees. It usually takes five to six years for the tree trunk to decompose [6]. Many crude palm oil mills harness the fiber and shell inside their low-pressure boilers, and the EFBs are either burned, causing air pollution, or sent to the plantation to decompose naturally.

Palm oil is dominant in the global economy, where palm oil is one of the seventeen main oils and fats on the international market. The oil palm fruit contains two different palm oils, such as crude palm oil (CPO) and palm kernel oil (PKO). CPO is extracted from mesocarp, whereas PKO is produced from seed or kernel. Palm oil is mainly utilized for the manufacture of margarine and compounds in cooking fats and oils and the manufacture of soap, cosmetics, candles, and detergents. The generation of PKO accounts for nearly 12% of its palm oil production [7]. The progress of the palm oil industry in Malaysia is the product of ideal weather, advanced technologies, and facilities for milling and processing, practical and good management skills, and research and development. The government of Malaysia is wholly committed to the industry's growth and supports the global expansion of palm oil production. As a result, palm oil is now readily embraced internationally, and palm oil has been exported to more than 140 countries in the world by Malaysia [1].

5. Waste Generation

The Malaysian palm oil industry has grown rapidly in the last decades. This growth has resulted in a rapid expansion of oil palm cultivation areas, as well as a rise in the number of operating palm oil mills. The oil palm industry generates a large amount of oil palm biomass waste in the field and at oil palm mills. Waste such as oil palm trunks (OPT) and oil palm fronds (OPF) are created during replanting and pruning at oil palm plantations after their economic life span has elapsed. In palm oil mills, converting fresh fruit bunch (FFB) into crude palm oil (CPO) has led to the production of high volumes of solid and liquid waste. The solid wastes from mills include empty fruit bunches (EFB), oil palm shells (OPS), and palm kernel shells (PKS). Meanwhile, the liquid waste is palm oil mill effluent (POME) [1,4].

For every kg of palm oil produced, about 4 kg of dry biomass is generated, with roughly a third of this represented by FFB derived wastes and the remaining two-thirds by trunk and frond material [11]. Oil palm trunk (OPT) is a solid waste that is abundantly available throughout the year. Huge volumes of OPT biomass wastes are collected during replanting, as trees that have outlived their useful lifespan are felled. It is considered the least expensive lignocellulosic raw material when compared to wood. In contrast, oil palm fronds (OPF) fibers are made up of varying sizes of vascular bundles. Vascular files are common in parenchymal tissue with a thin wall. Each buddle is composed of vessels, fibers, round gloves, phloem, and parenchyma tissue. The xylem and phloem tissues are easily distinguished, with the phloem separated into two distinct sections in each bundle [12].

Empty fruit bunches (EFB) are abundant as fibrous material of solely biological origin in a standard palm oil mill. EFB fibers are hard and solid multicellular fibers with a central section known as a lacuna that has porous surface morphology. The cross section of a fiber is a polygon with a compact bundle or vascular packet enclosed by dense cell layers. In monocytes, vascular fibers are normally covered by a few layers of thick cell walls, which provide tensile strength as well as side compression force. EFB contains no chemical or mineral additives, and depending on how it is handled at the mill, it is free of foreign elements such as gravel, wood residues, nails, waste, and so on. 1 kg of wet EFB is generated for every kg of palm oil. Besides that, EFB contained high amount of water which is over 60% due to biological growth and steam sterilization at the plant [7].

On the other hand, liquid waste is formed by extracting palm oil from a wet process in a decanter. This liquid waste is combined with waste from cooling water and sterilizer to yield palm oil mill effluent (POME). POME is a highly polluting wastewater generated by palm oil milling operations. POME is a viscous brown liquid with fine suspended solids that has a pH range of 4 to 5. POME is also a liquid with high concentrations of organic acids and a COD level greater than 20,000 mg/L. The relatively high content of potassium, nitrogen, phosphorus caused POME to have strong malodor [1,7]. Before it can be released into the waterway, it must be treated properly and effectively. POME is considered as the highest volume of wastes produced in palm oil mill. The higher the amount of FFB used, the more POME produced. For every tons of processed FFB, a standard palm oil mill produces 600-700 kg of POME [5].

In addition, oil palm shells (OPS) is a waste product derived from palm oil mills. The shape of OPS aggregate differs based on the extraction process or breaking of the nut. The shape of OPS may be irregular, circular, angular, flaky, or polygonal. An OPS is also lightweight, with a relatively smooth shell surface texture [13]. Furthermore, the OPS are an organic aggregate with numerous pores and a high water absorption rate. A study demonstrated that water absorption for 24 hours was 23.30% with OPS aggregate sizes ranging from 14 to 5 mm, in comparison with water absorption for 24 hours was 0.76% with granite [14]. The shell fractions remaining after the nut has been extracted after crushing in the palm oil mill is known as palm kernel shells (PKS). Kernel shells are a fibrous material conveniently treated in bulk from the production line to the final application. Dust-like fractions and tiny fibers are combined with large and small shell fractions. PKS has a poor water absorption rate as compared to OPS, with water absorption rates ranging from 11% to 13% [15].

6. Potential of Oil Palm Waste

Over time, the number of palm oil mills in operation has increased. In 1990, only 261 mills with a gross production capacity of 42.8 million tonnes of fresh fruit bunch (FFB) were in the process each year. This has resulted in soil and groundwater contamination and the release of harmful gasses leading to rapid environmental degradation. Palm oil mills generate considerable amounts of biomass waste and create a massive volume of waste [15]. The oil palm processing mills, on the other hand, contribute substantially to environmental deterioration on both the input and output sides of their operations. On the input side, crude palm oil (CPO) consumes vast amounts of water and electricity. In contrast, on the output side, processing practices accumulate large quantities of solid waste, wastewater, and air pollution. The solid wastes constitute mesocarp fruit fibers (MF), palm kernel shells (PKS), pressed fruit fibers (PPF), and empty fruit bunches (EFB). Liquid waste is formed by extracting palm oil from a wet process in a decanter. This liquid waste is combined with waste from cooling water and sterilizer to yield palm oil mill effluent (POME) [1].

However, due to biological growth and steam sterilization at the plant, EFB is saturated with water. Since EFB has a moisture content of about 67%, it must be pre-processed before being regarded as good fuel. EFB, in contrast to shells and fibers, are typically burned, creating air pollution, or returned to the plantations as mulch. EFB can be gathered easily and are available for use in all palm oil mills. Since fibers and shells are easier to handle and produce higher-quality fuels than EFB, it would be beneficial to use EFB for on-site energy demand while making fibers and shells available for off-site use generate more revenue than burning on-site [1].

Palm oil processing also produces Palm Oil Mill Effluent (POME), which is highly polluting wastewater. It is often disposed of in disposal ponds, resulting in contaminant leaching into groundwater and soil and the release of methane gas into the atmosphere. Instead, POME could be used to produce biogas via anaerobic digestion. Many palm oil mills have already implemented this method to meet water quality requirements for industrial effluent. The gas, on the other hand, is flared off [5]. Anaerobic digestion is commonly used as a primary treatment for POME in the industry. Liquid effluents from palm oil mills, for example, can be transformed anaerobically into biogas, which can then be used to generate electricity through gas-fired engines or gas turbines [1]. During POME digestion, the odor emitted into the ambient air lowered the air quality in the surrounding lagoons [7].

Traditionally, mesocarp fiber is combined with kernel shells and used as a solid fuel to power the mill. Mulching sends the extra fiber and EFB to the plantation [15]. Typically, the fibers recovered from pressed palm fruits are burned as sources of energy for palm oil mills. Carotenoids, vitamin E, and sterols were present in abundance in the fibers as well [13]. Because of the high volume of free fatty acid and peroxide values, oil palm fibers are ideal for non- edible applications. Nutraceutical products can be made from minor components such as phospholipids, vitamin E, and carotenes [13].

Since PKS contains palm oil residues, it has a substantially higher heating advantage than traditional lignocellulosic biomass. PKS is a high-quality biomass fuel with a uniform size distribution, simple operation, easy crushing, and minimal biological activity due to reduced moisture content instead of other industry residues [13]. Palm oil mills' press fiber and shell have historically been used as a solid fuel for steam boilers. The steam generated is used to power turbines that generate electricity. These two solid fuels will produce more than enough energy to fulfil the energy requirements of a palm oil mill [1]. Optimization of the utilisation of oil palm wastes into the value-added products is highly essential to ensure the wastes are well managed and not left abundant in the plant estates. However, it is important to ensure that the selected treatment process is effective to ensure mass production of the respective value added products.

Acknowledgments
The authors acknowledge the Ministry of Education Malaysia (FRGS19-091-0700) and International Islamic University Malaysia for providing financial support to this project.

References

[1] Liew WL, Muda K, Azraai M, Affam AC, Kheang S 2017 Agro-industrial waste sustainable management–a potential source of economic benefits to palm oil mills in Malaysia *Journal of Urban and Environmental Engineering* **11** 108-118

[2] Malaysian Palm Oil Council (MPOC) 2021 About the Palm Oil *Retrieved on 4 April 2021* http://mpoc.org.my/about-palm-oil/

[3] Aljuboori AH 2013 Oil palm biomass residue in Malaysia: availability and sustainability *International Journal of biomass & renewables* **2** 13-18

[4] Kurnia JC, Jangam SV, Akhtar S, Sasmito AP, Mujumdar AS 2016 Advances in biofuel production from oil palm and palm oil processing wastes: a review *Biofuel Research Journal* **3** 332-346

[5] Khalil HP, Amouzgar P, Jawaid M, Hassan A, Ahmad F, Hadiyana A, Dungani R 2012 New approach to oil palm trunk core lumber material properties enhancement via resin impregnation *Journal of Biobased Materials and Bioenergy* **6** 299-308

[6] Lim KO 2017 The future energy potential of replanting wastes in Malaysia *International Energy Journal* **10** 3-4

[7] Abdullah N, Sulaiman F 2013 The oil palm wastes in Malaysia *Biomass now-sustainable growth and use* **30** 75-93

[8] Abdul Khalil, H P S, Bhat, A H, Jawaid, M, Amouzgar, P, Ridzuan, R, and Said, M R 2010 Agro wastes: Mechanical and physical properties of resin impregnated oil palm trunk core lumber *Polymer Composites* **31** 638-644

[9] Khalid S 2021 Boustead Plantations posts profitable 4Q, thanks to higher palm product prices *Retrieved on 19 Mei 2021* https://www.theedgemarkets.com/article/boustead-plantations posts-profitable-4q-thanks-higher-palm-product-prices.

[10] Hai TC, Ng A, Prudente C, Pang C, Yee JT 2001 Balancing the need for sustainable oil palm development and conservation: The Lower Kinabatangan floodplains experience *In ISP*

National Seminar **1** 1-4

[11] Dungani R, Aditiawati P, Aprilia S, Yuniarti K, Karliati T, Suwandhi I, and Sumardi I 2018 Biomaterial from oil palm waste: properties, characterization and applications. *Palm Oil* **31** 1-6

[12] Sulaiman F, Abdullah N, Gerhauser H, Shariff A 2011 An outlook of Malaysian energy, oil palm industry and its utilization of wastes as useful resources *Biomass and bioenergy* **35** 3775-3786

[13] Abd Majid R, Mohammad AW, May CY 2012 Properties of residual palm pressed fiber oil *Journal Oil Palm Residues* **24** 1310-1317

[14] Mannan MA, Ganapathy C 2002 Engineering properties of concrete with oil palm shell as coarse aggregate *Journals of Construction and Building Materials* **1** 29-34

[15] Otti VI, Ifeanyichukwu HI, Nwaorum FC, Ogbuagu FU 2014 Sustainable oil palm waste management in engineering development *Civil and Environmental Research* **6** 121-125

IOP Conf. Series: Materials Science and Engineering 1192 (2021) 012029 doi:10.1088/1757-899X/1192/1/012029

Forensic analysis on printer inks via chemometrics

L H Mohd Zawawi[1], N F Mohamed Azmin[1,4], M F Abd. Wahab[1], S I Ibrahim[2] and M Y Mohd Yunus[3]

[1]Department of Biotechnology Engineering, International Islamic University Malaysia, Jalan Gombak, 53100 Kuala Lumpur, Malaysia
[2]Department of Electrical Engineering, International Islamic University Malaysia, Jalan Gombak, 53100 Kuala Lumpur, Malaysia
[3]Faculty of Chemical and Process Engineering Technology, Universiti Malaysia Pahang, Lebuhraya Tun Razak, 26300 Kuantan, Pahang, Malaysia

[4]Corresponding author's email: norfadhillah@iium.edu.my

Abstract. Printer inks are becoming necessary for utilization for wide range of purposes by society in current times with rapid development in technology and digital media area. Thus, forgery and counterfeiting becoming easier for the criminals. It is dangerous as some criminals will misused the technology by mean of addition and adulteration of parts of text or numbers on document as the inks and document can be made as an evidence in the trial court. Thus, the characterization and differentiation of the printed inks in the suspected documents (civil or criminal cases) may provide important information about the authenticity of the printer inks. The focus of this study to differentiate the chemical component of three different types of sample inks by incorporation of FTIR spectrophotometer with principal component analysis. The unique features of the ink samples were unmasked from the score plots of the principal component analysis. Thus, the graphical representation provided by the FTIR spectra with principal component analysis enabled the discrimination certain chemical in the printer inks.

1. Introduction

Ink cartridges and inkjet printers are important technology for everyone daily uses whether for office or personal purposes. Inkjet printers can be used to print out many different types of documents. Consequently, the negative parts of this technology are criminals are able to manipulate this technology to commit several crimes such as forgery and counterfeiting. The crime of forgery generally concerns the making of fake documents (certificates or identification cards), false banknotes, stamps and checks, the changing of an existing document (e.g., prescription) or the creation of a signature without authorization (false stamps, false signs) [1 need to find fwd ref]. These documents also can be used as evidences in a court for some particular cases. The verification of the authenticity of these documents are crucial because if these documents are lacking in evidence of authenticity, the trial judge have the authority to dismiss the document as irrelevant evidence. Therefore, forensic chemists and forensic laboratories are increasingly engaged in the verification and examination of the documents and inks. Paper and printer inks can be used as a sample by the chemist to examine the authenticity by providing details with regards to the different types of chemical compound of the ink and when its manufacture with its sources [2]. This information can be related to examine in case a particular ink is counterfeit or authentic.

The utilization of ink cartridges with printer will never fade away anytime soon in the current time or foreseeable future because it can translate as a medium between people and digital world. The interesting and challenging parts about inks and printers is that anyone with access to low-cost scanners, printers, inks and personal computers are capable of generating forgery, counterfeit and unauthorized duplication of some particular documents [3]. The immediate progress of the printing technologies has given high advantage towards the criminals. The information about chemical formulation of each manufactured ink is unknown, therefore there is quite a challenge and can be out into question to identify the brand of the printing ink cartridges [4]. This project focusing on to differentiate the chemical components in the three different types of cartridge inks by developing a modeling of the inks by using multivariate data analytical techniques with FT-IR spectroscopy and MATLAB. The modeling will be functioning as a database for the chemists and examiners to aid their verification and authentication process by extracting the particular information about the inks rather than making the modeling from the scratch.

This study focuses on the three different types of authenticity of printer inks. The identification and characterization of printer inks can be performed based on the properties of functional groups exists in the sample. The samples in then analyzed by the FTIR spectrophotometer. The chemical bonds in the samples will vibrate once the infrared radiation that interacts with the matter is absorbed which produce the infrared spectrum. The correlation shows the infrared band positions and chemical structure becomes the fundamental information for the analyst to identify the unknown molecules or even the structural identification. After that, the dataset obtain from the FTIR spectra will be further analyzed using principal component analysis. Principal component analysis is one of the chemometric methods to recognize the similarities and differences of the chemical information by providing the pattern of the sample [5,6]. In addition, significant information about the samples also can be extracted from the principal component analysis [7].

2. Materials and Method

Six different types of inks were used in this study. The authentic and claimed as authentic printer inks of the same brand were bought separately at the local stores in Kuala Lumpur whilst the counterfeit printer inks were bought from Shopee online stores. The color of the printer inks used in this project is a yellow color. The samples were divided into *Group 1* and *Group 2* as shown in Table 1. *Group 1* consists inks that were printed on papers using a personal office printer followed by FTIR analyses whilst *Group 2* consists samples of inks which were tested directly by the FTIR spectrophotometer. The brand of the personal printer model used in this project was Epson printer model L100. Samples in *Group 1* were printed out on A4 sized papers (70 gsm) from a brand named Topvalu Multipurpose. Then the printed papers were cut into rectangular blocks with a size of 3.81 cm height and 5.08 cm width.

Table 1. Two groups of coloured and black printer inks representing the authentic, claimed as authentic and counterfeit samples.

Type of Ink	Brand	Colour/Black	Group 1	Group 2
Authentic	Epson T6641 (Yellow)	Colour	Sample 1	Sample 7
	Epson T6641 (Black)	Black	Sample 2	Sample 8
Claimed as authentic	Epson Refill Ink Compatible with T664 Series	Colour	Sample 3	Sample 9
		Black	Sample 4	Sample 10
Counterfeit	Universal Refill Ink for All CISS or Ink Tank Printer	Colour	Sample 5	Sample 11
		Black	Sample 6	Sample 12

The analysis process carried out using FTIR spectrophotometer which is located in International Institute for Halal Research and Training (INHART) laboratory at IIUM Gombak. The range of spectra is 4000 cm^{-1} to 400 cm^{-1} per one scan per sample. The generated dataset from FTIR spectrophotometer will be exported from Excel to MATLAB software version MATLAB® R2018b 7.10.0.499 produced by MathWorks company prior to performing principal component analysis (PCA). The PCA reduces the dimensions of a high dimensional data matrix onto a simple and interpretable PCA plots. In Figure 1, the data matrix X comprises by n by p matrix where 'n' represents the absorbance values of the samples and 'p' represent the number of spectral ranges. The score plots obtained from principal component analysis were further analyzed for the spectral characterization and identification of samples.

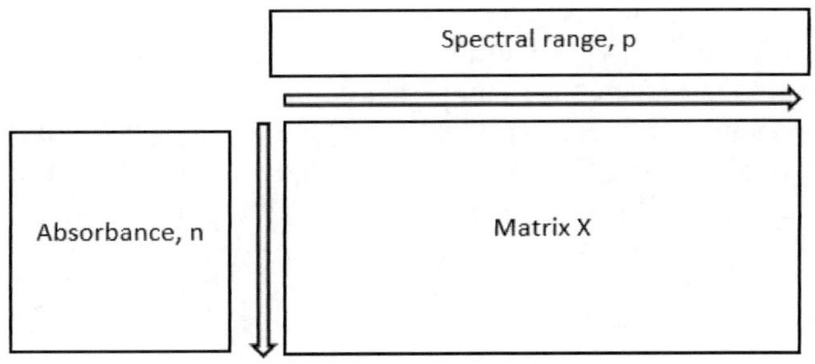

Figure 1. Two-dimensional data array, X.

3. Results and Discussions

The FTIR spectra illustrating the chemical structures of each type of printer inks used in this study were shown in Figure 2. The odd numbered samples represent the coloured printer inks (Samples 1, 3, 5, 7, 9 and 11) whilst the even numbered samples represent the black printer inks (Samples 2, 4, 6, 8, 10 and 12). Differences in the chemical arrangement and chemical composition in the inks result in distinctive spectra of each sample. When an organic compound interacts with infrared radiation, certain energy frequencies are absorbed while others are transmitted or reflected. The presence of functional groups in the substance determines whether the frequencies were absorbed or transmitted. To aid to the examination of the FTIR spectra, the spectra were assigned into four regions to determine the possible functional groups or chemical compounds in the inks (Figure 2).

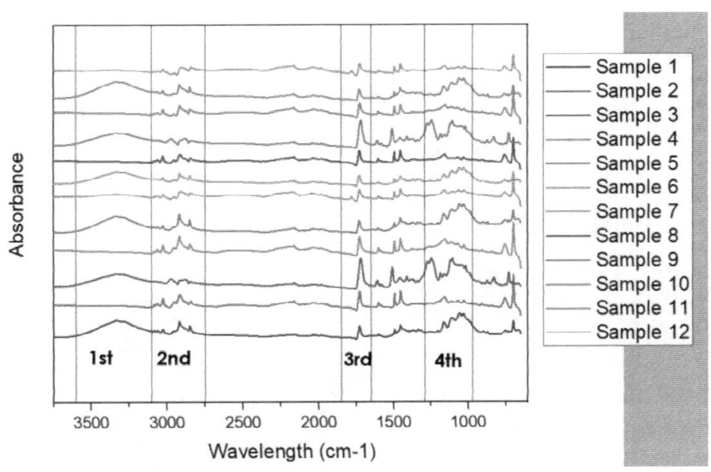

Figure 2. FTIR spectra for all printer ink samples and the four regions of analyses.

The first (1st) region at wavelength of 3600 – 3200 cm^{-1} suggested the presence of hydroxyl group (O-H) with possible presence of compounds of carboxylic acid, alcohol and aldehydes. Clearly, all coloured printer inks showed strong absorption band in the first region due to presence of O-H group. However, weak absorption band was displayed by all of black printer inks samples at the wavenumber

IOP Conf. Series: Materials Science and Engineering 1192 (2021) 012029 doi:10.1088/1757-899X/1192/1/012029

of 3600 – 3200 cm^{-1}. This suggested that O-H group is absent in the black printer inks. The second (2nd) region at wavelength of 3200 – 2750 cm^{-1} corresponds to the C-H stretch with possible presence of compounds of alkenes, unsaturated rings (-CH=CH-) or aromatic compounds i.e., the aliphatic groups (-CH, $-CH_2$, $-CH_3$). One of the rapid diagnoses for the identification of chemical structure in the FTIR studies is the presence of single bond, double bond or aromatic ring formed by the C-H molecular stretch. The third (3rd) region at wavelength of 1900 – 1700 cm^{-1} suggested the presence of ketones group i.e., C=O functional group with possible presence of acid halides, aldehydes, amides, anhydrides, and carboxylic acids. Finally, the fourth (4th) region at wavelength of 1350 – 950 cm^{-1} corresponds to the presence of and sulphur (S=O), fluorine (C-F) and ether compound (C-O-C) functional group.

The principal component analysis (PCA) was performed on the spectra of twelve printer inks. The eigenvalues from the PCA decomposition showed that the first three principal components (PC) explain 94.78 percent variability in the printer inks, as shown in the scree plot (Figure 3).

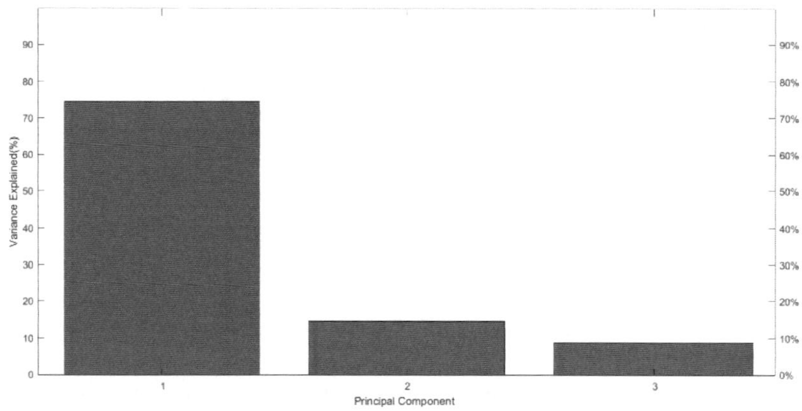

Figure 3. PC analysis of printer inks. Scree plot of the total variability explained by the first three PCs of the printer ink.

The score plot generated by performing PCA in the first region showed that PC1 successfully separates the black inks from the coloured inks due to the presence of the O-H functional group (Figure 4). The black inks were positioned on the right of PC whilst the coloured inks were positioned on the left of PC1.

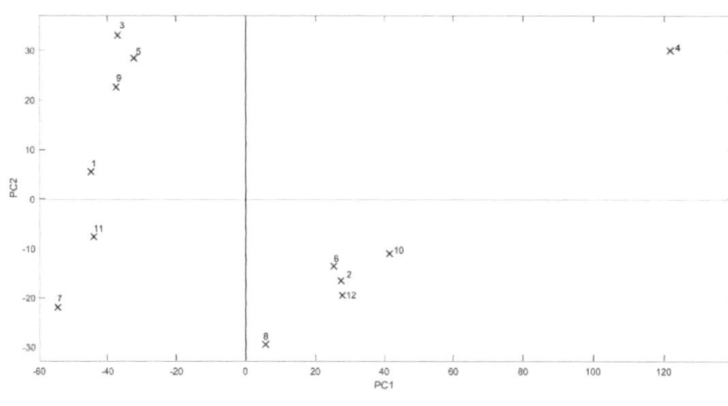

Figure 4. Score plot between first principal component and second principal component at region 3600 – 3200 cm^{-1}.

In the region 3200 – 3000 cm^{-1}, PC2 slightly separates the claimed to be authentic inks from the original and counterfeit inks as shown in Figure 5. The score plot divided into four quadrants which are quadrant 1 (located at positive region of principal component 1 and positive region of principal component 2), quadrant 2 (located at positive region of principal component 1 and negative region of principal component 2), quadrant 3 (located at negative region of principal component 1 and negative region of principal component 2) and quadrat 4 (located at negative region of principal component 1 and positive region of principal component 2).The samples in quadrants 1 and quadrant 2 the samples are displayed to have adsorption band below wavenumber of 3000 cm^{-1} while all the other samples in quadrant 3 and quadrant 4 shown to have peak above wavenumber of 3000 cm^{-1}. The score plot results show to distinguish between aliphatic and unsaturated compound in this project because any C-H stretch above 3000 cm^{-1} wavenumbers is the resulting from the C-H stretch of the aromatic, heteroaromatic, alkyne or alkene compounds meanwhile below than wavenumber of 3000 cm^{-1} indicates the compound is in aliphatic groups.

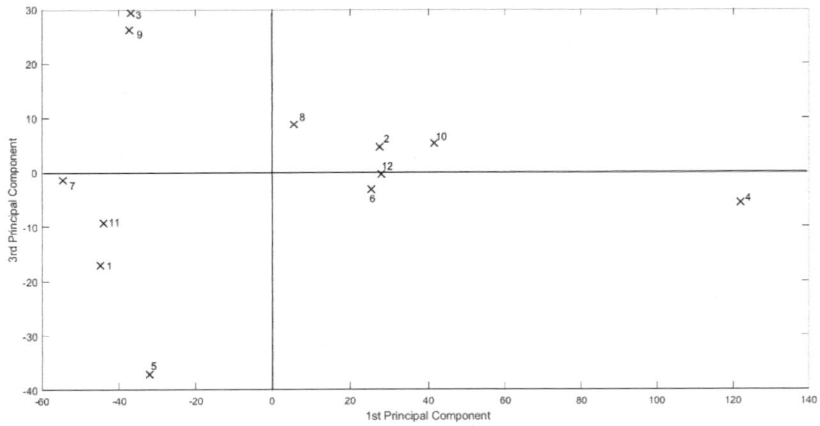

Figure 5. Score plot between first principal component and second principal component at region 3200 – 3000 cm^{-1}.

Figure 6 shows the score plot in the third region. The influence of ketone group or C = O in the spectrum compared to other functional groups such as O-H and C-H is the most intense due to the high intermolecular force or the difference in electronegativity shown by the oxygen with carbon bond. The wavelength of 1900 – 1700 cm-1, the present or absence of ketone functional group C=O can be detected. The other possible compounds are also aldehydes, carboxylic acid and esters. From the plot, only sample 4 were seen in the quadrant 4 meanwhile in the quadrant 3 there were only sample 5 and sample 11. In quadrant 1, there are 5 samples allocated which are sample 2, sample 6, sample 10 and sample 12. Sample 1, sample 3, sample 7 and sample 9 were in the quadrant 4. The score plot results show to distinguish between the samples with presence of ketone functional group and the sample with absence of ketone functional group.

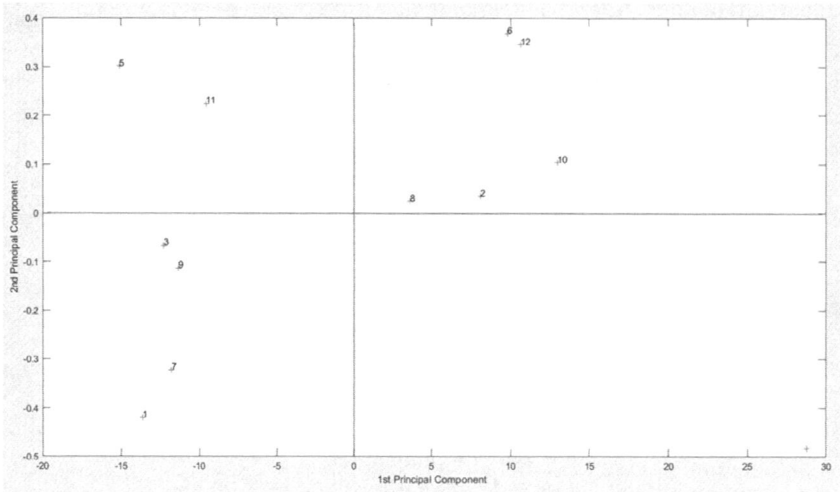

Figure 6. Score plot between first principal component and second principal component at region 1900 – 1700 cm^{-1}.

4. Conclusions

The implication of pairing between analytical method with chemometrics for the differentiation and characterization of several different types of printer inks were briefly discussed. The score plot of the principal component analysis enabled the discrimination of certain printer inks. The three proposed regions were significant to identify some of the compounds exist in the samples. The limitation of this study is that this study is applicable for same exact brands of samples used in this project only. Although there is no exact pattern to differentiate between counterfeit ink and authentic, some chemical component present in the printer inks can be identified and characterized. From forensic perspective, this situation may complicate the identification of the authenticity of printed ink deposited on a document. Thus, more research should be done to find out the other variables affecting the printer inks; hence producing a developing a printer inks database which can be used further for forensic purposes.

Acknowledgments

Authors wishing to acknowledge the Department of Biotechnology Engineering, Kulliyyah of Engineering IIUM for providing support to conduct this study.

References

[1] Kumar R and Sharma V 2018 Chemometrics in forensic science *TrAC - Trends Anal. Chem.* **105** pp 191–201

[2] Ferreira A, Navarro L C, Pinheiro G, do. Santos J A and Rocha A 2015 Laser printer attribution: Exploring new features and beyond *Forensic Sci. Int.* **247** pp 105–125

[3] Donnelly S, Marrero J E, Cornell T, Fowler K and Allison J 2010 Analysis of pigmented inkjet printer inks and printed documents by laser desorption/mass spectrometry *J. Forensic Sci.* **55** pp 129–135

[4] Zielinski A, Windson Isidoro Haminiuki C, Alberti A, Nogueira A, Demiate I and Granato D 2014 Comparative study of the phenolic compounds and the in vitro antioxidant activity of different Brazilian teas using multivariate statistical techniques *Food Research International* **60**

[5] Aston S O, Provost J and Masselink H 2008 *Jet printing with reactive dyes*. *Journal of the Society of Dyers and Colourists* **109** pp 147-152

[6] Martín-Alberca C, Zapata F, Carrascosa H, Ortega-Ojeda F E and García-Ruiz C 2016 Study of consumer fireworks post-blast residues by ATR-FTIR *Talanta* **149** pp 257–265

[7] Kamil M, Naeim M, Asri M, Nur W and Mat S 2015 Fourier Transform Infrared (FTIR) Spectroscopy and Principal Component Analysis (PCA) of Unbranded Black Ballpoint Pen Inks **6** pp 48–53

IOP Conf. Series: Materials Science and Engineering 1192 (2021) 012030 doi:10.1088/1757-899X/1192/1/012030

Thermoplastic sago starch nanocomposites wound dressing fortified with antibiotic-modified HNT

N Abuhamed[1], Z Ahmad[1], N Sarifuddin[1]

[1]Department of Manufacturing and Materials Engineering, Kulliyyah of Engineering, International Islamic University Malaysia, Gombak, 53100 Kuala Lumpur, Malaysia.

Email: zuraidaa@iium.edu.my

Abstract. Starches were reported to promote wound healing. However, the hydrophilicity of starch help absorbs the exudates from the wounds during the healing process, but it also enables a bacterial infection that slows the healing process. Halloysite nanotubes (HNT) are attracting many biological technologies because of their high loading capacity and biocompatibility. This paper investigates the modified HNT as a carrier for antimicrobials agent in wound healing materials. Halloysite was modified by dispersing it with chloramphenicol solution using a magnetic stirring method. Thermoplastic sago Starch (TPSS)/modified HNT (MHNTs) biocomposite films of different compositions (0.25, 0.5, 0.75 and 1 wt. % HNT) were then developed using the solution casting method. SEM revealed that modified HNT shows good dispersion on the TPSS matrix. With the introduction of modified HNT, the FTIR peaks of TPSS have altered at the peak of 3693.21 cm^{-1} and 1040.05 cm^{-1}. In addition, modified HNT reduced the water absorption rate of the TPSS films. Furthermore, modified HNT showed good resistance to bacterial culture and significantly reduced the biodegradability rate of TPSS compared to pristine HNT. From the findings, HNT can be a potential carrier for antibacterial agents to withstand bacterial attacks.

1. Introduction

Nowadays, there seems to be an increasing interest in biodegradable polymers as wound healing materials to reduce environmental pollution caused by synthetic dressing materials. In addition, the need to minimize the carbon footprint of synthetic polymers is driving new biomaterials developed from these renewable sources. Environmental issues and a lack of fossil resources have indeed prompted the growth of new biomaterials. Starch is a suitable choice for replacing toxic and non-biodegradable polymeric materials commonly used in the industry. In addition, the use of starch in non-food products is considered to be of innovative interest.

Starch has several benefits, including low cost, highly available, and absolute compostability with no harmful substances. This renewable resource is the primary ingredient for most biopolymers on the market. starch-based biopolymers are accounting for 85 % to 90 % of all biopolymers available.[1] Native starch, on the other hand, is brittle and has weak mechanical properties. [2, 3] Starch-based materials have several drawbacks, including low processability, long-term stability, and high water sensitivity. [1] As a result, sago starch must be thermoplasticized by adding plasticizers, including glycerol, to improve plasticity and minimize brittleness. Several studies have used biopolymers in medical applications because of their biocompatibility and environmental sustainability [4]. Furthermore, researchers have proven polysaccharides starches to be effective wound dressing materials [5][6]. Wound dressing compounds polysaccharides, such as sago starch, help to facilitate and promote healing. However, starch hydrophilicity absorbs the exudates from the wounds during the healing process, which most likely enables a bacterial infection that delays the healing process. As a result, an antimicrobial agent must be used to address this problem [5].

Halloysite nanotubes (HNT) were commonly used in drug delivery and tissue engineering scaffolds in the medical industry. However, halloysite has no antibacterial effect, and several studies have shown that HNT must be modified to provide an antibacterial effect [7][8]. In addition, the hollow structure of the halloysite makes it an excellent carrier for an antimicrobial agent. The need for HNT to be modified is to prevent bacterial infection that thermoplastic sago starch (TPSS) promotes during the wound healing process. Therefore, HNT is proposed to be modified with an antibacterial agent for wound dressing films. Henceforth this paper investigates the effect of modified HNT on thermoplastic sago starch nanocomposites.

2. Materials and Methods

2.1. Materials
Four main materials were used in this experiment: sago starch, glycerol, distilled water, and halloysite nanotubes (HNTs). In this study, the Malaysian sago starch powder (25% amylose) with an average diameter of 53.66μm, length of 388.76μm, and density of 0.571g/cm3 was used. Glycerol (95% purity) with a density of 1.232g/cm^3 was purchased from Merck Sdn. Bhd. Ultrafine grade HNTs were obtained from Imerys Tableware Asia Limited, New Zealand.

2.2. HNT modification
HNT was modified by dispersing 5 grams of HNT in 20 ml of distilled water and magnetically stirred in a hotplate stirrer for about 30 minutes at room temperature. Afterward, 2.0 ml of liquid antibiotics (chloramphenicol) were added to the HNT solution and mixed for 15 minutes. Finally, the solution of HNT was placed in an oven for 24 hours at 30 °C. After 24 hours of drying the HNT solution. Shear stress ball mill homogenization was used to reduce the HNT conglomerate to achieve good homogeneous dispersion of HNTs [9].

2.3. Film Preparation
The sago starch used in this experiment was previously oven-dried at 80°C for 8 hours and sieved using a mechanical shaker with a sieve size of 50μm. Based on the previous study, the optimum composition used for the preparation of the composite films was 6.5% sago starch, 3.5% glycerol, and 90% distilled water. Stock solution with the desired weight percentage of modified HNT according to Table 1. The solution contains modified HNT, sago starch, glycerol and distilled water while magnetically stirred at a gradually elevated temperature. The mixture was continuously stirred for 20min under the controlled temperature until it reached 75°C, where the starch mixture started to gelatinize around 70°C – 75°C. The mixture was then solvent casted on a smooth and flat surface and oven-dried for 20 hours at 40°C. The resultant composite film was peeled off the surface and stored in a sealed plastic bag for at least 40 hours at room temperature for further testing.

Table 1. Formulations for modified HNT/TPSS nanocomposites films

HNT (wt%)	Sago Starch (g)	Glycerol (g)	Distilled Water (ml)	MHNT (g)
TPSS	6.500	3.500	90.000	0
TPSS/MHNT 0.25	6.484	3.491	89.775	0.25
TPSS/MHNT 0.50	6.468	3.483	89.550	0.50
TPSS/MHNT 0.75	6.452	3.471	89.326	0.75
TPSS/MHNT 1.00	6.435	3.465	89.100	1.00

3. Materials Characterization and Evaluation

Morphological properties of wound dressing films were investigated using a JEOL JSM-IT100 scanning electron microscope (SEM) with a 10.0kV operating voltage. Thin-film specimens were sputter-coated with gold-palladium to avoid sample charging throughout imaging.

Meanwhile, the Nicolet iS50 FTIR spectrometer was used to determine the chemical group functionalities of the thin films. The films were tested using a 20-scan mode with a resolution of 4 cm^{-1} and a wavelength range of 650-4000 cm^{-1}.

Moreover, the water absorption of wound healing films was determined using the ASTM D570 method. Specimens were cut into 20 mm x 20 mm squares, dried for 24 hours at 50 ° C, and afterward weighted. The specimens were then submerged in distilled water for 24 hours at a temperature of 23±1 °C. After 24 hours of immersion, specimens were removed and gently wiped clean of any water collected on the specimen surface, followed by a weight measurement to the nearest 0.001g. The percentage of water absorption, W_a, was determined using equation (1). Where m_a represents the sample's weight after immersion and m_b represents the sample's weight prior to immersion.

$$Wa = \frac{ma - mb}{mb} \times 100 \quad (1)$$

Furthermore, the soil burial method was conducted to check the biodegradability of the materials [10]. In this method planting plastic pots filled with black soil that have an approximately 18% moisture content. Test samples were cut to 20 mm × 20 mm and placed in a net cloth, and buried in the soil. Specimens were gently and cautiously brushed to clear the sticking soil after the burial period of 10 days. After that, the samples were dried in a 40°C oven for 30 minutes.

The antimicrobial susceptibility of the prepared films was measured using the Kirby-Bauer method in an agar diffusion disk. Filter papers with a diameter of 1cm were dipped in different HNT/TPSS composite gel and then put on top of the agar disk surface, inoculated with E. coli bacteria. The disks were placed in an incubator at 37°C for 24 hours, after which the effects were observed. The region of the inhibition zone can be calculated using equation (2).

$$Area\ of\ inhibition\ zone\ (mm^2) = Area\ of\ clear\ zone\ (mm^2) - Area\ of\ well\ (mm^2) \quad (2)$$

4. Results and discussion

4.1. Surface morphology

Figure 1 (a) shows the SEM morphology of TPSS, which reveals a smooth and clear surface resulting from the effective integration of amorphous sago starch granules with glycerol and water due to starch plasticization. The surface of the TPSS is smoothed with the addition of glycerol. On the surface, there were very few starch particles. This is attributed to the mixing process, as starch was not dissolved in water for a sufficient time to blend appropriately [11]. Therefore, better outcomes could be achieved with a slower heating rate since the samples would have more mixing time before gelatinizing.

On the other hand, the surface morphology of TPSS with the addition of 0.25, 0.5, 0.75, and 1 wt.% of MHNT is showed in Figure 1 (b)-(e). The introduction of small amounts of MHNT to the matrix results in proper dispersion. However, at higher concentrations, MHNT causes certain agglomeration in the TPSS matrix, and only a few sago granules were often visible on the matrix surface. Higher HNT loading results in more aggregates in the matrix. This may be attributed to the modification of halloysite, which resulted in a weaker interfacial interaction with the polymer matrix. In addition, the modification of HNT with a strongly hydrophobic material (chloramphenicol) may explain the weaker interfacial interaction, which leads to weaker interfacial interaction between TPSS (hydrophilic) and modified HNT with hydrophobic substance [12].

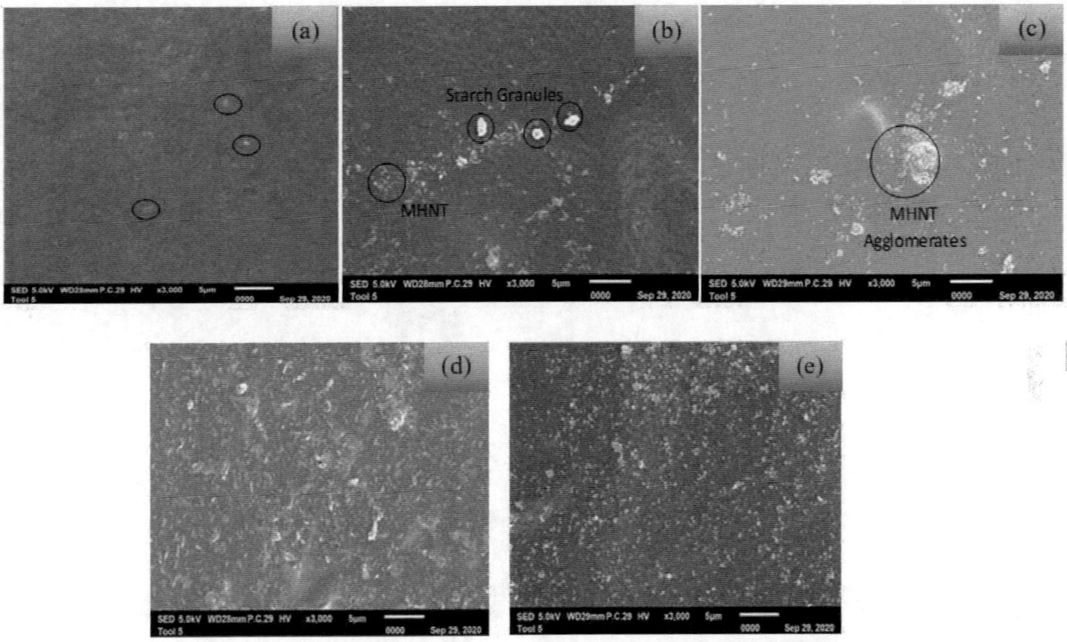

Figure 1. SEM micrograph of TPSS nanocomposite films at (a) 0 wt.% MHNT (b) 0.25 wt.% MHNT, (c) 0.5 wt.% MHNT, (d) 0.75 wt.% MHNT and (e) 1.0 wt.% MHNT

4.2. Chemical characterization

The OH stretching vibration and deformation and Si-O groups and stretchings can be seen in the FTIR spectra of MHNT in Figure 2. The antibiotics chloramphenicol did not show any alteration on the TPSS/MHNT peaks. According to early studies, the absorption peaks of substances encapsulated within the HNTare normally not revealed [13]. Chloramphenicol molecules are encapsulated within the tubes of MHNT, according to the current findings. These results are also backed up by antimicrobial and biodegradability studies. The inner-surface hydroxyl group's O-H stretching, which mildly changed from 3693.86 cm^{-1} for MHNT to 3691.88, 3692.45, 3692.97, and 3693.12 MHNT/TPSS at 0.25, 0.5, 0.75, and 1 wt. %, respectively. This is attributed to reactions between the inner surface hydroxyl groups of MHNT and the C-O-C group of starch or glycerol. With the introduction of more MHNT, the Si-O bonding increases. The notable peak of the TPSS at 1042.50 cm^{-1} had moved towards the peak of MHNT at 1041.17 cm^{-1}, as shown. For MHNT/TPSS at 0.25, 0.5, 0.75, and 1 wt. %, MHNT peaks generated at 1041.75 cm^{-1}, 1041.72 cm-1, 1041.69 cm-1, and 1041.57 cm-1, respectively. The Si-O bond represents the typical MHNT-matrix interconnection through bond water and CH2 stretching, which decreases the biocomposite's water absorption rate [14]. Previous studies have shown that modified antimicrobial HNT has a similar impact on TPSS FTIR spectra [14][15].

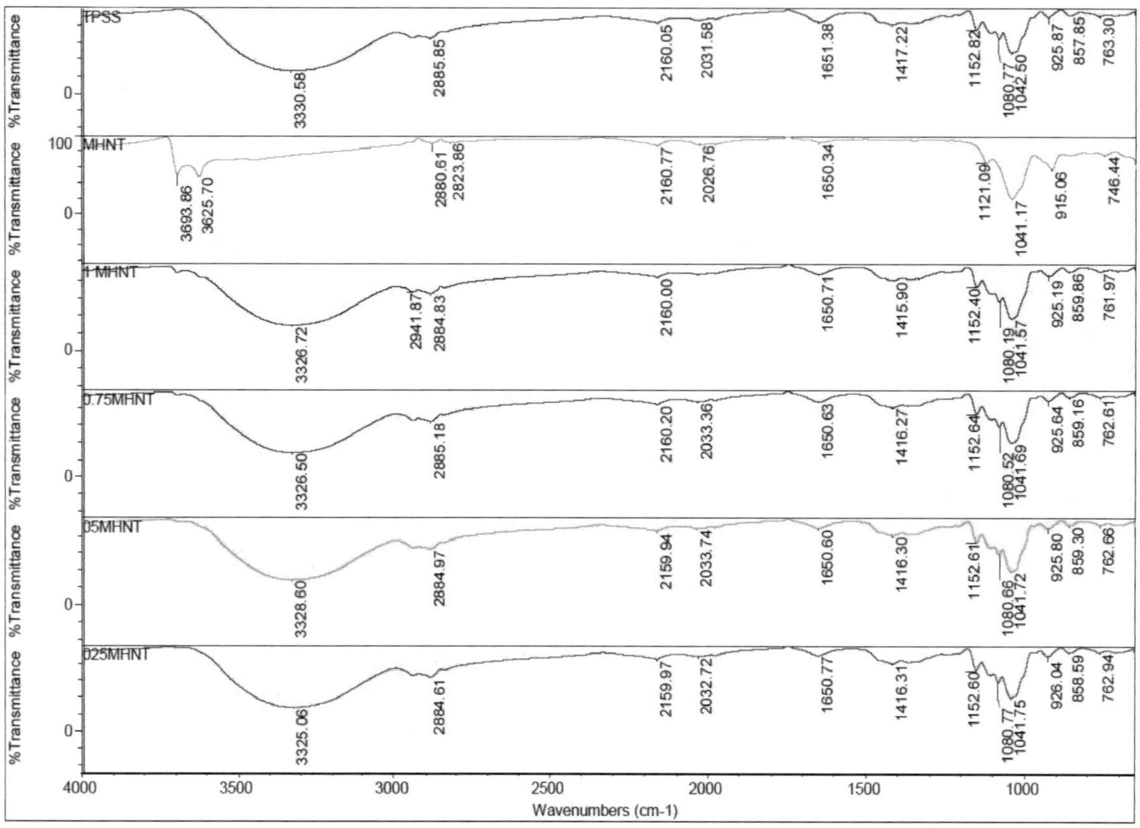

Figure 2. FTIR spectra of plasticized starch matrix (TPSS) containing 0, 0.25, 0.5, 0.75 and 1 wt.% of MHNT for wavenumber between 750 cm^{-1} to 4000cm^{-1}.

4.3. Water absorption

The addition of MHNT reduces the water absorption rate from 72.96% to 60.00%, 54.17%, 54.55%, and 53.13% when adding 0.25, 0.5.0.75, and 1 wt.% of HNT, respectively, as illustrated in Figure 3. When MHNT is mixed into the TPSS matrix, it reduces water absorption. The increase in crystallinity of the composites is responsible for this. HNT also creates a tortuous path that makes it difficult for water molecules to move through. Also, modified halloysite lumen was loaded with chloramphenicol, a highly hydrophobic antibiotic that makes the films containing MHNT have less capacity for water absorption [16].

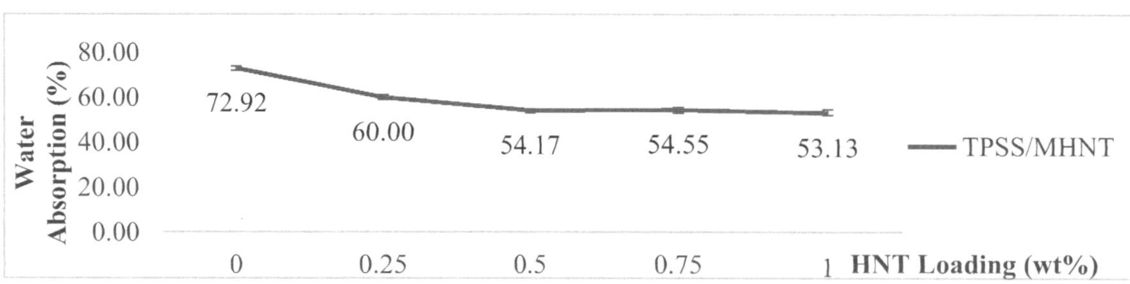

Figure 3. Water absorption for TPSS/HNT & TPSS/MHNT composites

The interaction between the nanocomposite films and the water molecules would be reduced as a result. The reduction is also due to the chloramphenicol antibiotics, a highly hydrophobic material being encapsulated in MHNT. Water absorption is essential for wound healing since it increases the films' ability to absorb wound exudates, which speeds up the healing process [15][16].

4.4. Biodegradation

Weight loss and crack formation were reduced with the addition of MHNT compared to composite films without MHNT. Films containing 0.25 and 0.5 wt. % MHNT developed cracks after 40 days, but they were less severe than the composites with only TPSS, as seen in Figure 4. The composites display no cracks or damage when the MHNT content is increased to 0.75 and 1 wt. %. The high amount of MHNT in the composite resulted in better antimicrobial properties and a lower degradability rate. The cracks will impact the weight loss and degradation because the surfaces would be exposed to the activity of microbes as a result of the cracks [17]. Higher levels of MHNT help slow down the rate of weight loss and degradation due to increased structural rigidity and the antibacterial property promoted by chloramphenicol.

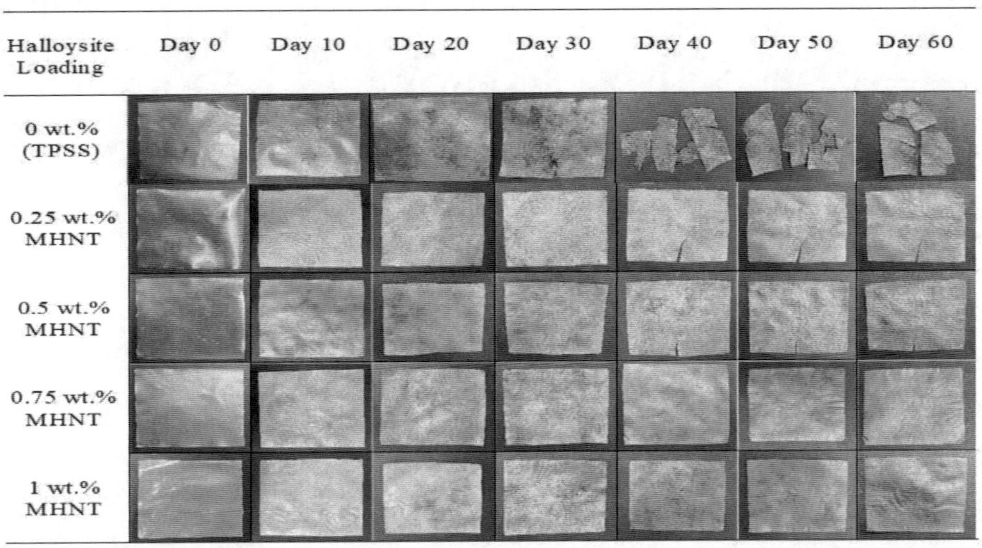

Figure 4. The degradation condition of the nanocomposites after certain burial days

4.5. Antimicrobial inhibition test

The disc diffusion method in Figure 5 revealed that halloysite is not an antimicrobial agent in nature. However, it showed strong resistance to bacterial culture after being modified with an antimicrobial material. Halloysite was modified in this study by solution mixing, and when combined with TPSS, a successful inhibition area was observed, as seen in Figure 5. With further increasing the MHNT in TPSS, a wider inhibition area is obtained. This demonstrates modified halloysite's ability to serve as an antibacterial container for wound dressings.

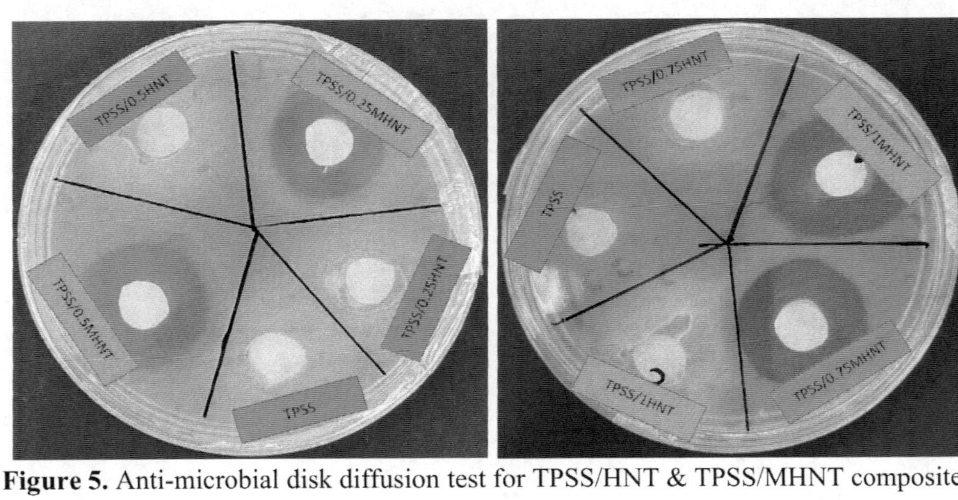

Figure 5. Anti-microbial disk diffusion test for TPSS/HNT & TPSS/MHNT composites.

5. Conclusion

The MHNT/TPSS nanocomposite films were successfully produced via the solvent casting method, revealing that HNT was well dispersed in the TPSS matrix at lower loading. Furthermore, based on the FTIR spectra obtained, it can be confirmed that the addition of HNT into TPSS films led to the formation of hydrogen bonding and Si-O, Al-OH, and Al-Si-OH bonding in the film matrix. TPSS with 0.25 wt. % MHNT was found to be the best biocomposite in this study to be used as a wound-healing material. This candidate had a higher water absorption rate compare to the other biocomposites with higher MHNT loading. Water absorption is crucial in wound healing because it helps the exudates from the wounds during the healing process. In addition, the biocomposite's biodegradability was high at 0.25 wt. % MHNT, compared to other higher loadings; as a result, this candidate is less damaging to the environment. Finally, the candidate demonstrated adequate antimicrobial capacity, suggesting that it could be used for wound healing applications.

6. References

[1] Xie F, Pollet E, Halley PJ, Averous L 2013 *Progress in Polymer Science* 1;38(10-11):1590-628

[2] Freire C, Podczeck F, Veiga F, Sousa J 2009 *European Journal of Pharmaceutics and Biopharmaceutics 72*(3) pp 587-594

[3] García NL, Famá L, Dufresne A, Aranguren M and Goyanes S 2009 *Food Research International 42*(8) pp 976-982

[4] Waghmare V S, Wadke P R, Dyawanapelly S, Deshpande A, Jain R and Dandekar P 2018 *Bioactive materials,3*(3) pp 255-266

[5] 5. Mandal A, Sekar S, Meera K M, Mukherjee A, Sastry T P and Mandal A B 2014 Physical Chemistry Chemical Physics 16(37) pp 20175-20183.

[6] Amal B, Veena B, Jayachandran V P and Shilpa J 2015 *Journal of Materials Science: Materials in Medicine 26*(5) p 181

[7] Kurczewska J, Pecyna P, Ratajczak M, Gajęcka M and Schroeder G 2017 *Saudi pharmaceutical journal 25*(6) pp 911-920

[8] Makaremi M, Pasbakhsh P, Cavallaro G, Lazzara G, Aw Y K, Lee S M and Milioto S 2017 *ACS applied materials & interfaces 9*(20) pp 17476-17488

[9] Deng S, Zhang J and Ye L 2009 *Composites Science and Technology 69*(14) pp 2497-2505

[10] Ahmad Z, Hermain H Y and Razak N H 2018 *Journal of Engineering Science and Technology* pp 86 – 93.

[11] Ahmad Z, Hermain H Y and Abdul Razak N H 2015 *Advanced Materials Research* Vol.1115 pp 394-397 Trans Tech Publications Ltd

[12] Makaremi M, Pasbakhsh P, Cavallaro G, Lazzara G, Aw Y K, Lee S M and Milioto S 2017 *ACS applied materials & interfaces* 9(20) pp 17476-17488

[13] Jang S H, Jang S R, Lee G M, Ryu J H, Park S I and Park N H 2017 *Journal of food science*, *82*(9), pp.2113-2120

[14] Abdullah Z W and Dong Y 2019 *Frontiers in materials 6* p 58

[15] Meira S M, Zehetmeyer G, Werner J O and Brandelli A 2017 *Food Hydrocolloids 63* pp 561-570.

[16] Barman M, Mahmood S, Augustine R, Hasan A, Thomas S and Ghosal K *International Journal of Biological Macromolecules 162* pp 1849-1861

[17] Ojijo V and Ray S S 2014 *Progress in Materials Science 62* pp 1-57

IOP Conf. Series: Materials Science and Engineering 1192 (2021) 012031 doi:10.1088/1757-899X/1192/1/012031

Effect of Stabilizers in the Synthesis of Silver Nanoparticles and Methylene Blue Oxidation

Dzilal Amir[1], Ricca Rahman Nasaruddin[1,2,*] , Nurul Sakinah Engliman[1,2], Sarina Sulaiman[1], Mohd Sufri Mastuli[3]

[1]Department of Biotechnology Engineering, Kulliyyah of Engineering, International Islamic University Malaysia Jalan Gombak, 53100 Kuala Lumpur, Malaysia
[2]Bioenvironmental Engineering Research Centre, International Islamic University Malaysia
[3]School of Chemistry and Environment, Faculty of Applied Sciences, Universiti Teknologi MARA, 40450 Shah Alam, Selangor, Malaysia

[*]Email of corresponding author: riccanasaruddin@iium.edu.my

Abstract. Metal nanocatalysts have received increasing attention in catalysis due to their higher reactivity and surface area-to-volume ratio at nano-size. Silver nanoparticle (AgNP) is among metal nanocatalysts that have been studied in various catalytic reactions (e.g., hydrogenation and oxidation). However, the high reactivity of AgNPs at nano-size caused instability and aggregation. Therefore, stabilizing molecules (or stabilizers) are always applied to maintain the nano size of AgNPs and prevent aggregation. Herein, the effects of different types and molar ratio of stabilizers-to-Ag precursor, to the synthesized AgNPs (i.e, size and concentration) were investigated. Two types of stabilizers, polyvinylpyrrolidone (PVP) and citrate were used in this study. The roles of stabilizers to the catalytic performance of synthesized AgNPs were then elucidated by using methylene blue oxidation as the model reaction. The UV-Vis absorption analyses showed that both stabilizers produced slightly different size and concentration of AgNPs based on the different wavelength and absorption intensity of the peak. We also found that the molar ratio of stabilizers-to-Ag precursor that produced better yield of AgNPs was 1:1 and 1:3 for PVP and citrate, respectively. Then, AgNPs stabilized by citrate was found having slightly higher catalytic activity in the methylene blue oxidation than AgNPs stabilized by PVP. This study provides insights to the roles of stabilizers for the synthesis of stable AgNPs with efficient catalytic reaction and can be used as guideline to other metal nanocatalysts.

1. Introduction

The advancement of nanotechnology has significantly increased the number of innovation and studies of nanomaterials, including metal nanocatalysts. Metal catalysts at nano-size are more reactive and thus nanocatalysts with higher reactivity and surface area-to-volume ratio are more favourable in catalysis as only lower amount is needed for higher catalytic activity. These metal nanocatalysts are not only applied for chemical transformations of value-added products but also to the catalytic removal of hazardous pollutants in the environment. For instance, a worrying amount of hazardous pollutants (especially, organic pollutants) are generated daily by many industries, thus require sustainable and efficient technology for their removal. Silver nanoparticle (AgNP) is among noble metals (e.g., platinum and gold) which have been applied as catalysts and it is preferable because it is cheaper than other noble

IOP Conf. Series: Materials Science and Engineering 1192 (2021) 012031 doi:10.1088/1757-899X/1192/1/012031

metals. AgNPs have been also reported as a good catalyst for the removal of organic pollutants (e.g., 4-nitrophenols and methylene blue) in wastewater [1].

Size of metal NPs is important to control their physical and chemical properties, including luminescence, conductivity, and catalytic activity. Thus, the ability to control the size of metal NPs during their synthesis is of crucial importance [2]. The size of metal NPs is usually depending on the synthesis parameters and conditions such as (e.g., stabilizers, reductant, temperature, pH, reaction time, etc). Among all, stabilizers are the most significant ingredients to modulate the size of the metal NPs during the synthesis and to maintain its stability by passivating and reducing the high surface charge of metal NPs, thus prevent further growth and aggregation [3]. Types of stabilizers used in the synthesis of metal NPs are polymers, dendrimers, anionic molecules and ligands. Polyvinylpyrrolidone (PVP), a water-soluble polymer and citrate, an anionic molecule are commonly used as stabilizers for the synthesis of metal NPs including AgNPs. In some studies, they were reported to serve as reductant, stabilizer as well as complexing agent [4].

The possibility of a precise control of size and mass production remains the most attractive goal for the synthesis of AgNPs [5]. Therefore, in this study, we investigated the effects of different type and amount of stabilizer (i.e., molar ratio of stabilizer-to-Ag precursor) to the synthesis of AgNPs and their catalytic activities. Methylene blue (MB) oxidation was used as a model reaction because methylene blue is the common dye in the wastewater. Therefore, its removal by catalytic oxidation is of high interest in wastewater treatment as the AgNPs could be incorporated into the existing materials (e.g., adsorbents and filter membranes) used in the wastewater treatment to improve the capability of those materials in removing the dyes in wastewater.

2. Materials and Methods

2.1. Preparation of chemicals

Silver precursor (silver nitrate, $AgNO_3$), polyvinylpyrrolidone (PVP) and trisodium citrate ($Na_3C_6H_5O_7$) and reductant ($NaBH_4$) were purchased from Sigma Aldrich while methylene blue (MB) and 50% hydrogen peroxide (H_2O_2) were purchased from local suppliers. For all experimental works, solutions were prepared using ultrapure water (denoted by UPW; 18.2 MΩ.cm @ 25°C). Beaker, volumetric flask, and other glassware were rinsed with UPW before use.

2.2. Synthesis of AgNPs with different type and amount of stabilizers

At a room temperature (~27 ± 1°C), 0.05 mL of 40 mM $AgNO_3$ solution was mixed with UPW with 40 mM PVP solution in a 15 mL vial. The volume of UPW and PVP was varied based on the different molar ratio of $AgNO_3$-to-PVP (1:1, 1:3, 1:5 and 1:7) which are summarized in table 1. The UPW was added to the mixture to make-up the total volume of 3 mL. The mixture was shaken using hand for 10 seconds and allowed to mix for 10 minutes. After 10 minutes, 0.02mL $NaBH_4$ (prepared by dissolving 0.04 g in 10 mL UPW) was added into the solution and the mixture was left for etching at room temperature for 70 minutes.

Table 1. The molar ratios and volume of reagents used for the synthesis of AgNPs

Molar ratio of $AgNO_3$:Stabilizers	Volume of reagents (mL)		
	$AgNO_3$ (40 mM)	Stabilizer (40 mM)	UPW
1:1	0.05	0.05	2.9
1:3	0.05	0.15	2.8
1:5	0.05	0.25	2.7
1:7	0.05	0.35	2.6

The colour of the solution turns into yellowish colour and the yellow colour became brighter after longer reaction time, indicating the formation of AgNPs. The as-synthesized AgNPs were then characterized by UV-Vis spectroscopy whereby 0.2 mL sample was diluted with 1.8 mL UPW. The above-mentioned steps were repeated for the synthesis of AgNPs stabilized by citrate. Of note, the concentration of the as-synthesized AgNPs is based on the mol of Ag precursor in the 3 mL total solution, which is 0.67 mM. Due to high sensitivity of AgNPs, all synthesis process and $AgNO_3$ solution were kept in the dark to avoid any photochemical reaction during experimental work. All the experiments were done in duplicates.

2.3. Methylene blue (MB) oxidation

The catalytic activity of as-synthesized AgNPs was determined using methylene blue (MB) oxidation with hydrogen peroxide (H_2O_2) as oxidizing agent. Briefly, in 7 mL vial, 0.2 mL of 0.05M MB was mixed with 0.2 mL AgNPs (0.67 mM). Then 1.7 mL of oxidizing agent (50% H_2O_2 solution) was added to the mixture and was shaken by hand to start the oxidation. The mixture was analysed using UV-Vis spectroscopy at 0, 20, 40 and 60 minutes. The blue colour of mixture due to the MB gradually changed reduce in intensity and finally to colourless after the oxidation completed. The catalytic tests were done in duplicates. Two control experiments were done which one of them was without oxidant and catalyst (also called as blank control) and another control was without the catalyst only. The blank control was measured by UV-Vis to determine the initial absorbance of the MB (A_o) before catalytic reaction occur. The catalytic tests were carried out at ambient temperature and under the present or normal light. The values of UV-Vis absorption were recorded and compared. The catalytic tests were done in duplicates.

3. Result and Discussion

3.1. Effects of stabilizers amount to the formation of AgNPs

The formation AgNPs can be observed by colour appearance and characterization by UV-Vis absorption peak. Colourless solution turns into yellowish brown after addition of $NaBH_4$ which gave a maximum absorbance at ~400 nm as shown in figure 1(a-b). It is well known that this yellowish brown color arises owing to excitation of surface plasmon resonance (SPR) vibration in the AgNPs [6].

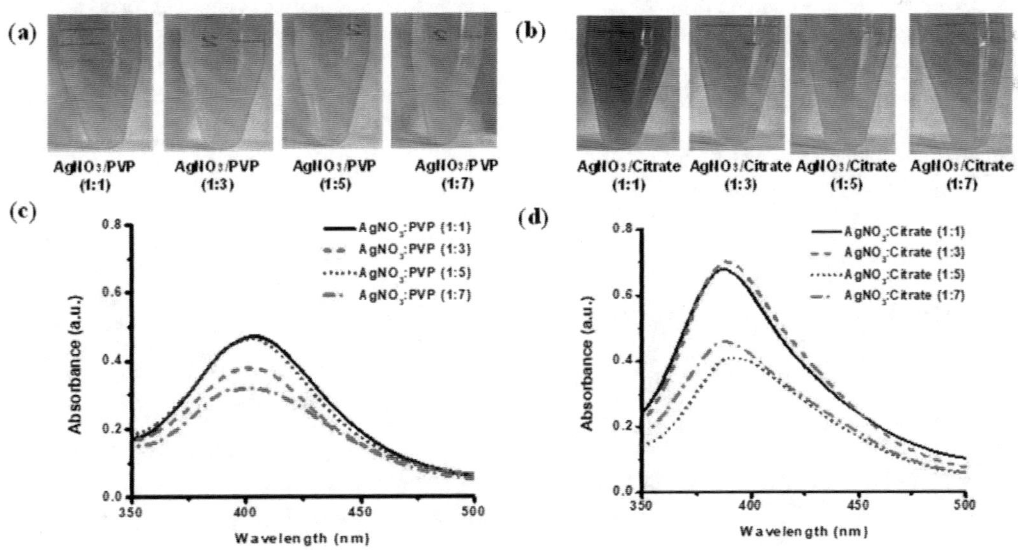

Figure 1. Color of (a) AgNPs-PVP and (b) AgNPs-Citrate solution and the UV-Vis absorption spectra for (c) AgNPs-PVP and (d) AgNPs-Citrate solution.

Figure 1(a-b) also shows that the colour of solution looks similar for both AgNPs stabilized by PVP (AgNPs-PVP) and citrate (AgNPs-Citrate), except for AgNPs-Citrate with 1:1 molar ratio of $AgNO_3$-to-Citrate. The colour is darker and more to brownish, probably due to formation of more and bigger AgNPs [7]. In general, the surface plasmon resonance of AgNPs produces a UV-Vis absorption peak at around 400 nm as can be seen in figure 1(c-d), but may slightly vary depending on the size of AgNPs [8].

Results in figure 1(c-d) also indicates that the highest peak of AgNPs synthesized by PVP and citrate were obtained when using $AgNO_3$-to-stabilizer molar ratio of 1:1 and 1:3, respectively. The highest peak at particular molar ratio of (AgNPs-PVP) and (AgNPs-Citrate), indicating those amounts are sufficient to form higher AgNPs than other molar ratio. As PVP has bigger structure than citrate based on molecular weight, the molar ratio of 1:1 (PVP) and 1:3 (citrate) is comparable to produce the highest amount of AgNPs. The higher molar ratio (i.e 1:5 and 1:7) for both PVP and Citrate produce lower AgNPs due to excessive amount of stabilizer, hence hinder the stabilizer to form complex with Ag^+.

3.2. Comparing the roles of different stabilizer to the formation of AgNPs

The UV-Vis absorption spectra of the AgNPs-PVP and AgNPs-Citrate solution were also compared with the control experiment whereby the AgNPs were produced without addition of stabilizers. However, AgNPs were still formed in the control experiments due to the presence of borohydride anions (BH_4^-) from the reductant ($NaBH_4$) which are not only capable to reduce the Ag^+ into Ag^0 but also to form complex and stabilize Ag^+ cations and AgNPs in the solution [9]. Based on the UV-Vis absorption spectra in figure 2(a), when other parameters and synthesis conditions were held constant, it can be seen that different stabilizers lead to the formation of AgNPs with different size and concentration which indicated by the different wavelength and intensity for the highest absorption peak. The highest absorption peak of AgNPs-PVP was found at higher wavelength (404 nm) than that of AgNPs-Citrate (388 nm), indicating the size of AgNPs-PVP could be slightly higher than AgNPs-Citrate. However, the intensity of the absorption peak of AgNPs-PVP is lower than that of AgNPs-Citrate, meaning there were more AgNPs species in AgNPs-Citrate solution than in AgNPs-PVP. The reason could be due to the more Ag-citrate complex can be formed with citrate anions which has more oxygens (in form of O^- and =O moieties) which are able to stabilize and reduce more Ag^+ cations, thus forming more AgNPs in the solution.

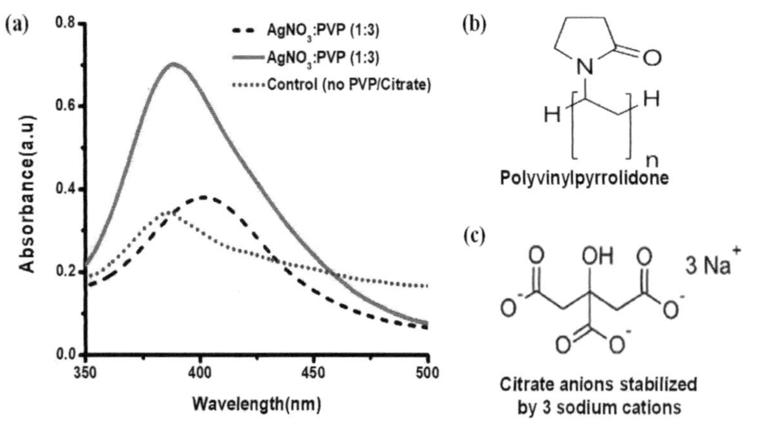

Figure 2. (a) UV-Vis absorption spectra of AgNPs-PVP, AgNPs-Citrate and control experiment (AgNPs-without stabilizer) and chemical structure of (b) (PVP) and (c) citrate

When compared with AgNPs synthesized without stabilizer (denoted by AgNPs-control), it can be seen that the absorbance peak for the AgNPs-control appears at the wavelength comparable to that of

AgNPs-citrate (388 nm). Interestingly, the absorption intensity of AgNPs-control is the lowest which suggest that there was insufficient reducing capability of $NaBH_4$ to reduce Ag^+ into AgNPs and both stabilizers, PVP and citrate are not only stabilizing the AgNPs, but also act as the reductant in the synthesis of AgNPs, leading to more formation of AgNPs in the solution.

3.3. Comparing the catalytic activity of AgNPs

The cationic MB dye is one of the water pollutants, which consumes the dissolved oxygen in water and endangers the aquatic system [10]. The catalytic ability of as-synthesized AgNPs was investigated with degradation of MB with presence of H_2O_2 for oxidation reaction figure 3. H_2O_2 generates OH radicals and oxidize the MB by generating discoloration. MB typically shows absorption maxima band at about 664 nm in aqueous solution due to π / π* and n / π* transitions and UV Vis absorption peak of MB is at 631nm wavelength from figure 3.

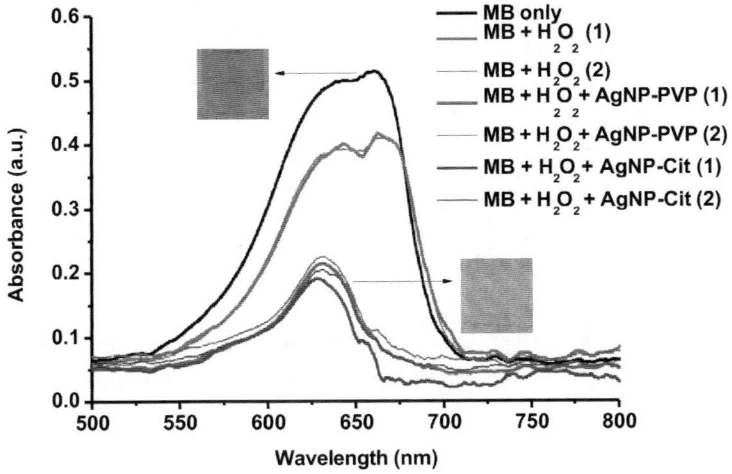

Figure 3. UV-Vis absorption spectra of methylene blue oxidation after 1 hour reaction for control experiments and AgNPs stabilized by PVP and citrate

After one hour of reaction, it was found that the UV-Vis absorption peak of MB decreased, and the blue color of MB solution reduced. MB degrade faster with catalyst (i.e AgNPs) as compared to H_2O_2 alone (without catalyst) and by comparing between two stabilizers, citrate performs better catalytic degradation as compared to PVP towards MB oxidation. One of the possible reasons is due bigger size and presence of alkyl chain of PVP that reduce accessibility to substrate (MB) hence reducing catalytic activity.

4. Conclusion

In the present study, stable silver nanoparticles were successfully prepared by a stabilizers PVP and citrate. The procedure was relatively easy, rapid, inexpensive, eco-friendly. Furthermore, this study demonstrated the effect of different molar ratio and type of stabilizers to produce AgNPs. It was found that the best molar ratio of stabilizers-to-Ag precursor for the highest yield of AgNPs synthesized using PVP and citrate was 1:1 and 1:3, respectively. It was also found that with similar molar ratio of stabilizers-to-Ag precursor, the yield and morphology of the Gaps such as size and shape could be different due to the different wavelength of the UV-Vis absorption peak. TEM analysis can be further done to confirm the different size and shape of the AgNPs produced by different stabilizers. Catalytic activity was demonstrated on MB to check the effectiveness of AgNPs as catalyst. The reduction of MB dye at was done at ambient conditions. The MB dye degraded within 60 min, signifying the usefulness of the synthesized AgNPs as a good catalyst for wastewater treatment. Different stabilizers also affect the catalytic activity as PVP is a polymer with a longer chain length reduced the accessibility of the substrate to the catalyst, thus leading to lower catalytic activity than AgNPs stabilized by citrate. This study provides a guideline in the selection of stabilizers for the synthesis of AgNPs nanocatalysts.

Acknowledgement
This is a part of Master study of Dzilal Amir. The principal investigator, Dr. Ricca Rahman Nasaruddin would like to acknowledge the financial support from Ministry of Higher Education Malaysia under research grant (FRGS-RACER/1/2019/TK02/UIAM//1) for the research works.

References

[1] Fundneider T, Acevedo Alonso V, Wick A, Albrecht D, and Lackner S, 2021, *Water Res.*, **189**, 116588
[2] Xue Y, Chen S, Yu J, Bunes B R, Xue Z, Xu J, Lu B, and Zang L, 2020, *J. Mater. Chem. C*, **8**, 10136
[3] Das S, Bandyopadhyay K, and Ghosh M M, 2021, *Inorg. Chem. Commun.*, **123**, 108319
[4] Zeng S, Vahidzadeh E, VanEssen C G, Kar P, Kisslinger R, Goswami A, Zhang Y, Mahdi N, Riddell S, Kobryn A E, Gusarov S, Kumar P, and Shankar K, 2020, *Appl. Catal. B: Environ.*, **267**, 118644
[5] Yaqoob A A, Umar K, and Ibrahim M N M, 2020, *Appl. Nanoscience*, **10**, 1369
[6] Saha J, Begum A, Mukherjee A, and Kumar S, 2017, *Sustain. Environ. Res.*, **27**, 245
[7] Deshpande J B, Chakrabarty S, and Kulkarni A A, 2020, *Chem. Eng. J.l*, 127753
[8] Ganash E A and Altuwirqi R M, 2021, *Materials*, **14**, 2326
[9] Iravani S, Korbekandi H, Mirmohammadi S V, and Zolfaghari B, 2014, *Res. Pharm. Sci.*, **9**, 385-406
[10] Atta A M, Moustafa Y M, Al-Lohedan H A, Ezzat A O, and Hashem A I, 2020, *ACS Omega*, **5**, 2829-2842

Evaluating the potential of chitin extracted from shrimp shell wastes as support material of gold nanoclusters (AuNCs) for catalysis

Nabila Azirah Zakaria[1], Ricca Rahman Nasaruddin[1,2] , Wan Mohd Fazli Wan Nawawi[1,2]

[1]Department of Biotechnology Engineering, Kulliyyah of Engineering, International Islamic University Malaysia, Jalan Gombak, 53100 Kuala Lumpur, Malaysia

[2]Bioenvironmental Engineering Research Centre, International Islamic University Malaysia

Email of corresponding author: riccanasaruddin@iium.edu.my/riccarahman16@gmail.com

Abstract. Thiolate-protected gold nanoclusters (AuNCs) have been intensively studied due to their appealing physicochemical properties that can be used in various applications, especially catalysis. However, thiolate-protected AuNCs in a free suspension form, tend to aggregate during catalytic reaction due to the high surface reactivity of its ultrasmall size, hence, reducing the rate of their catalytic reaction. Their recovery and reusability are also difficult. Therefore, several support materials have been studied to immobilize the thiolate-protected AuNCs for a better recovery, reusability and stability in solution and during a catalytic reaction. Herein, chitin extracted from shrimp shell waste was used as support material for the immobilization of the AuNCs protected by *p*-mercaptobenzoic acid ligands, using deposition-precipitation method. The free chitin and chitin immobilized with thiolate-protected AuNCs (abbreviated as AuNCs/Chitin) were characterized using FTIR. The catalytic properties of the AuNCs/Chitin were then evaluated in the hydrogenation of 4-nitrophenol in solution as the model reaction. It was found that the chitin extracted from SSW has a good potential to be support materials of thiolate-protected AuNCs because the shorter reaction time it took (12 minutes) as compared to the reaction without catalyst (60 minutes) and it can be recovered and reused twice.

1. Introduction

Gold nanoclusters (AuNCS) have received increasing attention due to their intriguing physicochemical properties which offer a myriad of applications (e.g., catalysis [1-4], and biomedical applications [5-10]). Unlike plasmon gold nanoparticles (Au NPs) which have continuous or semi-continuous energy levels [11], Au NCs have higher surface area-to-volume ratio, distinctive discrete electronic structure and redox behavior [12-15]. These different properties could cause better or different catalytic properties of AuNCs as compared to AuNPs. However, due to ultrasmall size, protecting ligands are compulsory in the synthesis of AuNCs to prevent their aggregation and maintain their monodispersity and ultrasmall size. Among various ligands, thiolates with sulfhydryl (-SH) functional group are commonly used to synthesize AuNCs due to the strong bonding between sulfur and gold (Au-S bond) [16].

Nevertheless, a study by Nasaruddin, et. al. (2018) AuNCs reported that the thiolate-protected AuNCs in a free suspension form, were still unstable and easily aggregated during a catalytic reaction (i.e., 4-nitrophenol hydrogenation in solution) [16]. Therefore, it is crucial to immobilize the thiolate-protected AuNCs to improve their stability during a catalytic reaction and to increase its recovery and reusability potential in catalysis.

The immobilization of thiolate-protected AuNCs is usually done using solid support materials (e.g., carbon-based materials, water-soluble and insoluble polymers, mesoporous silica, and metal oxide [17]). Natural biopolymer such as cellulose, chitosan and chitin could also be used as support material. We selected chitin in this study because it is the most abundant natural nitrogen-containing biopolymer and it has been used for various applications such as health supplements [18-20], agricultural [21, 22] and water treatment [23-25]. It's potential as support material of metal nanocatalysts is still rarely reported. Interestingly, chitin can be extracted from crustacean shells (e.g., crab shell and shrimp shell wastes), hence, it can be a sustainable support material for metal nanocatalysts and could further overcome the environmental concerns caused by the wastes [26].

In this study, we investigated the immobilization of AuNCs on chitin extracted from shrimp shell wastes (abbreviated as SSW) and evaluated their catalytic activity, recovery and reusability in the hydrogenation of 4-nitrophenol in solution, as the model reaction.

2. Experimental Section

2.1. Preparation of chemicals and materials

The grey shrimp was bought from the local supermarket (AEON Big at Wangsa Maju). The other chemicals such as hydrochloric acid (HCl), sodium hydroxide (NaOH) and 4-nitrophenol were obtained from the Department of Biotechnology Engineering (BTED), Kulliyyah of Engineering, International Islamic University Malaysia while gold precursor, tetrachloroauric(III) acid ($HAuCl_4$) and sodium borohydride ($NaBH_4$) were purchased from Sigma Aldrich. All the glasswares were rinsed with ultrapure water before use.

2.2. Chitin extraction from SSW

The extraction of chitin from the SSW was done based on the conventional method reported in Yang et. al., (2019) [4]. Briefly, the SSW was dried under sunlight for 10 hours to decrease its moisture content. After drying under the sunlight, the SSW was blended to be in a powder form and then was left to dry again. The dried SSW powder was then weighted to identify its moisture content. For the extraction of the chitin from the SSW, deproteinization was done by stirring one gram (1 g) of SSW powder with 20 mL of NaOH aqueous solution (5 wt%) at 80.0 ± 5 °C for 2 hours. The sample was then washed using distilled water, dried and weighted. Next, the dried deproteinized sample was stirred with 10 mL HCl aqueous solution (5 wt%) at room temperature for 1 hour for the demineralization process. The demineralized sample (extracted chitin) was then washed using distilled water, dried and weighted. The extracted chitin was then characterized using FTIR.

2.3. Synthesis of AuNCs

The thiolate-protected AuNCs used in this study were protected by thiolate ligands, *p*-mercaptobenzoic acid (*p*-MBA). They were synthesized using a reported carbon dioxide (CO)-reduction method developed by Xie and co-workers [27]. Briefly, 0.25 mL of aqueous solution of *p*-MBA (50 mM in 150 mM NaOH) and 0.25 mL HAuCl4 (40 mM) were mixed in 9.5 mL ultrapure water under stirring (500 rpm). After 2 hours, the colour of the solution turned to light yellow, indicating that the Au^{3+} ions had been reduced to Au^{+}-SR complexes. The pH of the solution was then brought to 11.6 by the addition of NaOH and CO was bubbled through the solution for 2 mins. The flask was sealed airtight and the synthesis reaction was done for 72 hours. The colour of the solutions was gradually turned into dark brown indicating the presence of thiolate-protected AuNCs. The final as-synthesized thiolate-protected

AuNCs with molarity of 0.5 mM (based on the mol of Au in the solution) were analyzed by UV-Vis and further used for the immobilization and catalytic tests.

2.4. Immobilization of AuNCs on extracted chitin

The immobilization of AuNCs was done by using deposition-precipitation method. Appropriate amount of dried chitin and deionized water (0.9 mL) was added simultaneously in a vial. It is important to make sure the pH of the solution is at pH 7.0. Then, 0.1 mL of as-synthesized thiolate-protected AuNCs solution were added afterwards. The mixture was stirred at 600 rpm for 1 hour and at room temperature. The change in colour of the mixture and the presence of precipitate were observed. The mixture was then centrifuged at 1000 rpm and the remaining liquid was taken out for UV-Vis absorption analysis. Meanwhile, the solid precipitates which were the immobilized AuNCs on chitin (abbreviated by AuNCs/Chitin) were then dried at room temperature overnight.

2.5. Hydrogenation of 4-nitrophenol (4-NP) in solution

The AuNCs/Chitin was tested for its catalytic activity through the hydrogenation of 4-nitrophenol in solution. In this experiment, 25 μL of 4-nitrophenol (4 mM) was added into 0.01 g of AuNCs/Chitin. Then 2 mL of $NaBH_4$ (0.25 M) was added into the solution to start the catalytic reaction. The yellowish colour of deprotonated 4-nitropheno (i.e., 4-nitrophenolate ions) gradually decreased to colourless, indicating the conversion of 4-nitrophenol to 4-aminophenol. After a catalytic reaction was done, the AuNCs/Chitin was recovered by centrifugation. Then, the recovered AuNCs/Chitin was tested again to study its recoverability and reusability using the same model reaction. The kinetic study was done using UV-Vis spectroscopy under kinetic mode by focusing on the absorbance of 4-nitrophenolate ions at wavelength 400 nm only. The UV-Vis absorption at 400 nm decreased with the time of reaction. According to Nasaruddin et al., (2018) since the high amount of $NaBH_4$ was added into the solution, the reaction could be regarded as the first order reaction and the slope of the decreasing UV-Vis absorption at 400 nm was taken as the apparent reaction rate constant, k_{app} (s^{-1}) [28].

3. Results and discussion

3.1. Characterisation of the extracted chitin

The average yield of extracted chitin was 38.96 ± 10.86 wt.%. From the FTIR spectra (Figure 1), the characterization peaks of the extracted chitin are at around 1010 cm^{-1} for -CO stretching band, 1428 cm^{-1} for $-CH_3$ deformation, 1556 cm^{-1} for amide vibrational band II, 1620 cm^{-1} for amide vibrational I with a single H-bond, 1652 cm^{-1} for amide vibrational I with double H-bond, 3256 cm^{-1} for -NH stretching band and 3431 cm^{-1} for -OH stretching band. The decreasing of peak intensity at around 1550 cm^{-1} and 1640 cm^{-1} (belong to amide groups of protein) between the shrimp shell waste sample and the deproteinized sample shows the practical removal of protein from the SSW sample. Moreover, the sharp peak at 872 cm^{-1} and at 1408 cm^{-1} (labelled by orange dashed line) that can be seen on deproteinized sample shows a vibration of carbonate ions ($-CO_3$) of $CaCO_3$ mineral. These peaks are more noticeable on the sample after deproteinization process as compared to the raw SSW because the protein matrix has been removed, thus exposing more minerals which further ease the demineralization process [29].

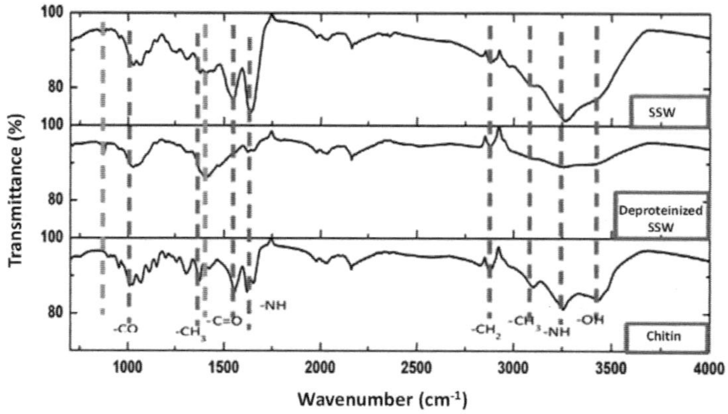

Figure 1. FTIR spectra shows the functional group of shrimp shell waste (SSW), deproteinized SSW and the final chitin after the demineralization of the deproteinized sample.

3.2. Immobilization of AuNCs on extracted chitin

The diluted thiolate-protected AuNCs solution was light brown in colour (refer to the inset image before centrifugation in Figure 2). As shown by the UV-Vis absorption spectra (black line; Figure 2), the thiolate-protected AuNCs shows featureless peak indicating the possibility of several species of AuNCs in the solution. During the stirring, immobilization happened, and AuNCs/Chitin was recovered by centrifugation. The immobilization of the thiolate-protected AuNCs could be seen indirectly when the colour of the remaining solution was colourless (refer to the inset image of after centrifugation in Figure 2). Due to the ultrasmall size of AuNCs, centrifugation of AuNCs solution at 1000 rpm did not result in precipitates and the color of the solution remained brownish. The UV-Vis absorption spectra in Figure 2 indicate that the peak for thiolate-protected Au NCs in the remaining solution almost disappeared after immobilization which could be an indirect evidence for the successful immobilization of thiolate-protected AuNCs on the surface of chitin.

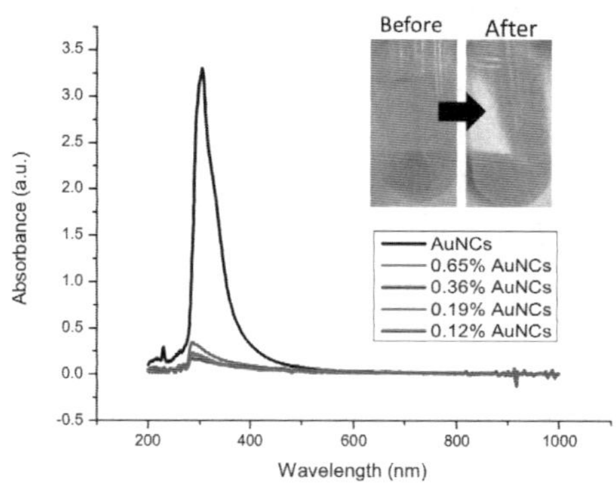

Figure 2. The UV-Vis absorption spectra of the thiolate-protected AuNCs and their remaining solution taken after their immobilization on chitin. Insert shows the image of the thiolate-protected AuNCs and chitin mixture before and after the centrifugation.

3.3. Hydrogenation of 4-nitrophenol by AuNCs/Chitin

The catalytic activity of the AuNCs/Chitin was determined by using hydrogenation of 4-nitrophenol (4-NP) in solution as the model reaction due to the simplicity of the process as the catalytic reaction can be observed by UV-Vis spectroscopy. Figure 3a shows the initial spectrum of 4-NP at 400 nm after the addition of $NaBH_4$. The absorption peak at 400 nm gradually decreased and the absorption peak at 296 nm appeared and gradually increased indicating the presence of the product, 4-aminophenol (4-AP).

Based on Figure 3a, the control reaction of hydrogenation of 4-NP (without any catalyst) took about 60 minutes to complete, meanwhile with the presence of Au NCs/Chitin as catalyst (Figure 3b), the reaction only took 12 minutes to complete.

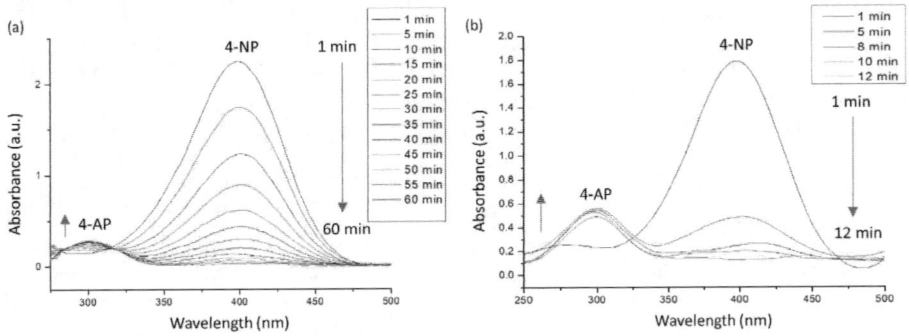

Figure 3. (a) The model reaction of hydrogenation of 4-nitrophenol (4-NP) which takes 60 minutes to complete. (b) The same reaction with the presence of Au NCs/Chitin as catalyst takes 12 minutes to complete. (Reaction conditions: 0.01 g AuNCs/Chitin; 5000:1 molar ratio of $NaBH_4$:4-NP; room temperature; in 3 mL cuvette; no stirring)

Kinetic mode analysis under UV-Vis absorption spectroscopy at 400 nm was done to show clear differences between the AuNCs/Chitin and the control experiments (i.e., reaction with the free suspension of thiolate-protected AuNCs, chitin only and without any catalyst). Based on the results in Figure 4a, the UV-Vis absorption peak of the reactant decreased with the reaction time. At the beginning of the catalytic reaction, the slope of the absorbance at 400 nm was stepper, showing a fast conversion of 4-NP to 4-AP. The Au NCs/Chitin shows a faster reaction rate compared to the control reaction without catalyst (inside the vial has 4-NP, $NaBH_4$ and ultrapure water only). However, the free suspension of thiolate-protected Au NCs only (labelled by AuNCs) had a faster reaction rate than AuNCs/Chitin. This result is expected because Au NCs solution is more reactive and has more active site exposure compared to Au NCs/Chitin. The immobilization may hinder some of the active sites due to the interface interaction between the thiolate-protected AuNCs and the surface of chitin. Nonetheless, the free suspension of thiolate-protected Au NCs cannot be recovered and reused, thus making it unfavourable for catalysis.

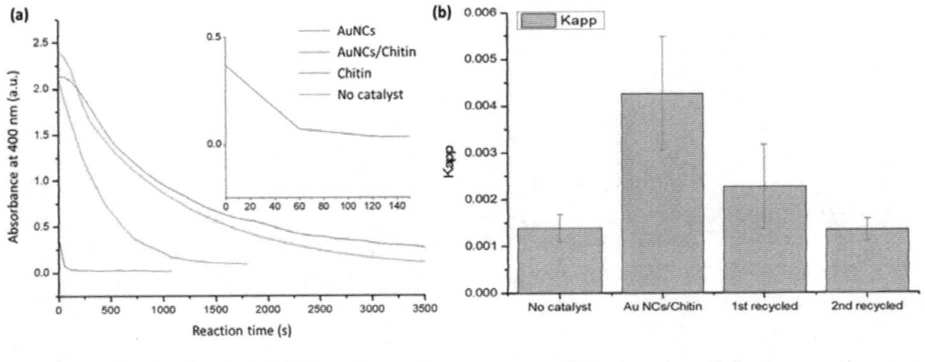

Figure 4. (a) Kinetic mode UV-Vis absorption spectra of 4-nitrophenol hydrogenation catalysed by AuNCs/Chitin, free suspension of thiolate-protected Au NCs, chitin and control experiment with no catalyst. (b) The k_{app} of hydrogenation of 4-nitrophenol of AuNCs/Chitin and the recovered Au NCs/Chitin.

Under kinetic mode, the apparent reaction rate constant (k_{app}) can also be determined by measuring the slope of the decreasing absorbance. From the graph in Figure 4b, the k_{app} for Au NCs/Chitin is about 3 times higher than that from the control reaction. The immobilized Au NCs/Chitin was recovered after being tested for the catalytic reaction to see the recoverability and reusability of the catalyst. Figure 4b also shows the k_{app} of the recovered and reused Au NCs/Chitin. The k_{app} of AuNCs/Chitin decreased from 0.0042 s^{-1} to 0.0022 s^{-1} in the first recycle and to 0.0014 s^{-1} in the second recycle which similar to the k_{app} of the reaction without any catalyst. The decreasing catalytic activity was probably due to some thiolate Au NCs that leached out during the catalytic activity and the recovery process [30].

4. Conclusion

Chitin extracted from shrimp shell wastes has shown a good potential as supporting materials for the immobilization of thiolate-protected Au NC for catalysis. The immobilization of thiolate-protected AuNCs on chitin allowed their recovery and they can be reused for 3 times in the hydrogenation of 4-nitrophenol with acceptable catalytic activity as compared to control condition (no catalyst at all). Overall, this project shows potential to use low cost chitin from shrimp shell waste as support materials to immobilize nanocatalysts. The study has significant impact for adding values to the shrimp shell waste and can be a solution for the solid waste management. The immobilization will also help to reduce wastage of gold nanocatalysts during catalytic reaction because it can be recovered and reused.

Acknowledgment

This is a part of the Final Year Project of Nabila Azirah Zakaria. The principal investigator, Nasaruddin RR would like to acknowledge the financial support from Ministry of Higher Education Malaysia under research grant (FRGS-RACER/1/2019/TK02/UIAM//1) for majority of the research works. Nasaruddin RR would also like to appreciate Nabila Azirah Zakaria for being a good supervisee. Although it is just an undergraduate final year project, Sr. Nabila had been dedicated to do her research works.

References

[1] Li G, Abroshan H, Liu C, Zhuo S, Li Z, Xie Y, Kim H J, Rosi N L, and Jin R, 2016, *ACS Nano*, **10**, 7998
[2] Ouyang R and Jiang D, 2015, *ACS Catal.*, **5**, 6624
[3] Deng H, Wang S, Jin S, Yang S, Xu Y, Liu L, Xiang J, Hu D, and Zhu M, 2015, *Gold Bull.*, **48**,161
[4] Li G and Jin R, 2013, *Acc. Chem. Res.*, **46**, 1749
[5] Zhang X-D, Chen J, Luo Z, Wu D, Shen X, Song S-S, Sun Y-M, Liu P-X, Zhao J, Huo S, Fan S, Fan F, Liang X-J, and Xie J, 2014, *Adv. Healthc. Mater.*, **3**, 133
[6] Setyawati M I, Kutty R V, Tay C Y, Yuan X, Xie J, and Leong D T, 2014, *ACS Appl. Mater. Interfaces.*, **6**, 21822
[7] Goswami N, Luo Z, Yuan X, Leong D T, and Xie J, 2017, *Mater. Horizons*, **4**, 817
[8] Chatterjee B, Sahoo A K, Ghosh S S, and Chattopadhyay A, 2016, *RSC Adv.*, **6**, 113053
[9] Zhou F, Feng B, Yu H, Wang D, Wang T, Liu J, Meng Q, Wang S, Zhang P, Zhang Z, and Li Y, 2016, *Theranostics*, **6**, 679
[10] Chen T, Xu S, Zhao T, Zhu L, Wei D, Li Y, Zhang H, and Zhao C, 2012, *ACS Appl. Mater. Interfaces.*, **4**, 5766
[11] Qian H, Zhu M, Wu Z, and Jin R, 2012, *Acc. Chem. Res.*, **45**, 1470-1479
[12] Jin R, 2010, *Nanoscale*, **2**, 343
[13] Tay C Y, Yu Y, Setyawati M I, Xie J, and Leong D T, 2014, *Nano Res.*, **7**, 805

[14] Zhu M, Lanni E, Garg N, Bier M E, and Jin R, 2008, *J. Am. Chem. Soc.*, **130**, 1138
[15] Luo Z, Nachammai V, Zhang B, Yan N, Leong D T, Jiang D, and Xie J, 2014, *J. Am. Chem. Soc.*, **136**, 10577
[16] Nasaruddin R R, Chen T, Li J, Goswami N, Zhang J, Yan N, and Xie J, 2018, *ChemCatChem*, **10**, 395
[17] Duan B, Liu F, He M, and Zhang L, 2014, *Green Chemistry*, **16**, 2835
[18] Duan B, Zheng X, Xia Z, Fan X, Guo L, Liu J, Wang Y, Ye Q, and Zhang L, 2015, *Angew. Chem. Int. Ed.*, **54**, 5152
[19] Ding F, Deng H, Du Y, Shi X, and Wang Q, 2014, *Nanoscale*, **6**, 9477
[20] Huang Y, Zhong Z, Duan B, Zhang L, Yang Z, Wang Y, and Ye Q, 2014, *J. Mater. Chem. B*, **2**, 3427
[21] Said M, Atassi Y, Tally M, and Khatib H, 2018, *J. Polym. Environ.*, **26**, 3937
[22] Peter M G, 1995, *J. Macromol. Sci., Part A*, **32**, 629
[23] Bhatnagar A and Sillanpää M, 2009, *Adv. Colloid and Interface Sci.*, **152**, 26
[24] Li X, Sun J, Che Y, Lv Y, and Liu F, 2019, *Int. J. Biol. Macromol.*, **121**, 760
[25] Sarode S, Upadhyay P, Khosa M A, Mak T, Shakir A, Song S, and Ullah A, 2019, *Int. J. Biol. Macromol.*, **121**, 1086
[26] Ng C-H, Hein S, Ogawa K, Chandrkrachang S, and Stevens W F, 2007, *Carbohydr. Polym.*, **69**, 382
[27] Luo Z, Nachammai V, Zhang B, Yan N, Leong D T, Jiang D-e, and Xie J, 2014, *J. Am. Chem. Soc.*, **136**, 10577
[28] Li J, Nasaruddin R R, Feng Y, Yang J, Yan N, and Xie J, 2016, *Chem. Eur. J.*, **22**, 14816
[29] Yang H, Gözaydın G, Nasaruddin R R, Har J R G, Chen X, Wang X, and Yan N, 2019, *ACS Sustain. Chem. Eng.*, **7**, 5532
[30] Nasaruddin R R, Yao Q, Chen T, Hülsey M J, Yan N, and Xie J, 2018, *Nanoscale*, **10**, 23113

Turning Coconut Residue into Hydrochar using Hydrothermal Carbonization

Myra Shahira Lau Abdullah[1] and NoorAshrina A Hamid[1]*

[1]School of Chemical Engineering, Engineering Campus, Universiti Sains Malaysia, Seri Ampangan, Seberang Perai Selatan, 14300 Nibong Tebal, Pulau Pinang, Malaysia

Corresponding author: chrina@usm.my

Abstract. The use of renewable and sustainable energy resources is critical, especially considering the scarcity of fossil fuels and the need to combat air pollution. Various types of biomass feedstock, including coconut husk, have been studied in the past to convert into useful forms of hydrochar. However, no one had attempted to use coconut residue to produce hydrochar through hydrothermal carbonization. The goal of this study is to investigate the properties of synthesised hydrochar from coconut residue under various operating conditions in order to determine the highest yield of hydrochar produced prior to methylene blue degradation. Hydrochar was created from coconut residue using hydrothermal carbonization (HTC). The hydrochar was synthesised over a range of residence times of 6 hours, 12 hours, 18 hours, 24 hours, and 48 hours at a constant temperature of 200°C. Prior to the degradation of methylene blue, the properties of hydrochar were investigated using elemental analysis. The oxygen-to-carbon (O/C) and hydrogen-to-carbon (H/C) ratios for all hydrochars were 0.36-0.16 and 1.78-1.17, respectively. For all concentrations, the highest rate of methylene blue removal was approximately 70-80 percent. Overall, residence time had a minor impact on the chemical properties of hydrochars.

1. Introduction

Hydrochar is a two-phased char that contains both solid and liquid phases, making it one of the most competitive biomass energy sources. It's a high-value-added carbonaceous material with a lot of oxygenated functional groups that's made from either dry or wet biomass utilising water as a solvent medium.[1,2]. Because carbon-rich solid hydrochar has a similar physical structure and chemical properties as lignite, it can be used as a fuel to replace traditional energy sources.

Hydrothermal carbonization is a low-temperature, wet thermal conversion technique that is gaining popularity for producing hydrochar.. Compared to common carbonization methods such as pyrolysis and gasification, HTC is operated at mild temperatures (180 – 350 °C) and autogenous pressures. In recent years, many substrates have been used as feedstock to obtain solid products namely municipal solid waste, biomass and sewage sludge. However, the reported physicochemical properties (yield, fixed carbon content, etc) of the derived hydrochar are different with varies in feedstock type and processing variables including carbonization temperature, residence time, pressure, solid load and solid/water ratio[1].

Over the last few years, countless different types of biomasses such as woody biomass [3,4], herbaceous biomass [5] and fruit biomass [6-8] were used for HTC. It has shown that hydrochar characteristics highly depend not only on operating conditions but also on the types of feedstocks [1, 9-

11]. It was discovered that lignocellulose material feedstocks will experience HTC reactions pathways such as hydrolysis, dehydration, decarboxylation, polymerization, aromatization, and solid-solid reactions because all lignocellulose materials are primarily composed of cellulose, hemicellulose, lignins, and some extractives with only minor differences in composition [6]. These indicate that interrelationships in hydrochar properties may be detected using different feedstock types.

Earlier, a study on HTC of kernel babassu coconut bran was done by Costa et al. [7]. In the synthesis process, the temperature was around 120 - 180°C with t_R of 48 h and biomass to ratio of 10%. Research showed that the content of C increases and that of O decreases when the temperature rises. A decrease in the O/C and H/C ratios is seen as a positive change, as it means less water vapor, smoke, and wasted energy during the combustion process [7]. An experimental study on carbonization of corn corb was done with operating conditions of 250°C with intervals of 0.5 to 6 hours and 10% biomass to water ratio [8]. It was discovered that as time progresses, O and H concentrations decline. This means that both elements are converted into liquid H_2O and gas H_2S, and dehydration is extensive. Hydrochar yield did decline slightly over time regardless of residence time as reported by Gao et al. [10].

Arellano et al. further investigated the potential of corn corb-based hydrochar as an adsorbate for methylene blue and discovered that the adsorption performance of HTC decreased as the temperature increased [8]. In contrast, hydrochar synthesised at low temperatures has a high capacity for methylene blue adsorption. At lower temperatures, carbonaceous materials have smaller pore sizes, resulting in low adsorption capacity. The presence of polar functional groups derived from corn corb, which is the source of hydrochar, may also contribute to the lower adsorption. This is due to the intermediate polarity of methylene blue, which causes repulsion [8].

Coconut residue is the selected biomass feedstock in this study whereas hydrochar was produced using it via hydrothermal carbonization technique. Several characterization has been done in order to determine the optimum parameters for hydrochar production. Also, the potential of hydrochar as an adsorbent for degrading methylene blue is also elucidated in this work.

2. Materials and Method

Coconut residue was obtained from a local market, whereas methylene blue was obtained from Sigma Aldrich and used without further purification.

2.1. Biomass Feedstock Preparation

The coconut residue was dried at room temperature until it reached a constant weight before being storedin a sealed container for use in the next hydrochar synthesis.

2.2. Hydrothermal Carbonization Experiment

Hydrothermal carbonization was carried out through Hydrothermal Autoclave Reactor. 5g of coconut residue was weighed into a 100mL Teflon vessel with 50 mL of deionized water (DI water) filled in the vessel. The Teflon vessel was sealed airtight in the autoclave reactor. Then, the reactor was heated for 6 hours with temperature of 200°C and heating rate of 5°C/min. The reactor was cooled down to room temperature once the reaction is completed. The hydrochar produced was recovered by filtration using 0.45μm filter paper. The hydrochar was then washed for several times with DI water to remove the water-soluble volatile matter and oven-dried at 100°C for 24 hours. The experiment was repeated for 9,12, 18, 24 and 48 hours.

2.3. Methylene Blue Solution Preparation

A stock solution with 1000 M of methylene blue was prepared by mixing 1 g of methylene blue in 1 L of distilled water. Concentration of methylene blue was analysed using UV- Visible spectrometer (Cary 60, Agilent Technologies, USA) at the characteristic wavelength (λmax = 668 nm) for calibration purpose.

2.4. *Characterization of Hydrochar*

2.4.1. *Determination of Hydrochar Yield*

Hydrochar yield was calculated using Eq. 1 [11],

Hydrochar yield (%) = (Mass of Dried Hydrochar)/(Mass of Dried Coconut Residue)×100% Eq. 1

Apart from that, determination of hydrothermal severity, R_0 is done for estimation of biomass degradation and conversion. The higher the R_0 indicates lower yield of hydrochar due to hydrolysis and degradation of hydrochar of cellulose and hemicellulose in biomass as suggested by Falco et al. [9]. The calculation is a function of combined effect of processing temperature, T and residence time, t_R based on Eq. 2 [9].

$$R0 = t_R \times \exp[(T-100)/14.75] \quad \text{Eq. 2}$$

2.4.2. *Determination of Hydrochar Chemical Composition*

The chemical composition of yielded hydrochar was analysed via CHN Elemental Analyzer (Perkin Elmer 2400 Series II) to determine the contents of carbon, hydrogen and nitrogen in hydrochar. Elemental O content was calculated by subtracting (100 – C (percent) – H (percent) – N (percent) – S (percent) – Ash (percent) on a dry-mass basis, with a maximum of 0.1g of each hydrochar required for the analysis.

2.4.3. *Determination of Hydrochar as Adsorbate for Methylene Blue Removal*

Hydrochar had a remarkable feature as an adsorbent for environmental contaminant due to substantial oxygen-containing functional groups on its surface as reported previously [13]. However, no previous studies on the adsorption of methylene blue by hydrochar produced from coconut residue have been conducted. As a result, an investigation into the rate of removal of methylene blue by hydrochar was carried out. In this study, four different concentrations of methylene blue were used, which were 50M, 100M, 150M, and 200M, respectively. All methylene blue concentrations were prepared using the dilution method in a volumetric flask to obtain a more accurate concentration of diluted methylene blue solution. The dilution calculation can be calculated using Eq. 3.

$$M_1V_1=M_2V_2 \quad \text{Eq. 3}$$

Where M_1 and V_1 is the concentration and volume of methylene blue stock solution while M_2 and V_2 is concentration and volume of desired concentration of methylene blue solution.

3. Results and Discussions

This study was solely focused on the effect of residence time, t_R namely 6, 12, 18, 24 and 48 hours, on the properties of hydrochar. The biomass-to-water ratio is set at 10%, and the temperature is set at 200°C. It was discovered that the coconut residue only decomposed into hydrochar after the sixth hour. This is because coconut residue requires a temperature greater than 200°C to break the bond and thus convert it to hydrochar.

3.1. *Hydrochar Yield*

According to Figure 1, it was observed that the yield of hydrochar was slightly reduced with longer residence time, t_R. This demonstrated that hydrochar did not significantly affected by residence time as the yield was only drop approximately 4.6% from 6-48 hours. Reason of this is that at higher residence time, t_R lighter organic compounds and permanent gases are formed contributed to decline of hydrocharyield and escalate the carbon's porosity [17].

This trend was also observed by Kalderis et al, who found that hydrochar produced after 6 hours of residence time showed only a slight decrease in hydrochar yield. [14]. This indicates that, the structural rearrangement only occurs in the first 6 hours and turn out to be a stable structure afterwards. Apart from that, hydrothermal severity R_0, was calculated and plotted in Figure 1 in the form of logarithmic. The purpose of defining R_0 was to evaluate the degree of degradation and conversion of biomass. Base from the figure, log R_0 was inversely proportional to hydrochar yield where log R_0 increased with residence time while the yield decreased which was same trend as suggested by Wang et al. 2018 [12]. Gao et al. reported that hydrochar yield is monotonically reduced with rising of severity factor which was due to hydrolysis and degradation of cellulose and hemicellulose in biomass [11]. As a conclusion, longer residence time, t_R insignificantly affected the hydrochar yield as compared to elevating the temperature which is parallel with Zhang et al. where hydrochar produced from HTC of corn corb with different residence time (0.5-6 hours) does not shows significant effect on the yield of hydrochar with only 1.1% reduction was observed [6]. Besides that, long residence time at lower temperature also does not indicate that biomass is fully carbonized. However, the rate of carbonization can be elevated at longer residence time which means complete carbonization can be achieved by elevating the residence time at lower temperature but may took days as compared to elevating the temperature.

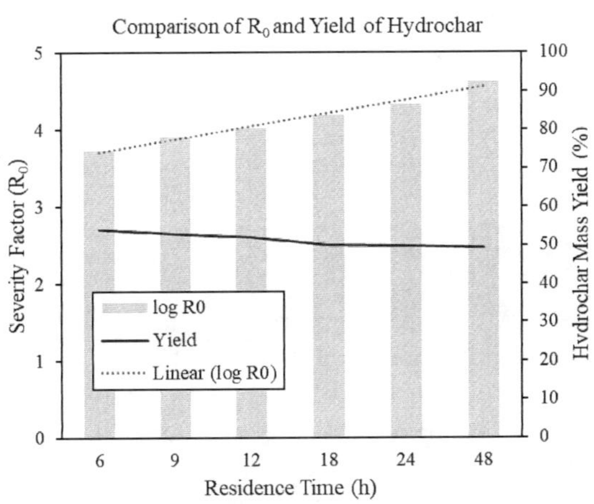

Figure 1. Severity factor vs. hydrochar mass yield at different residence times

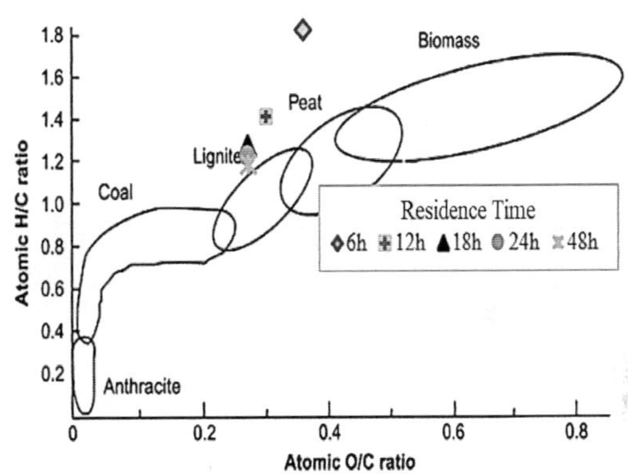

Figure 2. The O/C and H/C ratios for all hydrochars was plotted on Van Krevelen Diagram that adopted from Kambo & Dutta [2]

3.2. Chemical Composition of Hydrochar

Elemental analysis is a method to define the elemental composition of unknown substance. Determination of H and O content is crucial in order to obtain the elemental ratios of H/C and O/C. Both ratios are useful to identify the chemical transformation of coconut residue during hydrothermal carbonization (HTC). According to the results of table 1, the carbon content increases significantly from 6 to 18 hours and then slightly from 18 to 48 hours. The uptrend of carbon content is resulting from reactions occured during HTC which is dehydration and decarboxylation reactions [7-9]. The slight increment may due to current temperature was insufficient to cease the bond of macromolecules of biomass to be degraded [12]. Thus, higher temperature or longer residence time may be needed to increase the carbon content of hydrochar. Besides that, it was observed that oxygen content was decreased when residence time, t_R is increased. This phenomenon was also due to both dehydration and decarboxylation reactions during HTC as the combination of the two reactions brought to deoxygenating reactions [3]. As a result, the oxygen content of hydrochar decreased as the residence time increased.

Besides that, increment in nitrogen content was observed with increasing residence time. Supposedly, nitrogen content was decreased due to dissolution into processing liquid. Yet, the increment

IOP Conf. Series: Materials Science and Engineering 1192 (2021) 012033 doi:10.1088/1757-899X/1192/1/012033

of nitrogen content with residence time can be due to nitrogen content of coconut residue is removed by dissolution into hot water but it is retained in hydrochar due to longer reaction time. On the other hand, hydrogen content was inversely proportional to residence time. This can be explained by the occurrence of dehydration and decarboxylation reactions which removes both hydrogen (H) and oxygen (O) form coconut residue in the form of water [4].

Table 1. Elemental Composition of Hydrochar at Different t_R

	Residence Time, tR				
	6 h	12 h	18 h	24 h	48 h
C (%)	60.76±0.01	65.20±0.01	67.32±0.01	67.52±0.01	67.47±0.01
H (%)	9.00±0.01	7.53±0.01	7.02±0.01	6.83±0.01	6.57±0.01
N (%)	1.09±0.01	1.10±0.01	1.18±0.01	1.23±0.01	1.30±0.01
O (%)	29.15	26.17	24.48	24.42	24.66
H/C	1.78	1.39	1.25	1.21	1.17
O/C	0.36	0.30	0.27	0.27	0.27

O = 100 – (C+H+N)

Both H/C and O/C ratios of hydrochars decreased with increasing residence time. The decrease of H/C and O/C ratios was mainly because of dehydration and decarboxylation reaction during HTC [6, 10]. Reduction of H/C and O/C ratios indicate elimination of $-CH_3$ and CO_2 from biomass. High O/C ratio prompt to production of low carbonization rates of hydrochar and presence of polar functional group on hydrochars [10]. To visualized the degree of carbonization of hydrochar, a Van Krevelen diagram with H/C against O/C ratio was plotted (Figure 2). This diagram will function as an indicator where high ratio indicates presence of macromolecules (cellulose) while low ratio indicates presence ofmore condensed (aromatic) structure [3].

Corresponding to Figure 2, even though H/C was quite high, the result showed a decrease in H/C trend. Thus, lower H/C ratios and lower O/C ratios at longer residence times indicated an increase in condensed structure as well as the presence of macromolecules in the hydrochar. Aside from that, the coal band of all hydrochars except hydrochars at 6 hours was found to be like lignite, with H/C and O/C atomic ratios of lignite being typically 0.8-1.3 and 0.2-0.38, respectively, [18]. Hence, it was concluded that all hydrochars except the one produced at 6 hours were a low rank coal which is comparable to lignite. Thus, a good performance of co-combustion of hydrochars blended with coal can be achieved as these hydrochars has good compatibility with lignite. The trends of hydrochars produced is similar as the trend of hydrochar formed from Babassu kernel bran [7], corn corb [8] and bamboo [13] where all hydrochars shows reduction in both O/C and H/C atomic ratios due to deoxygenating reactions which is in parallel with results obtained from HTC of coconut residue.

3.3. Removal of Methylene Blue by Hydrochar

The results of the 50M hydrochar methylene blue removal produced during the 6-48 hour residence were studied. Figure 3 shows the results of dyes removed by hydrochar in 6 different time of residence from 6 - 48 hours with a 50 M methylene blue concentration, respectively. It was observed that the rate of removal was decreased with time which indicates that the methylene blue dye particles were capture by active site of hydrochar. Based on the plotted graph, the highest rate of removal of all concentration was hydrochar produced at 24 hours. This may due to the carbon content of hydrochar at 24 hours wasthe highest which was approximately 67.52%.

Although the hydrochar produced after 48 hours is longer than that produced after 24 hours, the rate of removal was lower. This could be because hydrochar produced at 24 hours was in powder form, whereas hydrochar produced at 48 hours was in small granular form, as was done for the study of hydrochar structure. As a result, the adsorbent capacity was reduced because the smaller surface area ofgranular hydrochar at 48 hours reduced the adsorbent capacity.

On the other hand, it was observed that the dye removal rate was significantly increased at first few hours before reach equilibrium after approximately 5 h. This phenomena suggested that the high rate of adsorption in the beginning as adsorbents site is all vacant initially while decreasing adsorption rate might ascribed to diminish in magnitude of adsorbent's vacant sites and dye concentration. Arellano et al. reported that, the mark down on rate of adsorption near the end of experiments denotes that monolayer formation of methylene blue onto adsorbent surface was possible to happened [8]. This occurrence may lessen the available active sites requisite for further adsorption after acquiring equilibrium.

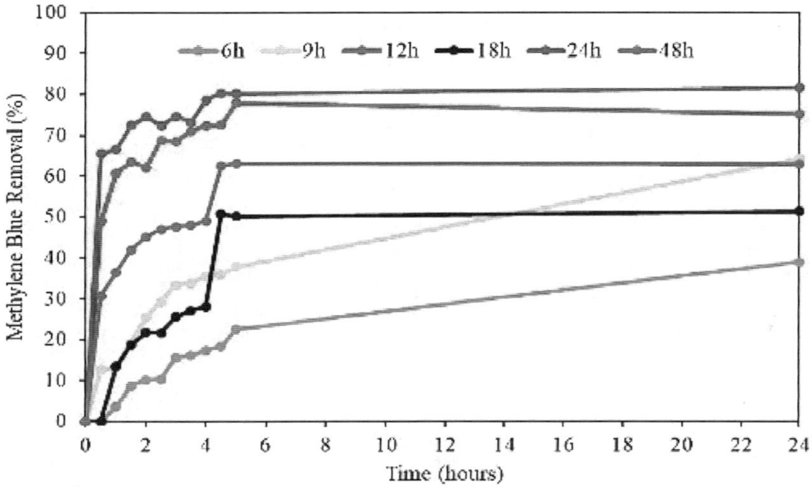

Figure 3. Methylene Blue Removal at 50M Concentration against Time

Conclusions

In this research study of production of hydrochar at different residence time, the average yield of hydrochar was approximately 50% where the mass yield of hydrochar did not show significant changes with longer t_R where the difference between mass yield of hydrochar with t_R at 6 hours and 48 hours was around 4.6%. However, the hydrothermal severity, R_0 shows an uptrend with increasing t_R which was accurate as R_0 increase would cause diminished on mass yield of hydrochar. In this study, the C content of hydrochars increased with longer t_R due to dehydration and decarboxylation reactions in the autoclave reactors. The C content was increased significantly at first 18 hours but then increased slightly after 18 hours may due to operating condition was insufficient to break the macromolecules. Overall, the highest MB removal was hydrochar produced at 24 hours which was approximately 80% with C content of 67.52%.

References

[1] Nzediegwu, C., Naeth, M. A., & Chang, S. X. (2021). Carbonization temperature and feedstock type interactively affect chemical, fuel, and surface properties of hydrochars. Bioresource Technology 330, 124976. https://doi.org/10.1016/j.biortech.2021.124976.

[2] Kambo, H. S., & Dutta, A. (2015). A comparative review of biochar and hydrochar in terms of production, physico-chemical properties and applications. Renewable and Sustainable EnergyReviews. https://doi.org/10.1016/j.rser.2015.01.050.

[3] Gao, P., Zhou, Y., Meng, F., Zhang, Y., Liu, Z., Zhang, W., & Xue, G. (2019). Preparation and characterization of hydrochar from waste eucalyptus bark by hydrothermal carbonization. Energy, 97(April 2019), 238–245. https://doi.org/10.1016/j.energy.2015.12.123.

[4] Kang, S., Li, X., Fan, J., & Chang, J. (2012). Characterization of hydrochars produced by hydrothermal carbonization of lignin, cellulose, d-xylose, and wood meal. Industrial and Engineering Chemistry Research, 51(26), 9023–9031. https://doi.org/10.1021/ie300565d.

[5] Reza, M. T., Wirth, B., Lüder, U., & Werner, M. (2014). Behavior of selected hydrolyzed and dehydrated products during hydrothermal carbonization of biomass. Bioresource Technology,169, 352–361. https://doi.org/10.1016/j.biortech.2014.07.010.

[6] Zhang, B., Heidari, M., Regmi, B., Salaudeen, S., Arku, P., Thimmannagari, M., & Dutta, A. (2018). Hydrothermal carbonization of fruit wastes: A promising technique for generating hydrochar. Energies, 11(8), 1–14. https://doi.org/10.3390/en11082022.

[7] Costa, R. S., Vieira, L. H. S., Ghosh, A., Santos, A. M. S., Ferreira, O. P., & Viana, B. C. (2019). Hydrothermal carbonization of waste babassu coconut biomass for solid fuel production. Revista Virtual de Quimica, 11(3), 626–641. https://doi.org/10.21577/1984-6835.20190048.

[8] Arellano, O., Flores, M., Guerra, J., Hidalgo, A., Rojas, D., & Strubinger, A. (2016). Hydrothermal carbonization of corncob and characterization of the obtained hydrochar. Chemical Engineering Transactions, 50, 235–240. https://doi.org/10.3303/CET1650040.

[9] Falco, C., Baccile, N., & Titirici, M. M. (2011). Morphological and structural differences between glucose, cellulose and lignocellulosic biomass derived hydrothermal carbons. Green Chemistry, 13(11), 3273–3281. https://doi.org/10.1039/c1gc15742f.

[10] Funke, A., & Ziegler, F. (2010). Hydrothermal carbonization of biomass: A summary and discussion of chemical mechanisms for process engineering. Biofuels, Bioproducts and Biorefining, 4(2), 160–177. https://doi.org/10.1002/bbb.198.

[11] Gao, Y., Wang, X., Wang, J., Li, X., Cheng, J., Yang, H., & Chen, H. (2013). Effect of residence time on chemical and structural properties of hydrochar obtained by hydrothermal carbonization of water hyacinth. Energy, 58, 376–383.https://doi.org/10.1016/j.energy.2013.06.023.

[12] Wang, T., Zhai, Y., Zhu, Y., Li, C., & Zeng, G. (2018). A review of the hydrothermal carbonization of biomass waste for hydrochar formation: Process conditions, fundamentals, and physicochemical properties. Renewable and Sustainable Energy Reviews, Vol. 90, pp. 223–247. https://doi.org/10.1016/j.rser.2018.03.071.

[13] Li, Y., Meas, A., Shan, S., Yang, R., & Gai, X. (2016). Production and optimization of bamboo hydrochars for adsorption of Congo red and 2-naphthol. Bioresource Technology, 207, 379– 386. https://doi.org/10.1016/j.biortech.2016.02.012.

[14] Kalderis, D., Kotti, M. S., Méndez, A., & Gascó, G. (2014). Characterization of hydrochars produced by hydrothermal carbonization of rice husk. Solid Earth, 5(1), 477–483. https://doi.org/10.5194/se-5-477-2014.

[15] Mohd Nor, N. 'Ain N., Abbasiliasi, S., Marikkar, M. N., Ariff, A., Amid, M., Lamasudin, D. U., Mustafa, S. (2017). Defatted coconut residue crude polysaccharides as potential prebiotics: study of their effects on proliferation and acidifying activity of probiotics in vitro. Journal of Food Science and Technology, 54(1), 164–173. https://doi.org/10.1007/s13197-016-2448-9.

[16] Ng, S. P., Tan, C. P., Lai, O. M., Long, K., & Mirhosseini, H. (2010). Extraction and characterization of dietary fiber from coconut residue. Journal of Food, Agriculture and Environment, 8(2), 172–177.

AUTHOR INDEX

Institute of Physics
Dirac House, Temple Back
Bristol BS1 6BE UK

CURRAN ASSOCIATES INC.
proceedings.com

ISSN: 1757-8981
ISBN 978-1-7138-4652-9
9 781713 846529